On Sets and Graphs

Eugenio G. Omodeo • Alberto Policriti
Alexandru I. Tomescu

On Sets and Graphs

Perspectives on Logic and Combinatorics

 Springer

Eugenio G. Omodeo
DMG/DMI (Department of Mathematics
and Geosciences, Section of Studies
in Mathematics and Information
Technology)
University of Trieste
Trieste, Italy

Alberto Policriti
DMIF (Department of Mathematics
Computer Science, and Physics)
University of Udine
Udine, Italy

Alexandru I. Tomescu
Department of Computer Science
Helsinki Institute for Information
Technology HIIT
University of Helsinki
Helsinki, Finland

ISBN 978-3-319-85536-3   ISBN 978-3-319-54981-1   (eBook)
DOI 10.1007/978-3-319-54981-1

Printed on acid-free paper

This Springer imprint is published by Springer Nature
The registered company is Springer International Publishing AG
The registered company address is: Gewerbestrasse 11, 6330 Cham, Switzerland

To Paola, Pietro, and Sara

To Giovanna

To Johanna and Aava

Foreword

Graphs are certainly the single most important data structure in computer science, mirroring somewhat the rôle sets play in mathematics. In fact, there is a very close and obvious relationship between graphs and sets, viz., each graph can be represented as a pair of sets, so very nearly we have something like a functor between graphs and sets. In programming, this symmetry is somewhat relaxed, since sets may as well be used for system construction, at least in the early design phases of software development (later sets vanish more and more into the processes and into the data structures actually implemented).

This book looks into the relationship between sets and graphs from different angles. One notes first that the set representation of graphs takes in fact for granted that the theory of sets is developed to such an extent that it may be usable for this representation. Looking into the development of set theory, however, one sees that there are many varieties of set theories around, so one probably wants to fix the ground before one can stand on it. In this monograph, this leads to the definition of sets that may be represented by graphs (in a curious reversal of rôles), demanding the question of characterizing those sets and those graphs for which such foundation issues arise. It gives rise to *set graphs*, roughly those graphs which underlie hereditary sets, which carry such an important burden in the foundation of set theory.

The authors examine the interplay of sets and graphs on a fairly fundamental level, formulating their set theory very carefully in such a way that some of the proofs become accessible to a proof assistant. They show that:

- one can find for every weakly extensional, acyclic digraph a Mostowski collapse by finite sets so that no two vertices are sent to the same set by the decoration,
- every graph admits an orientation which is weakly extensional and acyclic, so that one can regard its edges as membership arcs deprived of their natural orientation,

through the set-based proof-checker Referee (also known as the grown-up version of ÆtnaNova, it results from joint work of the two present Italian authors with Jacob T. Schwartz and D. Cantone). A brief introduction into Referee is given as well. Interestingly, Referee brings into the game the axiom of choice and not one of its—admittedly less well known—competitors like the axiom of determinacy. Technically, this happens through the arb operator familiar from languages like SETL.

One might ask whether this approach is a major route our discipline will be pursuing in the future: preparing the ground for machine-assisted proofs and then doing the proofs proper by a proof assistant. Well, probably. In some areas in logic and in complexity theory, proofs get so involved, so long, and so complicated that one scientist alone might no longer be able to conquer them. The analogy to programs comes to mind—programs tend to become so involved, so long, so complicated, and so many-faceted that a single programmer or a team of developers cannot really be its master any longer. But there is an important difference: programs are ultimately written for being executed on a computer, but a proof is intended to convey intellectual insight (and, of course, fun to its explorer), knowledge administering systems notwithstanding.

Since the universe of discourse is finite (but in the last chapter), combinatorial questions arise as well; the authors show that their approach renders counting certain graphs fairly straightforward and direct. They apply some Markov chain technique to the generation of random graphs, which at first sight looks like an unlikely match for the questions for which these chains are employed by the authors, but the technique soon turns out to an apt tool.

But one does not have to stop at finite objects, as the last chapter shows. Here the necessary set theory needs to be expanded, so that infinite sets can be handled. This excursion into infinity helps formulating and proving some properties which arise when well-foundedness of the basic relations is relaxed, and it gives surprising insights into Ramsey's theorem from finite combinatorics.

We should be glad that this book has been written. It helps us to understand the interplay of sets and graphs, in particular when wanting to look at graphs from an axiomatic set theory point of view. The book is also attractive because it formulates and solves some interesting combinatorial counting problems by directly manipulating the objects involved (rather than, say, setting up a generating function of some sort and manipulating it, which of course has also its merits). Finally, the use of the Referee system is not a mere add-on. It rather conveys some illuminating views into the inner working of some proofs requiring the power of set theory on both ends, viz., the tool, and the objects to be manipulated.

Professor Emeritus of Software Technology   Ernst-Erich Doberkat
and of Mathematics
Technische Universität Dortmund
Bochum, December 21, 2016

Preface

Will we say anything new on sets and graphs? Today's scholars are so closely acquainted with such entities—in theoretical computer science no less than in mathematics—that a book devoted to sets and graphs runs the risk of arousing, at best, an indulgent curiosity.

Why graphs? Without a good mastery of graphs, the design and analysis of computer algorithms would be unaffordable.

Why sets? Set Theory is commonly, though often implicitly, placed in the background of any other mathematical disciplines.

In particular, a graph, as customarily defined, is just a pair of finite sets: a set of vertices and a set of edges, of which the latter is included in the Cartesian square of the former. We see this reduction of graphs to sets neither as detracting from the autonomy of Graph Theory nor, in any deep or practical sense, revealing. One of the programmatic traits of this book is that graphs will endorse sets to the same extent to which sets will endorse graphs.

There is little controversy nowadays on the basilar role of sets and on the relevance of graphs for the development of algorithms. An obvious corollary then is: even algorithms are reducible, via Graph Theory, to Set Theory. As an indirect recognition of this, sets and associative maps (to wit, sets of ordered pairs) have become built-in types in many programming languages. But, in our daily experience, what abstractly speaking is just a set gets used in restrained ways; accordingly, in specific contexts, specialized concrete representations become preferable to others. As soon as concerns about representation arise, graphs pop up anew to the fore. This hints at why we will invest in cross-fertilization rather than giving priority to either graphs or sets in our investigation. Our undertaking, as we see it, is to exploit sets to shed light on certain issues that concern graphs while retaining our freedom to overturn perspective, whenever fit.

In the formal approach to Set Theory, a *set* is an object containing nothing but other sets as elements. This view brings uniformity into the foundations of mathematics and radically differs from the informal, naïve view according to which a set is a collection of elements whose nature we refrain from entering into. When we work in a universe where *everything is a set*, membership becomes the only relation we have to worry about and we find ourselves very close to Graph Theory.

This suggests a way in which graphs can represent sets: vertices will stand for sets and the edge relation will mimic the membership relation. This representation gives rise to interesting problems; in particular, it calls into play *hypersets*, whose notion somewhat generalizes the today prevailing conception of sets by permitting, e.g., membership to form cycles. By rooting sets into graphs, we are led to many combinatorial, structural, and computational questions: we will, in fact, study sets under the spotlights of combinatorial enumeration, canonical encodings by numbers, and random generation.

This book will privilege combinatorial questions over algorithmic issues in order to result in a more incisive manifesto. One of the exceptions will be our addressing the validity problem for set-theoretic sentences. With regard to solvable cases of that problem, the validity analysis is supported by graphs specially contrived to diagram collections of potential counterexamples to an alleged theorem. Such graphs also convey the causes of the combinatorial explosion that sometimes hinders logical analysis from going through.

Given a set, we can easily manage to build a graph that reflects the inner structure of that set. Doing this in the above-suggested manner will result in Directed edges. Can we conversely, but starting with a graph whose edges are undirected, find a set that conveniently represents it? Much of this book is devoted to an investigation on the graphs devoid of orientation that underlie sets, which we call *set graphs*. Through the study of their structure, we bring to light which graphs can be "implicitly" represented by sets. We elucidate the complexity status of the recognition problem for set graphs, characterize their class in terms of hereditary graph classes, and put forth polynomial algorithms for certain graph classes. Very little extant literature is concerned with the class of set graphs in its entirety, but many subclasses of it—suffice it to mention *claw-free* graphs here—have been studied since long. The set reading of graphs leads to simpler proofs of various classical results on claw-free graphs. We have taken advantage of the set-theoretic flavor of those proofs to formalize two of them with moderate effort in the set-based proof-checker Referee; the resulting new proofs are presented in this book in wealth of detail.

We will refrain, in general, from embarking on profound discussions about the infinite. Infinite sets do exist to us, mainly because we wish to adhere to the axioms of a standard set theory. Our hesitancy about the infinite is not grounded in philosophy; we simply decided to limit our focus to finite combinatorics.

In seeming conflict with the intention just stated, at a well-advanced phase of the exposition, we will not resist talking about a doubly stranded spiral formed by two infinite sets and describing it by means of slick set-theoretic formulae. An infinite graph will instantly show up that represents our spiral. Using it, we will shed light on a celebrated result, fundamental to *finite* combinatorics, namely, Ramsey's theorem.

A Word on the Audience for Whom This Book Is Intended

Much of the content of this book originates from papers recently published on scientific research journals, from which we selected topics which should be accessible with moderate effort also to nonspecialists, in particular to graduate students. The exercises put at the end of each chapter enable the reader to try her/his hand on the various topics so as to develop personally a deeper understanding of the subject matter.

This book assumes from the reader some familiarity with basic algorithm complexity and with standard programming techniques; anyhow, our algorithmic specifications will take the form of pseudo-code. To make our presentation as self-contained as possible on various topics we treat, we summarize some presupposed notions (e.g., NP-completeness) in panels which are spread all over the text.

The subject matter touches upon proof technology at some point; it is hence desirable that our reader has had some previous exposure to first-order predicate logic and formal methods. On the other hand, little knowledge of Set Theory and Graph Theory is assumed. For this reason, we try to present what is needed from those two areas of mathematics in a reasonably self-contained way, emphasizing concepts likely to be important in continuation of the work begun here, rather than technicalities. Foundational issues, for example, consideration of the strength or necessity of axioms or the precise relationship of our formal treatment to other weaker or stronger formalisms studied in the literature, are neglected.

Content of This Book

The introduction gives—mainly through examples—a rapid overview of the authors' way of combining the study of sets with the study of graphs.

Chapter 2 introduces two languages apt to support formal reasoning within Set Theory and Graph Theory, respectively. The primitive endowments of these languages rely on first-order predicate logic, and they are deliberately minimal; convenient syntactic extensions are then introduced conservatively. In the case of Set Theory, an inventory of sentences is produced from which the postulates of various specific axiom systems can be drawn: among those, the Zermelo-Fraenkel axioms; two axioms of opposite contents, namely, von Neumann's axiom of foundation and Aczel's anti-foundation axiom; and another incompatible pair of sentences, namely, Zermelo's infinity axiom (inessential for most of this book) and Tarski's axiom enforcing that only finite sets exist. In the case of graphs, the essential terminology is introduced, sometimes by resorting to the formal language in order to specify basic graph-theoretic notions such as acyclicity; we also hint at how significant graph theories often result from forbidding specific subgraphs.

Chapter 3 is also concerned with basics. It introduces two hierarchies of sets, one formed by the hereditarily finite sets, the other one—the celebrated von Neumann's *cumulative hierarchy*—also encompassing sets of infinite cardinality or rank. By virtue of the hierarchical construction of these set universes, membership is a well-founded relation over them, but this chapter also introduces two non-well-founded variants of the universe of hereditarily finite sets. The hierarchical hereditarily finite sets are intertwined with natural numbers, thanks to their anti-lexicographic order, first studied by Ackermann and also recalled in this chapter. Each hereditarily finite set is described in full by what we dub its *pointed membership graph*, whose acyclicity flags whether it is a hierarchical set (as opposed to a proper *hyperset*). A class of graphs named *membership graph* is also considered, which is broader than the class of pointed membership graphs; unlike in a *pointed* membership graph, there is no privileged vertex in a membership graph. Graphs closely akin to membership graphs, whose edges are also meant to mimic membership, though to a lesser degree of detail, are then exploited to solve some favorable cases of the *Entscheidungsproblem*, namely, to determine whether a set-theoretic formula subject to very stringent syntactical constraints is satisfiable or not.

The rest of the book is subdivided into two parts. Chapters 4 and 5 (along with their supplementary Appendix A) explain under what circumstances, and how, sets can conveniently model graphs; Chapters 6 through 8 investigate the converse issue: when is it convenient to represent sets by graphs? Notice that the use of graphs for handling the set-satisfaction problem mentioned at the end of the preceding paragraph pertains—at a relatively abstract level—to the latter circle of ideas.

Specifically, Chap. 4 undertakes the study of the class of *set graphs*, each of which results from a membership graph whose edge orientation has been forgotten. Set graphs hence have undirected edges; moreover, they are connected; their class includes graphs endowed with Hamiltonian paths and the so-called claw-free graphs. Given a graph with undirected edges, the problem of establishing whether or not it is a set graph is NP-complete; it amounts to finding an orientation of the edges such that the resulting directed graph neither forms cycles nor has distinct vertices endowed with the same children.

Chapter 5 presents recent proofs of two classical propositions concerning claw-free graphs. Since these proofs rely on the fact that claw-free graphs are set graphs, it is a straightforward task to develop them formally with the assistance of a proof-checker knowledgeable on sets, named Referee. Thus, by reporting on a concrete proof-checking experiment, this chapter offers a light introduction to proof technology. To convey the character of a proof script verifiable by means of our automated proof assistant, Appendix A shows excerpts of a formal proof of the fact that every connected graph has a vertex whose removal does not disrupt its connectedness.

Chapter 6 illustrates the usefulness of the set-to-graph correspondence by resorting to it for an explicit count of how many sets t of cardinality n enjoy this property: every element of t is also a subset of t. This result regards those hereditarily finite sets over which membership is well founded. The Ackermann encoding of their graphs is then revisited, to show that it can be obtained by means of a partition-refinement technique borrowed from algorithmics. A virtue of this technique is that it extends naturally to provide an ordering of non-well-founded hereditarily finite sets. This ordering can then be exploited for encoding non-well-founded hereditarily finite sets by dyadic rational numbers.

Chapter 7 studies how to generate a well-founded set of "size" n at random, so that each set of size n has equal probability to occur. Procedures of this kind can be of use for testing the correctness of algorithm implementations or for testing conjectures about data. Three general methods of generating combinatorial objects uniformly at random are described, two of which are based on the so-called combinatorial decomposition of the objects, while one is based on a Markov chain.

Chapter 8 broadens the scope of discussion to infinite sets and graphs. In its formulation dating back to Zermelo's original axiomatic system, the infinity axiom exhibits a single infinite set; as shown in this chapter, this axiom can be superseded by a sentence of greater syntactical simplicity, which brings into play two infinite sets twisted together. The simplest infinite graph underlying a pair of infinite sets that comply with this updated axiom can be used as a sort of abacus for speculating on finite combinatorics, in particular while addressing Ramsey's celebrated theorem.

Acknowledgments

We are grateful to Alberto Casagrande, Domenico Cantone, Giovanna D'Agostino, Agostino Dovier, Ernst-Erich Doberkat, Martin Golumbic, Alberto Marcone, Martin Milanič, Carla Piazza, and Romeo Rizzi for reading various parts of this book and to Salvatore Paxia, who enabled us to keep the Referee / ÆtnaNova proof-checking system alive. We also thank Alexandru-Alin Tudor for creating the three-dimensional images shown in Figs. 8.7 and 8.11.

This work was partially supported by the Academy of Finland under grants 250345 (CoECGR) and 274977, by the project *Specifica e verifica di algoritmi tramite strumenti basati sulla teoria degli insiemi* funded (2013) by INdAM/GNCS (Istituto Nazionale di Alta Matematica "F. Severi," Gruppo Nazionale per il Calcolo Scientifico) and by the project FRA-UniTS (2014) *Learning specifications and robustness in signal analysis*.

University of Trieste   Eugenio G. Omodeo
University of Udine   Alberto Policriti
University of Helsinki   Alexandru I. Tomescu
Helsinki/Trieste/Udine/Bucureşti/Berlin
November/December, 2016

Copyright Acknowledgments

The figures from Examples 4.3 and 4.4, Fig. 4.7 and some text excerpts from Chap. 4 have been reproduced with permission from Elsevier from [63]. Figures 4.3, 4.5, the figure from Exercise 4.5 and some text excerpts have been reproduced with permission from Elsevier from [64]. Some text excerpts have been reproduced with permission from Elsevier from [65].

The figure in Example 5.1, Figs. 5.2, 5.3, 5.4, 5.5 and some text excerpts from Chap. 5 have been reproduced with permission from Oxford University Press from [81]. The figures from Examples 5.3, 5.4, 5.5, Figs. 5.7, 5.8, 5.9, 5.11, 5.12, 5.19, 5.20, 5.21, 5.22, 5.23, 5.24, and some text excerpts from Chap. 5 have been reproduced with permission from Springer from [80].

Some text excerpts from Chap. 6 have been reproduced with permission from Elsevier from [94] and [99]. The figures in Examples 6.5, 6.6 and some text excerpts from Chap. 6 have been reproduced with permission from IOS Press from [25].

Figures 8.5 and 8.9 and some text excerpts from Chap. 8 have been reproduced with permission from Oxford University Press from [77]. Figure 8.19 and some text excerpts from Chap. 8 have been reproduced with permission from Cambridge University Press from [78].

Contents

Chapter 1
Introduction

> Graph Theory is a delightful playground for the exploration of proof techniques in discrete mathematics, and its results have applications in many areas of the computing, social and natural sciences.
>
> D. B. WEST, Introduction to Graph Theory.

This monograph revolves around the correspondences between sets and graphs and their applications to finite combinatorics, with an eye on proof methods and proof technology.

> Sets, as they are usually conceived, have *elements* or *members*. An element of a set may be a wolf, a grape, or a pigeon. It is important to know that a set itself may also be an element of some other set. [···] What may be surprising is not so much that sets may occur as elements, but that for mathematical purposes no other elements need ever be considered.
>
> P. R. HALMOS, Naive Set Theory.

Throughout, membership will be a *nested* relation. The cited passage by Paul Halmos points out that the restraint of using nothing but sets in the formation of sets does not in the least hinder generality, however disconcerting it may appear at first. The merits of nested usage of $\in$ will be a recurrent theme in this book.

This introductory chapter supplies informal and quite elementary reasons motivating our exploration of the correspondences between sets and graphs. It also surveys some key ideas behind the correspondences that will be set forth.

We will browse situations where graphs can help in the study of sets. Concerning combinatorial issues, graphs generally offer a better insight. We will, for instance, address counting problems of the kind: How many sets endowed with such and such features, and of a given "size," exist? A policy, to ease the solution of a problem of this nature, is to find a convenient representation of the sets of interest by means of graphs. The advantages thus gained are not entirely subjective, as we can illustrate by providing improved solutions to problems which had been formerly solved, as documented in the literature, but in more roundabout terms than ours.

© Springer International Publishing AG 2017
E.G. Omodeo et al., *On Sets and Graphs*, DOI 10.1007/978-3-319-54981-1_1

Sets must be allowed to be infinite, if their theory is to match its grand goals as a founding stone for mathematics; graphs, as mostly conceived in discrete mathematics, are finite. That the two types of entities are of comparable strength is hence implausible. However, to keep focused on finite combinatorics, this book will mostly deal with sets which are finite, seldom taking advantage of the fact that the most popular set theories, including one whose axioms will be recalled in Chap. 2,[1] do legitimize infinite sets.[2] A noticeable exception will be in Chap. 8, where the study of a "regular" form of infinity, describable by means of a pair of infinite sets as well as by an infinite graph, will culminate in a restatement of Ramsey's seminal theorem.

For the rest of this preamble, let us discuss how sets can be of help in the study of graphs. To find an answer, we must abandon the overly uniform view that a graph just consists of a set of vertices and a set of ordered pairs of those vertices: for, in the study of special classes of graphs, one may benefit from *ad hoc* representations.

At its simplest, the idea is already at work in the common case when one adopts a simplified representation of a graph whose edges are bidirectional. Naïve uniformity would impose that $\langle w, u \rangle$ and $\langle u, w \rangle$ jointly enter the edge-representing set when there is an edge between the vertices u and w; but, obviously, representing each edge by a two-vertex set $\{u, w\}$ instead of by the two opposite pairs $\langle w, u \rangle$, $\langle u, w \rangle$ implements a convenient data compression in this case.

Equally obvious thoughts follow from the remark that Graph Theory[3] is seldom interested in a graph per se but usually treats all isomorphic graphs on a par. For a representation method to be acceptable, it will hence suffice that it takes into charge at least one graph from each graph isomorphism class; multiple representations of the same class may be accepted, but then their multiplicity should be kept finite.

When, a few paragraphs ago, we mentioned studying special classes of graphs, we did not have in mind classes of isomorphic graphs: an example of what was intended is the class of all acyclic graphs, namely, those graphs (individuated to within isomorphism) in which no walk that respects the orientation of edges starts and ends with the same vertex. We will take into account many classes of

---

[1] Notice that we normally write "Set Theory" with uppercase initials only when referring to the field of mathematics encompassing intuitive approaches to sets, besides the study of variant axiomatic systems about sets. No matter how important, the Zermelo-Fraenkel theory, which we are referring to here, is but one of the many axiomatic systems which Set Theory considers worth of study. Other systems expunge infinite sets or enrich the set-universe with the aggregates sometimes called "hypersets."

[2] Here we cannot pass under complete silence that a field named combinatorial set theory, boldly addressing infinite combinatorics, has been flourishing for decades.

[3] In analogy with the above-traced distinction between Set Theory and set theories, we will distinguish between Graph Theory as a broad, largely semiformal field of study and the many graph theories formalizing multiple facets of the overall field.

graphs, some presumably familiar to the average reader (e.g., the acyclic graphs just mentioned) and some relatively exotic.

Typically, in this book, a graph will be represented by means of a set, equinumerous with the set of its vertices, which mimics edges by means of the membership relation. But how far-reaching can be such an edge-to-membership translation, given that membership is an asymmetric relation? Can we apply this translation to a graph whose edges are bidirectional? We will answer affirmatively, because we are prompt to accept that an undirected graph be represented by a set of sets which are in one-one correspondence $v \mapsto s_v$ with its vertices, provided this condition holds: $\{u, w\}$ is an edge if and only if either $s_u \in s_w$ or $s_w \in s_u$ holds. (Here, in watermark, readers may see a cause of multiple representations for the same graph.)

The proposed edge-to-membership translation turns out to be more effective in some cases than in others, as we will soon see, at p. 5, on a concrete example. Our translation will work most successfully for a class of graphs, devoid of orientation, which we will name set graphs. Among others, we will exploit it in connection with certain graphs, known since long as *claw-free graphs*, which form a subclass of our set graphs. This will give us the opportunity to exploit our edge-to-membership translation to get, in Chap. 5, novel and simpler proofs of a couple of important facts about claw-free graphs. Because of the simplicity of our new proofs compared to traditional ones and of their kinship with Set Theory, we can afford formalizing them in full with a proof checker which has built-in methods for reasoning about sets.

Alas, we must now point at a dissymmetry between membership, as it works over ordinary sets, and the edge relationship over the vertices of a graph. Circular walks in graphs must be admitted, if graphs are to nicely meet their broad spectrum of applications. Membership is a well-founded—and, consequently, an acyclic—relation over sets. How can sets be flexible enough for the edge-to-membership translation to be carried out with naturalness?

Here and there, this book will take into account "extraordinary" sets among which membership makes infinite descending chains of the form $\cdots \in x_3 \in x_2 \in x_1$. A specific axiom which von Neumann proposed in the 1920s forbids such chains; this did not prevent earlier studies on extraordinary sets, carried out in the framework of the preexistent Zermelo-Fraenkel axiomatic set theory, from having a continuation. In the 1980s, Aczel proposed a variant of the Zermelo-Fraenkel theory, where von Neumann's foundation axiom gets replaced by an antithetic axiom leading to a mature theory of non-well-founded sets. This book will mainly focus on well-founded sets; on occasions, though, we will suspend von Neumann's foundation axiom and bring Aczel's anti-foundation axiom into play, whenever extraordinary sets can enrich the interplay between sets and graphs.

We are now ready to enter the Introduction proper. What follows offers a more thorough overview of the subject matter of this book than what precedes. In order to keep the level of exposition as informal and light as possible, for the time being we will touch upon the various issues mainly through examples and images.

1.1 Tiny Motivating Examples

What do the structures shown in Fig. 1.1 represent? Anyone who has received a basic
training on graphs will readily see, here, two *complete graphs*: circles represent
vertices, the lines between them represent edges, and each vertex has any other
vertex as a neighbor. Presumably the line crossings, which are not marked by circles,
carry no meaning whatsoever, so one may want to strive, by means of a smarter
layout, to reduce their overall number. (This is said in order to stress the difference
between a graph as represented and the graph per se.)

Deeper examination of these two structures reveals that some properties, interesting for graph theory, hold for one graph but not for the other: for example, only
the graph on the right admits an *Eulerian circuit* (namely, a circular walk that visits
each edge exactly once); see Fig. 1.2.

On the other hand, both graphs can be oriented in a way devoid of cycles. Inspired
by the latter remark, a mathematically inclined reader may move on to prove that
every complete graph admits an *acyclic* orientation; that, to within isomorphism,
there is exactly one such orientation; and that, under that orientation, *no two vertices
have the same children* and *every grandchild of any vertex is itself a child* of that
vertex. If conversant with sets and with von Neumann's definition of the natural
numbers (which we will soon recall), one might then interpret the oriented edges
as membership and conclude that Fig. 1.1 simply set-theoretically "portrays" the
numbers 4 (as the set $\{0, 1, 2, 3\}$) and 5 (as the set $\{0, 1, 2, 3, 4\}$); see Fig. 1.3.

The graphs in Fig. 1.1, hence, offer a static view of two numbers; Fig. 1.3 shows
the same two numbers as resulting from a process of *counting*. One may contend
that merely redundant information has been added: how can it be of any use?
However, assigning an orientation to the edges of a graph so that they come to

Fig. 1.1 What do these
represent?

Fig. 1.2 An Eulerian circuit
of the graph on the right of
Fig. 1.3. Edges are labeled by
the order in which they
appear in the circuit

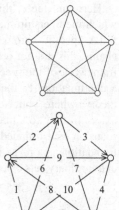

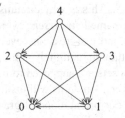

Fig. 1.3 Acyclic orientations of the graphs from Fig. 1.1

express membership amounts to imagining a recondite history of the construction of the graph, of which a suitable bottom-up decoration of the vertices by sets can offer successive snapshots. As this book will repeatedly illustrate (for less special graphs than those considered so far, of course), the successive stages of the history of a graph can sometimes be revealing of certain properties it enjoys—e.g., existence of particular paths across it—which either do not vary or evolve in an easily traceable manner, all over its construction.

As another quick example, consider a directed graph whose vertices are $0, 1, \ldots, n$, and in which $i > j$ holds for each edge $\langle i, j \rangle$ (a condition ensuring acyclicity). It can be shown that there exist distinct sets $s_0, s_1, \ldots, s_n$ such that $s_j \in s_i$ holds if and only if $\langle i, j \rangle$ is an edge. Such s_i's are readily found when each pair $\langle j + 1, j \rangle$ with $0 \leq j < n$ is an edge: simply put, in this case, $s_i = \{s_j : \langle i, j \rangle$ is an edge$\}$ for each i, so that $s_0 = \emptyset$, $s_1 = \{\emptyset\}$, either $s_2 = \{\{\emptyset\}\}$ or $s_2 = \{\emptyset, \{\emptyset\}\}$ holds, and so on.

How far-reaching is the example just seen? Can we, say, adapt it to the graph devoid of orientation resulting from the previous one when every edge $\langle i, j \rangle$ gets superseded by the bidirectional edge $\{i, j\}$? As said in the preamble to this Introduction, we will answer affirmatively by accepting that an undirected graph be represented via a one-one association $v \mapsto s_v$ of sets to its vertices ensuring that $\{u, w\}$ is an edge if and only if either $s_u \in s_w$ or $s_w \in s_u$ holds.

To show that the proposed edge-to-membership translation works better in some cases than in others, suffice it to point back to the example discussed two paragraphs ago: there we could swiftly exhibit the correspondence $i \mapsto s_i$ only after insisting that $\langle n, n-1 \rangle, \ldots, \langle 1, 0 \rangle$ be edges. The ongoing will make this point clearer and clearer.

Entia non sunt multiplicanda praeter necessitatem.[4]

(A formulation of Occam's razor principle.)

Is membership a cycle-free relation over sets? We have taken·this for granted in the foregoing because this is the prevailing view today, but we must warn the reader that niche set theories admitting the contrary exist; moreover, they can ease the modeling of circular phenomena such as the behaviors of finite automata of

---

[4]Entities should not proliferate beyond necessity.

various kinds. This is why circumscribed parts of this book will consider peculiar aggregates, named *hypersets*, which can have membership cycles in they inner structure. In those digressions, even the digraph of Fig. 1.2 will be capable of being decorated (though, definitely, not in a bottom-up fashion) by means of hypersets.

Ordinary sets are supposed to be *extensional*—i.e., distinct sets cannot have the same members—and this will justify our forbidding that two vertices have the same children in our orientations of graphs. Hypersets of the kind we will consider are subject to an equality criterion stronger than extensionality, which, while preventing needless proliferation of them, also contributes to making their exploitation more effective—in particular hypersets can be used, rather straightforwardly, to minimize automata. As will be explained in due course, the only way of interpreting the digraph of Fig. 1.2 as representing membership over a family of hypersets will be by a degenerate assignment sending all vertices to the same hyperset Ω: that unique object which solves the equation $x = \{x\}$. This example points to the crucial feature differentiating cyclic graphs representing sets from the acyclic ones: to wit, the enhanced extensionality criterion for judging which vertices is equivalent.

1.2 Why Sets as Graphs?

This book will mostly deal with hereditarily finite sets, that is, with sets that are finite, and so are their elements, and the elements of their elements, and so on. To simplify our exposition in this section, by set we mean a *well-founded* set, namely, one among whose elements, elements of elements, etc., the membership relation[5] does not form cycles or infinite descending chains of the form

$$\cdots \in x_3 \in x_2 \in x_1 .$$

The collection, which we will denote HF, of all hereditarily finite sets is the union

$$\mathsf{HF} = \bigcup_{n \in \mathbb{N}} \mathsf{HF}_n$$

of the *levels* HF_n—whose subscripts range over all nonnegative integers—determined by the inductive rules

$$\mathsf{HF}_0 = \emptyset,$$

$$\mathsf{HF}_{m+1} = \mathcal{P}(\mathsf{HF}_m),$$

where $\mathcal{P}(s)$ is the power set operator yielding the set of all subsets of the operand s. For example, $\mathsf{HF}_1 = \{\emptyset\}$ and $\mathsf{HF}_2 = \{\emptyset, \{\emptyset\}\}$.

---

[5]The precise extent of the set to which we are restricting membership will be clarified very soon, with the definition of $\mathsf{trCl}\,(\cdot)$.

Under the view that sets can have only other sets as elements, a most natural question is: If we interpret the membership relation among sets as the arc relation among vertices corresponding to each set, what graphs do we get? Is there any further insight by interpreting a set as a graph? Are there new problems to be studied for these graphs?

First, we must "unravel" a set to obtain the collection of its elements, of the elements of its elements, and so on. This set is called the *transitive closure* of a set and is formally defined as:

$$\mathsf{trCl}\,(x) = x \cup \bigcup_{y \in x} \mathsf{trCl}\,(y),$$

so that clearly $\mathsf{trCl}\,(\emptyset) = \emptyset$. Having thus embedded the "full history" of a set x inside $\mathsf{trCl}\,(x)$, let us now consider the graph (V_x, E_x) which has

- $V_x = \mathsf{trCl}\,(x) \cup \{x\}$,
- $E_x = \{\langle u, w \rangle : u \in V_x \wedge w \in V_x \wedge w \in u\}$.

This graph, representing x, will be called the *pointed membership graph* of x; it is *pointed* in the sense that one of its vertices, namely, x, plays a special role—this vertex, called *point*, is easy to detect. The arcs outgoing from the point go toward the elements of x. If we remove x from this pointed graph, then we obtain the *membership graph* of x, which still captures the history of x but may forget which ones, among vertices, are the elements of x. We dwell on this in Example 1.1.

Example 1.1 Let x be the set $\{\emptyset, \{\{\emptyset\}\}, \{\emptyset, \{\emptyset\}\}\}$. By expanding the definition of transitive closure, we obtain $\mathsf{trCl}\,(x) = \{\emptyset, \{\emptyset\}, \{\{\emptyset\}\}, \{\emptyset, \{\emptyset\}\}\}$. Below on the left, we draw the pointed membership graph of x and on the right its membership graph.

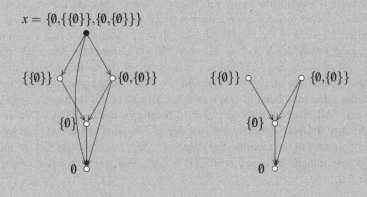

$$x = \{\emptyset, \{\{\emptyset\}\}, \{\emptyset, \{\emptyset\}\}\}$$

(continued)

Example 1.1 (continued)

  Passing now to consider the set $y = \mathsf{trCl}\,(x)$, we observe that $y = \mathsf{trCl}\,(y)$, that is, y has as elements all elements in its own "full history." Its pointed membership graph is shown below. Notice that x and $y = \mathsf{trCl}\,(x)$ have the same membership graph.

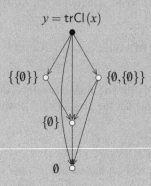

As illustrated just now in Example 1.1, any set y which equals $\mathsf{trCl}\,(x)$ for some set x has all sets in its own "full history" as elements; in formulae, $y = \mathsf{trCl}\,(y)$. Sets satisfying this equation are called *transitive*, and they will play a special role in this book. From a graph-theoretic point of view, in the pointed membership graph of such a y, there is an arc from y to each element of $\mathsf{trCl}\,(y)$; it is hence immaterial whether we work with its pointed or unpointed membership graph.

We are now ready to start giving a flavor of the problems where the graph-theoretic insight pays off. As seen in Example 1.1, the sets x and $\mathsf{trCl}\,(x)$ have the same transitive closure, which amounts to having the same membership graph. The simple question arises: how many sets have the same transitive closure as a given set x? We can easily answer this question in graph-theoretic terms as follows.

Consider a *source* of the membership graph of x, namely, a vertex v with no *in-neighbors*; in formulae, $N^-(v) = \emptyset$, where $N^-(v)$ denotes the set of all vertices that have an outgoing arc to v. Since v is a vertex of the membership graph of x, it must be an element of x, else we would not find it in the graph. Therefore, any set with the same membership graph as x has all sources of its membership graph as elements. All other elements are free to belong or not to the set. Denote by n the cardinality $|\mathsf{trCl}\,(x)|$ (namely, the number of elements of $\mathsf{trCl}\,(x)$) and by s the number of sources of the membership graph of x. We thus conclude that the number of sets whose transitive closure is $\mathsf{trCl}\,(x)$ is 2^{n-s}. We give a concrete example in Example 1.2.

Example 1.2 Let x be the set $\{\emptyset, \{\{\emptyset\}\}, \{\emptyset, \{\emptyset\}\}\}$ from Example 1.1, and recall its pointed and unpointed membership graphs drawn therein. Its membership graph has $n = 4$ vertices, out of which $s = 2$ are sources, namely, $\{\{\emptyset\}\}$ and $\{\emptyset, \{\emptyset\}\}$. Thus, any set with the same transitive closure as x must contain these two sources as elements. The set x also contains $\emptyset$, while $\mathsf{trCl}(x)$ additionally contains both $\emptyset$ and $\{\emptyset\}$. In total, there are $2^{n-s} = 2^{4-2} = 4$ sets with the same transitive closure as x. We draw them here below, this time omitting the vertex labels.

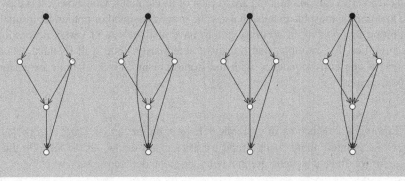

According to its definition, a pointed membership graph seems to be just a syntactic rewriting of a set x, from a sequence of nested uses of the set-constructor $\{\cdot, \ldots, \cdot\}$ to a description in terms of vertices V_x and arcs E_x. In fact, the set E_x is fully described by V_x, and there seems to be no gain in further introducing "redundant" arcs between elements of V_x.

However, the attentive reader will have noticed that in Example 1.2, we have in fact omitted the *vertex labels* from the drawings of the four pointed membership graphs! Yet, this posed no ambiguity in reconstructing the *sets* to which the pointed membership graphs correspond. Since the arc relationship mimics the restriction of membership to a transitive set of sets, we can retrieve the set $\mathsf{m}\,v$ corresponding to each vertex v of the graphs from Example 1.2 by setting $\mathsf{m}\,v = \{\mathsf{m}\,u : u \in N^+(v)\}$, where $N^+(v)$ denotes the set of *out-neighbors* of v, namely, the set of vertices to which there is an outgoing arc from v. Since our sets are well founded, membership graphs are acyclic, and thus the values $\mathsf{m}\,v$ can be constructed "bottom-up." For example, we can easily check that the second graph from the left in Example 1.2 is the pointed membership graph of the set x, as drawn in Example 1.1. The sets thus produced are called *vertex decorations*; the function m sending each vertex v of an acyclic graph to its decorating set $\mathsf{m}\,v$ will be called the *decoration* of the graph.

Through the book, we will repeatedly put to work the process of decorating a graph with a collection of sets. It can, and will, be applied to arbitrary graphs and even, counterintuitively, to graphs endowed with cycles (in what precedes, acyclicity was the only—but seemingly essential—graph property entering into a decoration). To further puzzle the reader, in Example 1.3, we show a different graph whose decoration leads to the same set x as the pointed membership graph seen in Example 1.1. This brings up the question: what is the additional property enjoyed by membership graphs that makes them "canonical representatives"? In other words, for what class $\mathscr{C}$ of graphs the above decorating function is bijective?

The sought-for property is clearly the one that distinct vertices have pairwise distinct sets of out-neighbors. This graph property will be called *extensionality*. All membership graphs (pointed or not) are in fact extensional, since each vertex is a set, and each vertex has an outgoing arc to each of its elements. Conversely, it is easy to see that non-isomorphic extensional acyclic graphs (pointed or not, under a suitable definition of *point* of an arbitrary graph) have distinct sets of vertex decorations. This implies, for example, that transitive hereditarily finite well-founded sets are in one-to-one correspondence with the isomorphism classes of finite extensional acyclic graphs.

Example 1.3 Much as in Example 1.1, let x be the set $\{\emptyset, \{\{\emptyset\}\}, \{\emptyset, \{\emptyset\}\}\}$, whose pointed membership graph we draw again below, on the left. On the right, we draw an acyclic graph having the same decorating sets.

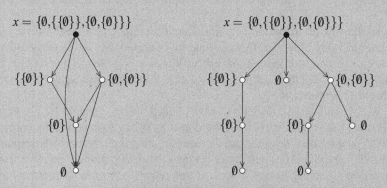

The graph on the left is extensional; the one on the right is not, because it has three *sinks*, namely, three vertices with the null set as out-neighborhood.

A fact worth of notice is that we had to consider *isomorphism classes* of extensional acyclic graphs, in the one-to-one correspondence cited above, because the vertex names in any such graph are immaterial when it comes to decorating it with sets. Since this is important for the rest of this section, let us pause to explain it a bit further in Example 1.4. Unlike the membership graph with which we started—whose vertices describe, by themselves, the entire structure—here the arcs describe the structure in full.

Example 1.4 Let $G = (V, E)$ be the extensional acyclic graph drawn below.

We did not specify "who" the three elements of V are; thus, we did not specify what are the arcs making up E (and consequently we did not show the names of the vertices in the above drawing). Assume however that $V = \{1, 2, 3\}$, where **1**,**2**, and **3** stand for distinct arbitrary names. We can see below that there are six different ways of constructing the set E of arcs so that the resulting graph is isomorphic to the one above. The decorations of all of these six graphs will be the same.

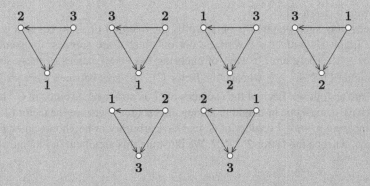

As it turns out in general, for any isomorphism class of an extensional acyclic graph with n vertices, $n!$ is the exact number of extensional acyclic graphs with vertex set $\{1, 2, \ldots, n\}$ that are isomorphic to any fixed representative of the class.

Our discussion has so far established that it is immaterial whether we regard hereditarily finite well-founded sets truly as sets or as special (to wit, as extensional and acyclic) graphs. In Example 1.2 we have seen how a graph-theoretic insight (that of *source* of a graph) helped in solving a simple counting problem.

Let us now look at another counting problem on sets, which we not only address with a graph-theoretic insight but fully solve via graphs. This problem is "given n, find the number of transitive sets with n elements." We will be satisfied if we manage to express the answer to such a problem as a recurrence relation depending on n and possibly other parameters. This problem was first solved in 1962 [92] in set-theoretic terms, using a bijection between sets and numbers (Ackermann's encoding, which we will also define in Sect. 3.3) and an encoding of sets by particular tuples of sets of numbers, which we will also study, in Sect. 6.3.1.

We will offer another solution below, which, besides being exclusively graph theoretic, requires very little additional machinery. This is in fact inspired from the count of acyclic graphs with n vertices, historically found later than the number of transitive sets (1970–1973) [101, 102]. We begin by reviewing the count of acyclic graphs with vertex set $\{1, 2, \ldots, n\}$, as presented, e.g., in [45]. We will then see that the same strategy also works for sets.

Let $a_{n,s}$ denote the number of acyclic graphs with vertex set $\{1, 2, \ldots, n\}$, with the additional property that there are exactly $s \in \{1, \ldots, n\}$ sources. The desired number is, thus, $a_n = \sum_{s=1}^{n} a_{n,s}$. We argue below that:

$$a_{n,s} = \binom{n}{s} \sum_{i=1}^{n-s} (2^s - 1)^i \, 2^{s(n-s-i)} a_{n-s,i}, \qquad (1.1)$$

where $a_{n,n} = 1$, for all $n \geq 1$.

Indeed, any such graph G is obtained by the addition of s sources to a generic acyclic graph G' with $n - s$ vertices, out of which i are sources, for some $i \in \{1 \ldots, n - s\}$. There are $\binom{n}{s}$ ways of choosing the vertex names of these sources from among $\{1, 2, \ldots, n\}$, whence the factor $\binom{n}{s}$. The arcs between these s sources and G' are as follows. Each of the i sources of G' is no longer a source in G'; hence, it must have at least one in-neighbor among the s sources, whence the factor $(2^s - 1)^i$. The remaining $n - s - i$ vertices of G' can have arbitrary in-neighbors among these s sources, whence the factor $2^{s(n-s-i)}$. We illustrate this argument in Example 1.5.

Example 1.5 Let G be the acyclic graph with vertex set $\{1, 2, 3, 4, 5\}$ drawn below.

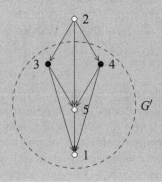

(continued)

Example 1.5 (continued)

Graph G has $s = 1$ sources, so G has been counted by $a_{5,1}$. Following the decomposition (1.1), we see that G was obtained by choosing the name of its source to be named 2 and adding it to the graph G' made up of the vertices $\{1, 3, 4, 5\}$. Graph G' has been counted by $a_{4,2}$, because it has four vertices and $i = 2$ sources (it is immaterial whether the names of its vertices are $\{1, 3, 4, 5\}$ or $\{1, 2, 3, 4\}$). We see that the unique source 2 of G has arcs toward both sources of G', shown in black. Any arc from 2 toward the non-source vertices of G', namely, 5 and 1, is allowed; in this case, we have the arc $\langle 2, 5 \rangle$.

Let now e_n denote the number of extensional acyclic graphs with vertex set $\{1, 2, \ldots, n\}$. As observed at the end of Example 1.3, each isomorphism class of an extensional acyclic graph with n vertices will be accounted for in e_n exactly $n!$ times. Thus, the number of transitive sets with n elements will simply be $e_n/n!$.

As above, let $e_{n,s}$ denote the number of extensional acyclic graphs with vertex set $\{1, 2, \ldots, n\}$ that have exactly $s \in \{1, \ldots, n\}$ sources. If in the recurrence for $a_{n,s}$, we added all s sources at the same time, the extensionality property now mandates that we make a more accurate count, by adding only one source at a time. We will see below that:

$$ e_{n,s} = \frac{n}{s} \left((2^{n-s} - (n-1))e_{n-1,s-1} + \sum_{i=0}^{n-s-1} \binom{s+i}{i+1} 2^{n-1-(s+i)} e_{n-1,s+i} \right), \quad (1.2) $$

where $e_{1,1} = 1$, and we interpret $e_{n,0}$ as 0, for all $n \geq 2$. Indeed, such a graph G is obtained by the addition of a source v to a generic graph G' with $n - 1$ vertices, in two possible ways.

First, G' can have $s - 1$ sources, and thus v must have out-neighbors only among the non-source vertices of G' (so that G has s sources in total). The out-neighborhood of v must also be different from that of the $n - 1$ vertices of G'. Thus, there are $(2^{n-s} - (n-1))e_{n-1,s-1}$ ways to add v to a generic G'.

Second, G' can have $s + i$ sources, for $i \in \{0, \ldots, n-s-1\}$, in which case v must have outgoing arcs toward exactly $i + 1$ sources of G' (so that G has s sources in total). There are $\binom{s+i}{i+1}$ ways of choosing these $i + 1$ out-neighbors of v. Additionally, vertex v can have arbitrary arcs toward the remaining $n - 1 - (s + i)$ vertices of G', since its out-neighborhood thus becomes different from that of the other vertices of G'. Hence, there are $\binom{s+i}{i+1} 2^{n-1-(s+i)} e_{n-1,s+i}$ ways to add v to a generic G'.

In (1.2) we multiply by n because the new source v can have any name among $\{1, 2, \ldots, n\}$, and we divide by s because each such G is obtained in this manner s times, by the addition of each of its s sources. We illustrate this argument in Example 1.6 below.

Example 1.6 Let G be the extensional acyclic graph with vertex set $\{1, 2, 3, 4, 5, 6\}$ drawn below on the left.

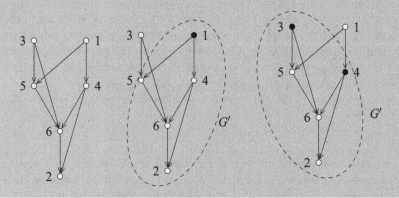

Graph G has two sources; thus, G has been counted by $e_{6,2}$. It can be obtained in two ways, by the addition of each of its sources to a graph with five vertices:

- The first possibility is drawn above in the center: we add source 3 to a graph G' with one source, vertex 1 (shown in black). Source 3 chooses its out-neighbors among the non-source vertices of G', making sure their set is different from the out-neighborhood of all of the five vertices of G'.
- The second possibility is drawn above on the right: we add source 1 to a graph G' with two sources, vertices 3 and 4 (shown in black). Source 1 chooses exactly one among vertices 3 and 4 as out-neighbor (4 in this case). The out-neighborhood of source 1 becomes thus different from that of any other vertices of G', and it is free to choose its remaining out-neighbors among the other three vertices of G'.

In Sect. 6.1, we will study several other strategies for counting transitive sets as extensional acyclic graphs.

In Table 1.1, we show the first ten values of the recurrences a_n and e_n. If one studies closely the values from this table, a certain correlation between a_n and e_n appears. In fact, [112] proved that the proportion between e_n and a_n converges, and this limit is

$$\lim_{n \to \infty} \frac{e_n}{a_n} = 0.326210 \ldots$$

From a practical point of view, this means that in a probabilistic sense, acyclic and extensional acyclic graphs with vertex set $\{1, 2, \ldots, n\}$ are almost the same: if we pick such a random acyclic graph, there is a probability of about 1/3 that it is also extensional.

Table 1.1 The number of acyclic graphs (column a_n) and extensional acyclic graphs (column e_n) with vertex set $\{1, 2, \ldots, n\}$. Column $e_n/n!$ contains the number of transitive sets with n elements

n	a_n	e_n	$e_n/n!$
1	1	1	1
2	3	2	1
3	25	12	2
4	543	216	9
5	29281	10560	88
6	3781503	1297440	1802
7	1138779265	381013920	75598
8	783702329343	258918871680	6421599
9	1213442454842881	398362519618560	1097780312
10	4175098976430597632	13663013921119014400	376516036188

Since in fact any transitive set with n elements is represented by $n!$ such extensional acyclic graphs (recall Example 1.4), then we can actually generate uniformly at random a transitive set in this manner: generate an acyclic graph with vertex set $\{1, 2, \ldots, n\}$ (using, e.g., [61, 62, 106]), check whether it is extensional, and if so, return its decoration with sets, as constructed, e.g., in Example 1.3.

In Chap. 7, we will study this random generation problem more carefully, focusing on generating directly transitive sets (or extensional acyclic graphs), without passing through acyclic graphs first. It is useful here to briefly mention one of these methods for generating transitive sets with n elements (Sect. 7.2). This goes as follows. Start with an arbitrary extensional acyclic graph, and repeat the following simple operations—choose a random pair $\langle u, v \rangle$ of vertices, and:

- if $\langle u, v \rangle$ is an arc of the graph, remove it, if the resulting acyclic graph stays extensional
- if $\langle u, v \rangle$ is not an arc of the graph, add it as arc, if the resulting graphs stay extensional acyclic

If we manage to prove some "special" properties about this process and if it is repeated "long enough," we have that the final graph is generated uniformly at random.

The reason why this example is relevant is that these two operations are purely graph theoretic. In fact, it would be cumbersome to express, let alone implement, them on sets, for removing or adding an element to a set x means that all other sets that have a membership path to x need to change. A computer representation of sets based on graphs is not only more efficient, but it is also in a way more abstract: it allows special useful operations among sets that would be cumbersome to mimic in set-theoretic terms. After all needed operations have been done on this graph abstraction, then the set decoration technique instantiates the set encoded by the graph.

Another example where such abstract reasoning on graphs, rather than sets, is particularly useful is Sect. 3.5. Here we will discuss algorithms to decide whether a given formula about sets (e.g., conjunctions of literals $x = y$, $x \in y$, $x = \emptyset$, where variables $x, y, \ldots$ stand for sets) is satisfiable. We will see that when the formulas are simple enough, the most natural strategy is to start building a graph by interpreting all literals of the form $x \in y$ as arcs of a graph and, then, to reduce the satisfiability problem to that of checking certain properties of this graph, for example, acyclicity. If we also need to produce a satisfying assignment of sets for the variables $x, y, \ldots$, then it basically suffices to decorate its vertices by sets (more details on such construction can be found in Example 1.8). In Example 1.7 below, we briefly illustrate this idea on a concrete set formula.

Example 1.7 Consider the formula

$$a \neq b \wedge a \notin b \wedge b \notin a \wedge (\forall x \in a)(\forall y \in b)(x \in y \vee y \in x).$$

In order to satisfy the formula, we can reason *graph-theoretically* as follows: let us try to build a graph collecting the two membership graphs of a and b. We do this by starting to put two vertices "on the board," say v_a for a and v_b for b. We illustrate this construction step by step, showing on the right the set decorations attached to each vertex and highlighting in black the problematic cases.

Two nodes are necessary because of literal $a \neq b$, and moreover, they cannot be connected by arcs because of $a \notin b$ and $b \notin a$. Using two nodes, however, is not sufficient, since the graph is not extensional: a decoration by sets would result in $a = \emptyset = b$:

$$v_a \circ \qquad \bullet\, v_b \qquad\qquad a = \emptyset \circ \qquad \bullet\, \emptyset = b$$

Hence, we must add at least one out-neighbor to either v_a or v_b, say an out-neighbor u_a of v_a. At this point, v_b and u_a have the same, empty, out-neighborhood. Thus, the literal $a \notin b$ is no longer satisfied:

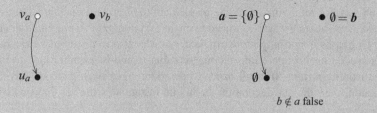

$b \notin a$ false

(continued)

Example 1.7 (continued)

Let us try to fix this by adding this time an out-neighbor of v_b, say u_b. Now, the subformula $(\forall x \in a)(\forall y \in b)(x \in y \lor y \in x)$ imposes an arc between u_b and u_a, say the arc $\langle u_b, u_a \rangle$. Notice, however, that at this point, u_b and v_a have the same out-neighborhood, which this time results in the literal $a \notin b$ being no longer satisfied:

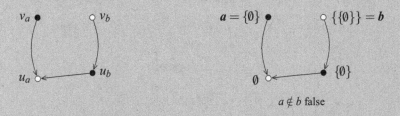

$a \notin b$ false

We must continue to add vertices and enrich our model. One strategy is to add again an out-neighbor for v_a, say w, and, because of the subformula $(\forall x \in a)(\forall y \in b)(x \in y \lor y \in x)$, say the arc $\langle w, u_b \rangle$. We are still not done because w and v_b have the same out-neighborhood and the literal $b \notin a$ becomes false. One way to continue is to add the arc $\langle w, u_a \rangle$, indeed producing a satisfying set assignment for the formula.

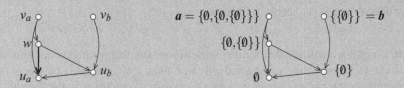

Another, infinitary, strategy is not to add any arc between the out-neighbors of v_a and, likewise, any arc between the out-neighbors of v_b. This results in a membership graph of a and b with a bipartite undirected graph underlying it. As above, the literals $a \notin b$ and $b \notin a$ will prevent our construction from ever breaking down, and we must alternately add one out-neighbor to v_b and one to v_a. We depict below the first stages of this construction, highlighting in black the problematic vertices of each stage.

(continued)

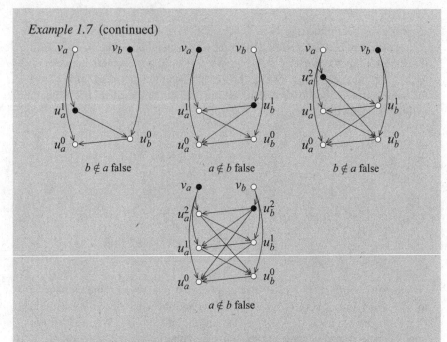

Example 1.7 (continued)

This scenario ultimately leads to a model in which both v_a and v_b are decorated by infinite sets. Adding further conjuncts to our formula, we can in fact force the membership graph of any of its models to have a bipartite undirected underlying graph and thus to be satisfied only by infinite sets. This will be discussed at length in Chap. 8.

In Chap. 8 we will see that for certain set-formulae, this method is not immediately viable, because they are satisfied only by *infinite* sets. A simple example of this phenomenon can be easily built by adding a couple of conjuncts to the formula of Example 1.7. This will be described in full details in Sect. 8.2.1. The perspective of sets as graphs is used somehow originally in Chap. 8. On the one hand, it provides a means to describe and depict infinite sets satisfying low complexity formulae— illustrating a peculiarity of set-theoretic satisfiability with respect to purely logical satisfiability. On the other hand, it provides a technique to *glue* together infinitely many satisfying assignments, thereby allowing to fully describe the entire *spectrum* of a formula (the collection of its models) in a single sentence. This last result is grounded on the decidability proved in [74, 75].

1.3   Why Graphs as Sets?

The most straightfoward way of instituting a correspondence between E and $\in$ consists in representing a directed graph as follows. Choose a collection of sets, one for each vertex of the graph, so that the membership relation between them mimics the edge relation between the corresponding vertices of the graph; to enforce that all sets associated with the vertices are distinct, add a "private" element to each set. The resulting family of sets can be constructed so that they are well founded if the graph has no cycles—see Example 1.8—and they are hypersets in the contrary case.

Example 1.8 Let $G = (V, E)$ be the acyclic graph drawn below

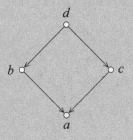

We see that b and c have, in G, the same out-neighborhood: $N_G^+(b) = N_G^+(c) = \{a\}$; in this sense, they *collide*. In order to enforce that the vertex out-neighborhoods are pairwise distinct sets, we can add a "private" vertex denoted $\emptyset_v$ as out-neighbor of each vertex v. We thus obtain the richer graph

$$G' = (V \cup \{\emptyset_v : v \in V\}, E \cup \{\langle v, \emptyset_v \rangle : v \in V\})$$

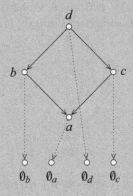

At this point, it does hold that all original vertices of G have pairwise distinct sets of out-neighbors in G'. Stated formally,

$$N_{G'}^+(u) \neq N_{G'}^+(v), \text{ for all distinct } u, v \in V.$$

(continued)

Example 1.8 (continued)

Thus, we can *decorate* each $v \in V$ with the set of all sets decorating the out-neighbors of v (by momentarily assuming that $\emptyset_v$ decorates each new vertex $\emptyset_v$). Since G was an acyclic graph, such decorations can be constructed in a bottom-up manner, as shown below:

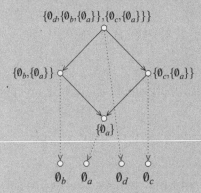

By means of additional vertex and arc insertions into G', one can enforce that *all* vertices of the resulting graph have pairwise distinct out-neighborhoods. A way, one among many, of achieving this is to add all possible arcs between the $\emptyset_v$ vertices so that they induce an acyclic graph (recall Fig. 1.3) and to finally add a sink $\emptyset$ that is an out-neighbor of all vertices $\emptyset_v$. We draw this construction below:

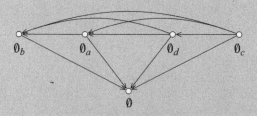

As a matter of fact, the $\emptyset_v$ vertices now have pairwise distinct sets of out-neighbors, and their out-neighborhoods are also distinct from the ones of all $v \in V$, because only they have $\emptyset$ as an out-neighbor.

As for undirected graphs, one enjoys greater freedom in representing them as sets. A naive way is to replace every undirected edge between two vertices x and y by a pair of opposite *arcs* between x and y and to then exploit the above representation technique. The particular graph gotten in this way will be representable only by means of hypersets, though. To act less naively, one will choose, for each edge of the given graph, only one of its two possible orientations and then will resort to the above representation. In order to model graphs by well-founded sets, it will suffice to impose that the resulting orientation be overall acyclic: any undirected graph can, in fact, be so oriented.

An even more natural way of representing undirected graphs as sets, which will also provide valuable insight into the represented graphs, goes as follows. Every undirected graph admits an acyclic orientation in which distinct vertices other than the sinks have distinct sets of out-neighbors. We call such an orientation *weakly extensional*. For example, the graph G' from Example 1.8 is weakly extensional. Look now at Example 1.9, bearing in mind the graphs G and G' from Example 1.8. The orientability property just stated ensures that the resulting orientation, thanks to its weakly extensionality, can be decorated as we have done with G' before. In brief, every finite undirected graph is isomorphic to the undirected graph underlying the (directed) membership relation among a collection of well-founded sets. Noteworthily, the sets in this collection are hereditarily finite.

Example 1.9 Let G be the undirected graph drawn below

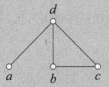

A weakly extensional acyclic orientation of G is the following graph, whose sinks are a and c:

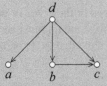

Assuming that a and c get decorated by special sets $\emptyset_a$ and $\emptyset_c$, respectively, its vertex decoration obtained in the same manner as in Example 1.8 is

(continued)

Example 1.9 (continued)

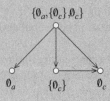

A simple strategy to obtain a weakly extensional acyclic orientation of a undirected graph G is to perform a depth-first visit of it and to orient its edges "forward" along the visit. More precisely, let $v_1, \ldots, v_n$ be the order in which vertices are first encountered during the visit of G: we simply orient each edge $\{v_i, v_j\}$ with $i < j$ as $\langle v_i, v_j \rangle$. For the time being, we explain the correctness of this procedure through the concrete case study, Example 1.10, below. This orientability property is discussed in greater detail in Chap. 5, where we give an alternative inductive proof.

Example 1.10 Let G be this undirected graph:

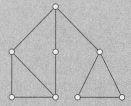

Below, we label the vertices in the order in which they get discovered by a depth-first visit of G starting at the topmost vertex in this drawing. We also highlight, for every vertex v, that edge $\{u, v\}$ across which we arrived at v.

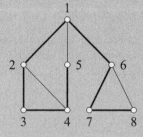

(continued)

Example 1.10 (continued)

By general considerations related to the depth-first character of the visit, the set of highlighted edges must form a forest, made up of as many trees as there are connected components in the given graph. We have in fact gotten only one tree—to be named T_G—for our G is connected. By further general considerations, there can be no *cross-edge* between different subtrees of T_G rooted at the same vertex v (the two endings of such an edge should in fact belong to the same subtree). This well-known property can easily be checked, visually, on the tree above.

The following orientation D of G is the one obtained by orienting each edge from lower to higher index:

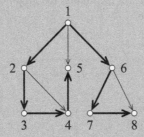

The acylicity of D plainly follows from its construction. Notice in particular that the two sinks of D are the ones labeled 5 and 8.

Suppose now that there is a collision in D between nonsink vertices x and y. Observe that it cannot hold that one is an ancestor of the other in T_G because otherwise there would be a directed path in D among them, which cannot allow collisions to happen. Therefore, x and y belong to different subtrees rooted at some vertex v of T_G. However, since x and y are not sinks, they have at least one common out-neighbor in D, which is also a common descendant of them in T_G, violating the fact that there are no cross-edges in T_G.

The representation of an undirected graph by means of a weakly extensional acyclic orientation is in general more parsimonious that the one obtained by introducing a private element for every vertex. However, it is as parsimonious as its number of sinks. The most parsimonious one, namely, an orientation with exactly one sink, is in fact an extensional one. As we will discuss in Example 1.11, the set decorations of the vertices of such an orientation will then form a *transitive* well-founded set.

In Chap. 4, we will see that it is computationally difficult (i.e., NP-hard) to find an extensional acyclic orientation of an arbitrary undirected graph and, for this reason, so is finding a weakly extensional acyclic orientation with the minimum number of sinks. However, some undirected graphs—which we will call *set graphs*[6]—do admit such an orientation. Among these undirected graphs are the ones which have a Hamiltonian path, namely, a path visiting every vertex exactly once (see Sect. 4.2).

Another historically well-studied class of undirected graphs that are set graphs is the one of *claw-free graphs*. We will provide a proof of this fact in Sect. 4.2 and later, in Chap. 5, will present an experimental verification of that proof by means of an automated deduction system based on Set Theory. Claw-free graphs are the undirected graphs that do not have any induced subgraph isomorphic to the small graph named *claw*. The claw, drawn on page 103, consists of a vertex and three other vertices adjacent only to it. For example, the undirected graph from Example 1.11 below is claw-free (thanks to the edge lying between *c* and *d*).

Example 1.11 Consider again the undirected graph *G* of Example 1.9, which we lay out differently here:

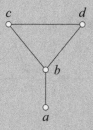

An extensional acyclic orientation of *G* is the following graph *D*:

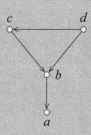

Since *D* is extensional, it has a unique sink and thus we do not need a special set for it. To obtain the sets decorating all vertices, use the simple rule "each vertex is decorated with the set of all sets decorating its out-neighbors":

(continued)

---

[6]Throughout most of this book, the term *graph* means directed graph. However, in keeping with the standard graph-theoretic terminology, in Part II, we will use *graphs* for undirected graphs and *digraphs* for directed graphs.

Example 1.11 (continued)

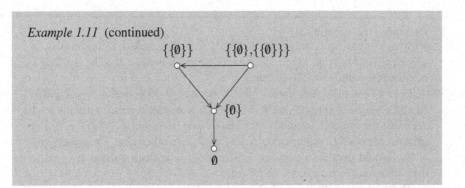

Not only claw-free graphs admit a highly parsimonious representation via a transitive well-founded set, but working with their extensional acyclic orientations is revealing of their structure. First, this will allow us to easily prove the following property, initially discovered in the 1970s:

> Every connected claw-free graph with an even number of vertices
> admits a perfect matching [108, 111].

(A *matching* in an undirected graph G is a collection of edges of G that do not share vertices. A matching is called *perfect* if every vertex of G belongs to one edge of the matching.)

We will discuss this proof at length in Chap. 5, where we will also formalize it in terms suitable for our proof checker knowledgeable on sets. In Example 1.12, we illustrate it through a concrete example.

Example 1.12 Let G be the undirected connected claw-free graph drawn below on the left, and let D be the extensional acyclic orientation of it drawn on its right:

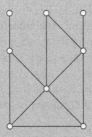 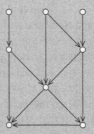

The proof that G has a perfect matching proceeds by induction, starting with the trivial base case of an undirected graph with 2 vertices and 1 edge. The inductive case uses a set-theoretic concept of *rank* of a hereditarily finite set, characterizable as the deepest level of nesting of $\{\cdot, \ldots, \cdot\}$ in its

(continued)

Example 1.12 (continued)
writing. For example, the set $\{\{\emptyset\}, \{\{\emptyset\}\}\}$ from Example 1.11 has rank 3. In an extensional acyclic graph, the rank of a vertex is defined to be the length of a longest path from it to the sink.

Let r be the maximum rank of the vertices of D. In our case, $r = 4$ and in fact all sources of D have rank 4. We can now consider a vertex y of rank $r-1$ such that the in-neighborhood of y is not empty; in formulae, $N^-(y) \neq \emptyset$. The claw-freeness of G implies that $N^-(y) \leq 2$; for, otherwise, y together with $N^-(y)$ would contain a claw in G. The proof now distinguishes two cases, depending on the cardinality of $N^-(y)$.

The first case, illustrated below on the left, is $N^-(y) = \{x\}$. From the choice of the ranks of y and x and the fact that D has a unique sink, it follows that $G - \{x, y\}$ is connected; plainly, it is also claw-free and it has an even number of vertices. Hence, we can apply the inductive hypothesis to it and obtain a perfect matching, as shown below in the center. Together with the edge $\{x, y\}$, this forms a perfect matching for the entire G, shown on the right.

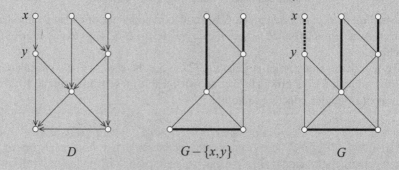

$$D \qquad\qquad G-\{x,y\} \qquad\qquad G$$

The second case, illustrated below on the left, is $N^-(y) = \{x, w\}$. Again, given the uniqueness of the sink of D, we can apply the inductive hypothesis to $G - \{x, w\}$, and obtain a perfect matching M for it, shown below in the center. In M, y must be matched with another vertex, say z, i.e., $\{y, z\} \in M$. Since G is claw-free, at least one of the edges $\{x, z\}$, $\{w, z\}$ must exist in G, otherwise the set $\{w, x, y, z\}$ would induce a claw. Assume for definiteness that $\{x, z\}$ is an edge. Then the perfect matching for G, shown below on the right, is $(M \setminus \{\{y, z\}\}) \cup \{\{x, z\}, \{w, y\}\}$.

(continued)

Example 1.12 (continued)

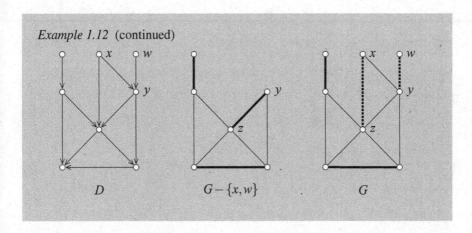

D $G - \{x, w\}$ G

The attentive reader may have noticed that in the proof illustrated in Example 1.12, we have made no use of the extensionality of the orientation of the claw-free graph. What we did use was uniqueness of the sink and the set-theoretic insight behind the concept of rank. This strategy can be reused in many cases when the properties to be proved can benefit from a decomposition of the graphs based on ranks. Another result about claw-free graphs provable by this strategy is:

The square of every connected claw-free graph is vertex pancyclic [60].

The *square* of an undirected graph is obtained from G by additionally connecting any two vertices that have a common neighbor. An undirected graph is *vertex pancyclic* if for every vertex v, it contains cycles of all possible lengths passing through v. A weaker version of this property, namely, that the square of every connected claw-free graph has a Hamiltonian cycle, will also be discussed and formalized with our proof checker in Chap. 5.

Part I
Basics

Chapter 2
Membership and Edge Relations

This book tackles fundamental graph-theoretical notions with a deductive methodology which we will approach from two standpoints, one rooted in Set Theory and one, more directly, in Dyadic Logic. Neither standpoint appears superior to the other one in all circumstances; actually, both are viable for a formally rigorous approach, as this chapter will begin to explain.

This chapter provides, along with the next, the basic terminology that will be used throughout the book. We refrain from being overly technical, and introduce unproblematic notions and terms in a colloquial way. A focus of our discourse will be on axioms for Set Theory, because these are an underpinning—crucial, though often left as implicit—of the machinery which we will play with elsewhere in the book. Our aim is not to uphold or study a specific axiomatization of Set Theory, but to give enough information to convince the reader of the viability of a fully formal approach; in fact, concerning non-secondary issues such as whether infinite sets should enter the game or not, or acyclicity vs. non-well foundedness of membership, we will indicate antithetical options without enforcing any particular choice beforehand.

Recall: We use upper case terms (e.g., Set Theory or Logic) for the informal notions and lower case correspondents (e.g., set theory or logic) for specific axiomatic counterparts of those notions.

2.1   Edge/Membership Languages and Their Structures

Every formal language exploited in this book will have a *signature* consisting of:

- nonlogical relation symbols, typically E for the edge relationship and/or ∈ for membership;
- the logical symbol of equality, =, sometimes regarded as dispensable inasmuch as expressible by other means;
- constants, for example, ∅;
- function symbols, for example, the operators ∩, \, ∪, and $\mathcal{P}$.

Along with these, the usual endowment of first-order logic—propositional connectives (¬, → , etc.), an infinite provision of individual variables, and quantifying signs (∀ and ∃)—will always be at hand. The conjunction connective, namely, ∧ (the dual of disjunction, ∨), will often be replaced by & (better readable); also, the biimplication connective ↔ will sometimes be written as ⇔.

"*Relator*" and "*functor*" are synonymous with "relation symbol" and "function symbol," respectively. Each relator and functor has an *arity*, namely, a positive integer, associated with it: this indicates its number of operands (see below). A relator or functor of arity one or two is called *monadic* or *dyadic*, respectively.

Our typical relators—E, ∈, and =—are dyadic and get used as "*infixes*," namely, in intermediate position between their operands. A dyadic or monadic functor which gets used according to some special notation (e.g., as an infix or as a construct subject to some priority conventions regarding its syntactic interplay with other functors and relators) is often called an *operator*.

Some set-theoretic semantics will be associated with the relator ∈ whenever this symbol will enter into play, not a fixed semantics, though. We will in fact agree upon some collection of set-theoretic axioms—sometimes duly indicated, often clear from the context—when using ∈ in formulae. Such axioms will also give meanings to various set operators.

2.1.1   Terms, Formulae, and Sentences

Definition 2.1 (Well-formed terms and formulae)   *Terms* and *formulae*, the main syntactic categories of our languages, are recursively defined as follows:

1. every individual variable is a term;
2. every constant is a term;
3. $f(t_1,\ldots,t_n)$ is a term when f is a functor of arity n and $t_1,\ldots,t_n$ are terms;
4. $R(t_1,\ldots,t_n)$ is an (atomic) formula when R is a relator of arity n and $t_1,\ldots,t_n$ are terms;
5. $\neg\varphi$ and $(\varphi \to \psi)$, etc. are formulae when φ and ψ are formulae;
6. $\forall v\,\varphi$ is a formula when v is an individual variable and φ is a formula.

(The formation rules 2., 3., and 4. depend, of course, on the signature of the edge/membership language under consideration.)

Fig. 2.1 Higher priority
indicates stronger "cohesive
power"

Construct			Priority
=	∈	E	5
¬		∀ν	4
	∧		3
	∨		2
→	↔	↮	1

This is an abstract characterization of the well-formed expressions of any of
our languages. It is tenable (but, for most of our purposes, immaterial) that every
formula should be seen, rather than as a string, as a directed, ordered tree each
of whose vertices has a label which can be: a variable, a constant, a functor, a
relator, the negation connective ¬, the implication connective →, or a *universal
quantifier* ∀ν. Any vertex which carries a certain label σ is called an *occurrence*
of that σ.

The above formation rules 1–6 enforce, among other constraints, that the
children[1] of any occurrence of σ comply in number with the arity of σ. In particular:
any vertex devoid of children must be labeled with either a variable or a constant;
moreover, any occurrence of ¬, or of a quantifier, must have exactly one child and
that child must be the root of a formula.

Concretely, strings (over an alphabet comprising parentheses) well represent
terms and formulae; and we will indulge to many "pretty printing" devices, such
as the use of parentheses of different sizes and shapes, recourse to operator
precedences in order to use parentheses sparingly (see Fig. 2.1), and macros for
derived constructs such as *existential quantifiers* (of the forms $\exists v$ and $\exists! v$, where
v is a variable), *restricted quantifiers* (of the forms $\forall v \in t$ and $\exists v \in t$, where v
is a variable, and t is a term where v does not occur), and customary propositional
connectives (∧ for conjunction, ∨ for disjunction, ↔ for biimplication). Convenient
abbreviation rules for new symbols are:

$$s \ni t \;\leftrightarrow_{\text{Def}}\; t \in s,$$

$$s \neq t \;\leftrightarrow_{\text{Def}}\; \neg(s = t), \qquad\qquad\qquad s \notin t \;\leftrightarrow_{\text{Def}}\; \neg(s \in t),$$

$$\varphi \wedge \psi \;\leftrightarrow_{\text{Def}}\; \neg(\varphi \rightarrow \neg\psi), \qquad\qquad \varphi \vee \psi \;\leftrightarrow_{\text{Def}}\; (\neg\varphi) \rightarrow \psi,$$

$$\varphi \leftrightarrow \psi \;\leftrightarrow_{\text{Def}}\; (\varphi \rightarrow \psi) \wedge (\psi \rightarrow \varphi), \qquad \exists v\, \varphi \;\leftrightarrow_{\text{Def}}\; \neg\forall v\, \neg\varphi,$$

$$\forall v, w\, \varphi \;\leftrightarrow_{\text{Def}}\; \forall v\, \forall w\, \varphi, \qquad\qquad \exists v, w\, \varphi \;\leftrightarrow_{\text{Def}}\; \exists v\, \exists w\, \varphi,$$

$$\forall v \in t\; \varphi \;\leftrightarrow_{\text{Def}}\; \forall v\, (v \in t \rightarrow \varphi), \qquad \exists v \in t\; \varphi \;\leftrightarrow_{\text{Def}}\; \exists v\, (v \in t \wedge \varphi),$$

$$\exists! v\, \varphi \;\leftrightarrow_{\text{Def}}\; \exists w\, \forall v\, (v = w \leftrightarrow \varphi), \qquad \exists! v \in t\; \varphi \;\leftrightarrow_{\text{Def}}\; \exists! v\, (v \in t \wedge \varphi);$$

In the bottommost left of these, w stands for a variable not occurring in $\exists v\, \varphi$.

---

[1]In this introductory phase, we refer by the colloquial word "child" (of a vertex) to what will be
later introduced with the more official name of *out-neighbor*.

Call *scope* of the occurrence of a monadic symbol (typically a quantifier) the formula rooted in the child of that occurrence. An occurrence of a variable v within a formula ψ is said to be *bound* if it belongs to the scope of some occurrence of the corresponding quantifier $\forall v$; if this is not the case, that occurrence is called *free*.

A formula inside which no variable has free occurrences is called a *sentence*. A standard way of converting a formula φ into a sentence is to prefix a list $\forall v_1 \cdots \forall v_m$ of quantifiers, where $v_1, \ldots, v_m$ are the distinct variables which occur free in φ (sorted, for definiteness, according to some specific criterion). The resulting

$$\varphi^\forall \leftrightarrow_{\mathrm{Def}} \forall v_1 \cdots \forall v_m \, \varphi,$$

is called the *universal closure* of φ; an example is

$$\left(\forall z(z \in x \leftrightarrow z \in y) \rightarrow x = y\right)^\forall \leftrightarrow_{\mathrm{Def}} \forall x \, \forall y \, [\forall z(z \in x \leftrightarrow z \in y) \rightarrow x = y] \, .$$

A *prenex* formula is defined to be one whose syntactic form is

$$Q_1 v_1 \cdots Q_n v_n \, \chi \, ,$$

where (1) each Q_i is either one of the quantifying signs $\forall, \exists$, (2) the variables $v_1, \ldots, v_n$ (with $n \geqslant 0$) are distinct, and (3) no quantifiers occur in χ. The quantifier-free part χ of such a formula is called its *matrix* and the quantifier list $Q_1 v_1 \cdots Q_n v_n$ its *prefix*. There are well-known techniques (cf., e.g., [19, Section 6.3.1]) for rewriting any sentence as a logically equivalent prenex sentence.

Only sentences are eligible as axioms; when an axiom is stated with free variables, its universal closure is understood. Thus, e.g., Panel 2.1 suggests that

$$\forall x, y \left((x \, \mathrm{E} \, y \vee y \, \mathrm{E} \, x) \rightarrow x \, \mathrm{E} \, x\right) \text{ and } \forall x, y, z \left(x \, \mathrm{E} \, y \rightarrow (y \, \mathrm{E} \, z \rightarrow x \, \mathrm{E} \, z)\right)$$

be adopted as the axioms of a theory of *preorders* and that

$$\forall x, y \, (x \, \mathrm{E} \, y \rightarrow y \, \mathrm{E} \, x) \text{ and } \forall x, y, z \left(x \, \mathrm{E} \, y \rightarrow (y \, \mathrm{E} \, z \rightarrow x \, \mathrm{E} \, z)\right)$$

be adopted as axioms for a theory of *equivalence* relations.[2]

---

[2]In the common situation when one wants the domain of the equivalence relation to be the entire universe of discourse, one will also postulate *full reflexivity*: $\forall x \, (x \, \mathrm{E} \, x)$.

Panel 2.1 Equivalence relations and partitions

An *equivalence relation* E enjoys the two properties

$$x \,\mathrm{E}\, y \;\rightarrow\; y \,\mathrm{E}\, x \qquad (symmetry),\text{ and}$$
$$x \,\mathrm{E}\, y \;\rightarrow\; (y \,\mathrm{E}\, z \;\rightarrow\; x \,\mathrm{E}\, z) \qquad (transitivity).$$

These entail that when x is in the domain D of E, that is, when $x \,\mathrm{E}\, y$ holds for some y (which, by symmetry, amounts precisely to saying that x is in the image of E), then $x \,\mathrm{E}\, x$. The latter property, formally statable as

$$(x \,\mathrm{E}\, y \;\vee\; y \,\mathrm{E}\, x) \;\rightarrow\; x \,\mathrm{E}\, x \qquad (reflexivity),$$

taken together with transitivity alone, characterizes *preorders*. Hence, an equivalence relation is a preorder of its domain.

A family $\mathscr{A}$ of nonvoid pairwise disjoint sets is said to be a *partition* of $\bigcup \mathscr{A}$, and its elements—and, occasionally, also the set $\mathcal{P}(\bigcup \mathscr{A}) \setminus \bigcup \mathscr{A}$—are called *blocks* of $\mathscr{A}$. The family $\mathscr{E}(D)$ of all equivalence relations on a fixed domain D is in one-to-one correspondence $E \mapsto D/E$ with the family of all partitions of D: in this correspondence the blocks of D/E, called the *E-classes*, are taken so that

$$x \,\mathrm{E}\, y \text{ holds iff } x \text{ and } y \text{ belong to the same block of } D/E.$$

The preorder $\sqsubseteq$ on $\mathcal{P}(\mathcal{P}(D))$ which is defined by the condition

$$\mathscr{A}_0 \sqsubseteq \mathscr{A}_1 \text{ iff for every } a \text{ in } \mathscr{A}_1, \text{ there is a } B \text{ in } \mathscr{A}_0 \text{ such that } a = \bigcup B,$$

when restricted to the partitions of D, partially orders them; indeed,

$$\text{if } D/E_0 \sqsubseteq D/E_1 \text{ and } D/E_1 \sqsubseteq D/E_0, \text{ then } E_0 = E_1.$$

It is *inductive*, in the sense that when $\mathscr{C}$ is a subfamily of $\mathscr{E}(D)$ such that either $E_0 \sqsubseteq E_1$ or $E_1 \sqsubseteq E_0$ holds for every pair E_0, E_1 in $\mathscr{C}$, then there exist an S and an I in $\mathscr{E}(D)$ such that for every E in $\mathscr{E}(D)$

$$C \sqsubseteq E \quad \text{for every } C \text{ in } \mathscr{C} \text{ iff} \quad S \sqsubseteq E \qquad (supremum),\text{ and}$$
$$E \sqsubseteq C \quad \text{for every } C \text{ in } \mathscr{C} \text{ iff} \quad E \sqsubseteq I \qquad (infimum).$$

The relation $\mathscr{A}_0 \sqsubseteq \mathscr{A}_1$ is read as "$\mathscr{A}_0$ is *finer* than $\mathscr{A}_1$" or as "$\mathscr{A}_1$ is *coarser* than $\mathscr{A}_0$"; obviously $D/E_0 \sqsubseteq D/E_1$ holds iff $E_0 \subseteq E_1$.

It is quite common that an equivalence relation E gets defined on the domain D of a function $r(\cdot)$ by the condition $x \,\mathrm{E}\, y$ iff $r(x) = r(y)$; when moreover $r(r(x)) = r(x)$ holds for every x, i.e., r associates the same element of B to all members of each block B of D/E, then r is called a *choice of representatives*.

2.1.2 Interpretations vs. Models

The most parsimonious languages of the kind described above are endowed with a single nonlogical relation symbol: we will denote[3] them as $\mathscr{L}_E = \{E, =\}$ and $\mathscr{L}_\in = \{\in, =\}$.

Formally speaking, (finite directed) graphs are the finite *interpretations* for the language $\mathscr{L}_E$. That is, no constraint other than finiteness on the interpretation of $\mathscr{L}_E$ is needed when we want to just talk about graphs.

The situation would not differ with $\mathscr{L}_\in$, unless special semantics were attached to $\in$, but if we are to use the symbol $\in$ in accordance with mathematical practice, interpretations must be subtler objects. Formally speaking, (finite) interpretations of $\mathscr{L}_\in$—interpreting $\in$ as the familiar membership—are nothing but the *transitive closure* of a set, namely, the collection of its elements, elements of the elements, and so on, to be introduced right before Definition 2.4. Rigor calls for a precise first-order axiomatization, such as the one set forth below, which specifies *models*. Mostly, in this book, we will simply work with finite transitive sets whose existence can be proved in *any* standard axiomatization of Set Theory. The only noticeable variant to this rule will refer to a membership relation which is *not well founded*, as discussed toward the end of Sect. 2.2.

Remark 2.1 We often indicate by $x \mapsto \boldsymbol{x}$ a function assigning an entity $\boldsymbol{x}$ of the "universe of discourse" to each free variable x of a formula φ belonging either to $\mathscr{L}_E$ or to the language $\mathscr{L}_\in$. Thus, typically, boldface letters designate the vertices of a graph when φ belongs to $\mathscr{L}_E$; in the case when φ belongs to $\mathscr{L}_\in$, they stand for sets.

On the syntactic plane, we indicate that $x_1, \ldots, x_k$ are the free variables of φ by writing $\varphi = \varphi(x_1, \ldots, x_k)$; on the semantic plane, $\varphi(\boldsymbol{x}_1, \ldots, \boldsymbol{x}_k)$ designates the truth value which φ assumes under the assignment $x_1 \mapsto \boldsymbol{x}_1, \ldots, x_k \mapsto \boldsymbol{x}_k$ of values to its free variables. Likewise, when $t = t(x_1, \ldots, x_k)$ is a term, $t(\boldsymbol{x}_1, \ldots, \boldsymbol{x}_k)$ designates a specific entity of the universe of discourse. ⊣

2.2 Set Theories

2.2.1 Axioms

A somewhat rudimentary—yet fully functional, for our purposes—axiomatic set theory is the one consisting of the following axioms. Even narrower

---

[3]As is customary, we will be using the set-theoretic language also as a convenient *meta*language. As an outcome, the first and second occurrence of "=" in "$\mathscr{L}_\in = \{\in, =\}$" differs in meaning.

Panel 2.2 Equality and membership
The equality symbol appears in both $\mathscr{L}_E$ and $\mathscr{L}_\in$. Precise formal treatment of equality would require the introduction (given for granted here) of either logical axioms or specific inference rules that embody full reflexivity, symmetry, and transitivity of the equality relation symbol $=$. One should also implement that substitution of equals for equals does not affect the truth value of the atomic formulae involving any relator other than $=$. For example, regarding E, a suitable equality axiom will take the form

$$\forall z \big((x = z \wedge x \, E \, y \rightarrow z \, E \, y) \wedge (y = z \wedge x \, E \, y \rightarrow x \, E \, z) \big);$$

and, similarly, regarding $\in$.

It is common that nodes x and y in a graph differ (so that they make the atomic formula $x = y$ false) but nevertheless entertain the same edge relationship with all other nodes. A peculiar feature of sets, instead, is that the analogue of this never happens: each set is *uniquely characterized* by the collection of its elements. This fact, usually stated in set theories as the *axiom of extensionality*, will allow us, when convenient to do so, to formally eliminate the symbol $=$ from our basic language $\mathscr{L}_\in$. In essence, this will be achieved by rewriting each atom of the form $x = y$ as $\forall z(z \in x \leftrightarrow z \in y)$—hence, at the cost of introducing extra universal quantifiers.

collections of axioms turn out to be suitable for some applications. A much deeper discussion on axiomatic systems for Set Theory can be found in the first chapter of [54], a text to which we refer the reader for any further reference on this matter.

- Axiom of *extensionality*, providing the semantic link between membership and equality (see Panel 2.2):

$$\forall x, y [\forall z(z \in x \leftrightarrow z \in y) \rightarrow x = y].$$

- Axiom of *foundation*, stating that every set—unless empty—has a $\in$ minimal element:

$$\forall x \big[\exists z(z \in x) \rightarrow \exists y \big(y \in x \wedge \forall w(w \in y \rightarrow w \notin x) \big) \big].$$

- Axiom of *pairing*, the first and most basic device needed for yielding doubleton and singleton sets, as well as ordered pairs:

$$\forall x, y \exists z(x \in z \wedge y \in z).$$

Panel 2.3 Historical notes on Set Theory
Set Theory was initially proposed as a study of infinite sets, its birthplace
being a series of papers published between 1874 and 1884 by Georg Cantor. In
these, Cantor proved the uncountability of real numbers, introduced cardinal
and ordinal numbers, and formulated the celebrated continuum hypothesis.

During the so-called foundational crisis in mathematics, one of the representative contradictions appearing was Bertrand Russell's 1901 antinomy
concerning a set formed precisely by all sets that do not belong to themselves.
Actually, it is the naturalness involved in the spontaneous concept of set that,
unless properly tamed, leads to Russell's and to other similar antinomies.

David Hilbert acknowledged that, on the one hand, Set Theory had pointed
out the necessity to perfect logical theory and that, on the other hand, Set
Theory itself, once established axiomatically, can lie at the foundations of
mathematics.

The axiomatization of Set Theory which has now become standard is the
one presented by Ernst Zermelo in 1908, with later emendations and additions
due to Hermann Weyl, to Abraham Fraenkel and Thoralf Skolem (1922/1923),
and to John von Neumann (1925). This theory is commonly referred to as the
Zermelo-Fraenkel set theory, or **ZF**.

- Axiom of *union*, the device needed to construct the monadic union of a given set:

$$\forall x \exists y \forall z, w\, (z \in x \wedge w \in z \rightarrow w \in y).$$

- The *comprehension* axiom scheme. This is an infinite collection of axioms which
ensure, for every formula φ ($= \varphi(z)$) and every set x, the existence of the subset
of x consisting of exactly those elements z of x which satisfy φ:

$$\left(\forall x \exists y \forall z (z \in y \leftrightarrow \varphi \wedge z \in x)\right)^{\vee}$$

(here y does not occur free in φ, whereas z typically does).

- The *replacement* axiom scheme. This is an infinite collection of axioms which
ensure, for every ψ ($= \psi(z, w)$) which specifies a mapping and every set x, the
existence of a set comprising all elements which have a counterimage in x:

$$\left(\forall x \exists y \left[(\forall z \exists! w\, \psi) \rightarrow \forall z(z \in x \rightarrow \exists w(w \in y \wedge \psi))\right]\right)^{\vee}$$

(here y does not occur free in ψ, whereas z and w typically do).

The full power of the collection of axioms specifying a theory often results from
the combined action of individual axioms. As an example regarding the theory
at hand, consider the axiom of union: taken alone, this statement guarantees the
existence of a set *including* (perhaps strictly) the familiar monadic union of a set;

taken in conjunction with comprehension and extensionality, it allows us to prove existence and uniqueness of *monadic union* $\bigcup x = \bigcup_{y \in x} y$. Other examples are the proof that to any set x there corresponds a *successor* $S(x) =_{\text{Def}} x \cup \{x\}$—the axioms involved in this proof are pairing, union, extensionality, and comprehension—the proof that there exists (by comprehension), and is unique (by extensionality), a set devoid of elements, namely, the *empty* (or "*null*") set, for which we introduce and use the constant $\emptyset$.

To begin seeing the constant $\emptyset$ and the operator $S(\cdot)$ at work, let us state the axiom of *infinity*, to which we will resort almost nowhere (until, in Chap. 8, we will discuss at length various formulations of it):

$$\exists v \left[\emptyset \in v \wedge \forall y \bigl(y \in v \to S(y) \in v \bigr) \right].$$

Panel 2.4 reminds the reader that Zermelo's original formulation of this axiom referred to a simpler way of defining the successor operation; it also highlights that the infinitude of the universe of sets is not a consequence of the axiom of infinity: the axioms of extensionality and pairing, plus any provision that bring the null set into play (e.g., the axiom of comprehension) suffice for that purpose.

Panel 2.4 Zermelo's numerals and the successor theory
The theory of "natural numbers with successor," cf. [33], regards an injective, acyclic function $S(\cdot)$ in whose domain there is only one individual, zero, which is not an S-image (a "successor"). This theory can be autonomously instructed on first-order logic, as follows, by means of two axioms plus denumerably infinitely many instances of an axiom scheme:

$$\forall x, y \left[S(x) = S(y) \to x = y \right],$$
$$\forall y \left[y \neq 0 \leftrightarrow \exists x \left(y = S(x) \right) \right],$$
$$\forall x \left[x \neq \underbrace{S(S(\cdots S(x) \cdots))}_{\text{one or more}} \right].$$

The privileged interpretation is, of course, a structure graphically representable as

$$\overset{0}{\bullet} \longmapsto \overset{1}{\bullet} \longmapsto \overset{2}{\bullet} \longmapsto \cdots \overset{n}{\bullet} \longmapsto \overset{S(n)}{\bullet} \longmapsto \cdots$$

and isomorphic to $\mathbb{N}$. This is the simplest way of modeling the axioms; however, extending this with any number of additional, independent chains

$$\cdots \longmapsto \bullet \longmapsto \cdots \overset{m}{\bullet} \longmapsto \overset{S(m)}{\bullet} \longmapsto \cdots$$

isomorphic to $\mathbb{Z}$ would be perfectly compatible with the axioms.

(continued)

Panel 2.4 (continued)

To carve inside Set Theory an equivalent of the above specification of $S(\cdot)$, observe that specific set-theoretic axioms ensure the existence of sets of the form $\{\cdots\{\varnothing\}\cdots\}$, where nesting is arbitrarily deep but remains anyway finite; an inductive use of extensionality also gives us that

$$\underbrace{\{\cdots\{\ \varnothing\}\cdots\}}_{i\ \text{times}} \neq \underbrace{\{\cdots\{\ \varnothing\}\cdots\}}_{j\ \text{times}} \text{ holds when } i \neq j$$

(e.g., $\varnothing$ and $\{\varnothing\}$ differ, because one has no elements, whereas the other has one; consequently $\varnothing$, $\{\varnothing\}$, and $\{\{\varnothing\}\}$ differ, and so on). This, by itself, shows that the universe of sets is infinite; moreover, it brings us rather close to what we aim at, if we simply define $0 =_{\text{Def}} \varnothing$ and, for all y: $S(y) =_{\text{Def}} \{y\}$. This was Zermelo's implementation of natural numbers [114] (to which von Neumann opposed an alternative standard which today predominates: after the latter, one defines the successor of y to be $y \cup \{y\}$). Now consider the issue: do the sets $\{\cdots\{\varnothing\}\cdots\}$ collectively constitute a set ζ? By postulating that

$$\exists z \left[\varnothing \in z \wedge \forall y \left(y \in z \rightarrow \{y\} \in z \right) \right],$$

Zermelo enforced an affirmative answer to this question: it will suffice, in fact, to use comprehension to intersect all z's that satisfy the statement of the infinity axiom just seen, and thus cut off the smallest superset ζ of $\{\varnothing\}$ which is closed under the operation $y \mapsto \{y\}$. With the infinity axiom available, we can hence surrogate within the realm of sets the theory of natural numbers with successor simply by restricting the quantifiers of its (previously autonomous) language to the set ζ.

Another axiom which will play a marginal role in this book is the *power set* axiom

$$\forall x \exists y \forall z [\forall w(w \in z \rightarrow w \in x) \rightarrow z \in y],$$

which in combination with comprehension and extensionality enables one to prove existence and uniqueness of the set, $\mathcal{P}(x)$, consisting precisely of the subsets of x. Its main interest, in the economy of this book, is that in Chap. 5, we will adopt the following as a plain (though by no means a standard) characterization of *finitude*: a set x is finite if and only if every nonnull set belonging to $\mathcal{P}(\mathcal{P}(x))$ has an inclusion-minimal element. We note in passing that, guided by this notion (and after [109]), one could enforce that only finite sets exist by adopting as an axiom

$$\forall k \left[\exists f \left(f \in k \right) \rightarrow \exists a \in k \ \forall b \in k \left(\left(\forall d \in b \ d \in a \right) \rightarrow b = a \right) \right].$$

Of course, this and the infinity axiom seen above are at odds, so as most one of the two could be adopted. Although we will devote little space to infinite sets in this book, we have reasons to refrain from postulating finitude.

Remark 2.2 A subtle axiom, much debated in the history of Set Theory, is the so-called *axiom of choice*. In its original formulation due to Zermelo, this axiom ensures that to any set p whose elements are nonnull, there corresponds a (choice) function f_p sending each $s \in p$ to an $x \in s$—for an equivalent statement, see Exercise 2.7. A stronger version of this axiom, known as the *universal choice axiom* [36], asserts the existence of a *global* function g which "glues together" all f_p's: in fact, g simultaneously sends each nonnull s in the universe of sets to an $x \in s$.

In the economy of this book, we consider the existence of a universal choice function as being fairly unproblematic. In fact, we will introduce in Chap. 5 a specialized such function named **arb**, which will also witness the well foundedness of membership over the universe of sets.

2.2.2 Extended Syntax

We will often introduce, quite freely, set-theoretic operators whose intended semantics can, nowadays, be regarded as classical. To make our formal treatment mercilessly rigorous, we should take into account each operator *per se* and prove it to be definable from our basic collection of axioms. Anyhow, after each addition of a new operator, the overall expressive power of our theory does not change. Below we mention a few operators that we will most often use throughout this book:

- dyadic union: $x \cup y$;
- set difference: $x \setminus y$;
- dyadic intersection and symmetric difference:
  $x \cap y =_{\text{Def}} x \setminus (x \setminus y)$, $x \bigtriangleup y =_{\text{Def}} (x \setminus y) \cup (y \setminus x)$;
- the "with" operator: $(x \text{ with } y) =_{\text{Def}} x \cup \{y\}$ (the abovementioned $S(x)$ is a special case of usage of this operator);
- the monadic union operator $\bigcup x$;
- the power set operator $\mathcal{P}(x)$.

Many more derived operators and relators can, and will, be introduced via abbreviations. Among the least problematic of them:

- the inclusion relator $\subseteq$;
- k-element collection (where k can be any nonnegative integer): $\{S_1, \ldots, S_k\}$;
- (Kuratowski) ordered pair: $\langle x, y \rangle = \{\{x\}, \{x, y\}\}$;
- Cartesian product: $x \times y =_{\text{Def}} \{\langle z, w \rangle : z \in x, w \in y\}$.

2.2.3   Setformers: A Perspicuous Set-Abstraction Construct

Set formers have several crucial advantages as language elements. First of all, they give us very powerful means for defining most mathematical objects of strategic interest. [···] A second advantage of set formers traces back to the fact that the human mind is 'perception dominated', in the sense that we all depend heavily upon many innate perceptual abilities, which operate rapidly and subconsciously, and by which the conscious (and reasoning) abilities of the mind are largely limited. [···] Where direct perception fails, we must fall back on more tortuous processes of reconstruction and detection, slowing progress by orders of magnitude. Hence the importance of notations, diagrams, graphs, animations, and scientific visualization techniques generally [···] From this point of view, much of the importance of set theory and its set-former notations lies in the fact that their syntax reveals various simplifications and relationships with which the mind operates comfortably.

(Jacob T. Schwartz, [105, p. 14])

The right-hand side of the abbreviating definition, seen last, of Cartesian product deserves a quick digression: a perspicuous device has shown up here, which is not a native construct of predicate calculus, but can be introduced as a conservative extension in any suitably rich set theory. The origin of set abstraction terms of this kind, dubbed *setformers* in the ongoing, can be traced back to *nineteenth*[th]-century logicians (Gottlob Frege and Giuseppe Peano, cf., Fig. 2.2), but their syntax lacks

----

·4 ε

Soit a une Cls; $x\varepsilon a$ signifie «x est un a».

ε est la lettre initiale du mot ἐστί.
Exemples : $9\,\varepsilon\,N^2$ $13\,\varepsilon\,N^2 + N^2$ $2^{61} - 1\,\varepsilon\,Np$
Sur la possibilité de remplacer le signe ε par une autre con-
vention voir F1897 note à la P2.
P signifie «proposition». Ce signe n'est pas un symbole de logique, car il ne se trouve pas dans les formules ; il est une simple abréviation.
Les P catégoriques ne sont pas l'objet du calcul logique.

·5 $\mathfrak{z}$

Soit p une P contenant une lettre x; la formule $x\,\mathfrak{z}\,p$ représente «la classe des x qui satisfont à la condition p».

On peut lire le signe $\mathfrak{z}$ par le mot «qui».
Exemple : $1\,\varepsilon\,x\mathfrak{z}\,(x^2 - 3x + 2 = 0)$

«l'unité est une racine de l'équation entre parenthèses».
Autres ex. :

§quot P1·0 §Dvr P1·0 §mp P2·6 §ϑ P·0 §Med P1·0 §λ P1·0 §q′ P4·0...

Dans la formule $x\,\mathfrak{z}\,p$, la lettre x est apparente.
Les deux signes $x\varepsilon$ et $x\mathfrak{z}$ représentent des opérations inverses.

Fig. 2.2 An excerpt of a writing of the year 1900 in which Peano introduces the class-forming construct as a sort of inverse of membership

a definite standard despite them being in widespread use today. At its simplest, the general syntax we adopt for setformers goes as follows:

Definition 2.2 (Syntax of setformers) A *setformer* is a term of the form

$$\{e : v_0 \, c_0 \, t_0, \, v_1 \, c_1 \, t_1, \ldots, v_n \, c_n \, t_n \mid \varphi\},$$

where (1) each c_i stands for either $\in$ or $\subseteq$, (2) $e \, (= e(v_0, \ldots, v_n))$ and $\varphi \, (= \varphi(v_0, \ldots, v_n))$ are a set term and a condition inside which the pairwise distinct variables v_i may occur free, and (3) none of the terms t_j is allowed to involve any of the variables $v_j, \ldots, v_n$ whose subscripts exceed $j - 1$.

The formulae $v_i \, c_i \, t_i$ occurring between ": " and " | " are called *iterators*; each of them renders its v_i a bound variable through the entire setfomer.

We can write, for example,

$$\mathcal{F}(x) = \{y : y \subseteq x \mid \mathsf{Finite}\,(y)\},$$
$$\bigcup x = \{z : y \in x, \, z \in y \mid \mathtt{true}\},$$

in order to form the set of all finite subsets of x and, respectively, the set of all elements of elements of x, dubbed the *unionset* of x. Similarly, in a context where V and A collect, respectively, vertices and arcs of a graph, we can ban arcs of the form $\langle u, u \rangle$ by requiring that

$$A \subseteq \{\langle u, w \rangle : u \in V, \, w \in V \mid u \neq w\}.$$

Taking advantage of setformers, we will frequently introduce abbreviations such as the one shown in the following, somewhat self-referential, example. Suppose that by means of the notation $\mathsf{In\_Graph}(G)$, we want to state that G is an ordered pair whose first component is a set V and whose second component is the converse of the membership relation $\ni$ among elements of V, represented as a set of ordered pairs. Here is the symbolic specification we propose:

$$\mathsf{In\_Graph}(G) \; \leftrightarrow_{\mathrm{Def}} \; \exists V, E \, [G = \langle V, E \rangle \wedge E = \{\langle x, y \rangle : x \in V, \, y \in V \mid x \ni y\}].$$

It is time to clarify beyond doubts what is the designation of a setformer (cf. [105, pp. 13–16]):

Definition 2.3 (Semantics of setformers) The law

$$\forall w \left(w \in \{e : v_0 \, c_0 \, t_0, \, v_1 \, c_1 \, t_1, \ldots, v_n \, c_n \, t_n \mid \varphi\} \leftrightarrow \right.$$
$$\left. \exists v_0, \ldots, v_n \left(\left(\bigwedge_{i=0}^{n} v_i \, c_i \, t_i \right) \wedge \varphi \wedge w = e \right) \right)$$

conveys the meaning of the setformer $\{e : v_0 \, c_0 \, t_0, \, v_1 \, c_1 \, t_1, \ldots, v_n \, c_n \, t_n \mid \varphi\}$ introduced by Definition 2.2.

This makes it evident that the iterators $v_i\, c_i\, t_i$ bind the variables v_i in a way closely akin to existential quantifiers.

Example 2.1 The above-seen specifications of $\mathcal{F}(x)$ and $\bigcup x$, by the law laid down with Definition 2.3, yield:

$$\forall w\big(w \in \mathcal{F}(x) \leftrightarrow \exists y\,(y \subseteq x \wedge \mathsf{Finite}\,(y) \wedge w = y)\big),$$
$$\forall w\big(w \in \textstyle\bigcup x \leftrightarrow \exists y \in x\, \exists z \in y\,(w = z)\big).$$

Example 2.2 Let us define, in the light of Kuratowski's definition of ordered pair:

$$\mathbf{img}\,(R) \leftrightarrow_{\mathrm{Def}} \{\,y : p \in R,\; x \in \textstyle\bigcup p,\; y \in \bigcup p \mid p = \langle x, y\rangle\,\}.$$

Thus, by Definition 2.3 giving us that

$$\forall w\big(w \in \mathbf{img}\,(R) \leftrightarrow \exists p \in R\, \exists x \in \textstyle\bigcup p\, \exists y \in \bigcup p\,(p = \langle x,y\rangle \wedge w = y)\big),$$

$\mathbf{img}\,(R)$ will designate the set of all 2nd components of pairs belonging to R.
Note that this formula can be rewritten as

$$\forall w\big(w \in \mathbf{img}\,(R) \leftrightarrow \exists p \in R\, \exists x\, \exists y\,(p = \langle x,y\rangle \wedge w = y)\big),$$

where two existential quantifiers are unrestricted; nevertheless, we are required to restrict x and y by means of iterators in the setformer.

Remark 2.3 In the general setformer $\{\,e : v_0\, c_0\, t_0, v_1\, c_1\, t_1, \ldots, v_n\, c_n\, t_n \mid \varphi\,\}$ introduced above, either the subpart "$e :$" or the "subpart $\mid \varphi$" (but not both) can be omitted in certain cases:

- The condition φ (together with the "$\mid$" that precedes it) is often omitted, in which case the understanding is that φ is a true sentence, e.g., $\emptyset = \emptyset$ (this convention legitimizes, in particular, the definition of Cartesian product seen short ago).
- The term e (together with the "$:$" that follows it) is often omitted, in which case the understanding is that e coincides with the variable v_0.

It thus makes sense to specify the *powerset* operator, to shorten the above-seen specifications of $\mathcal{F}(x)$ and unionset, and to specify the dual of unionset, monadic *intersection*, as follows:

$$\mathcal{P}(x) = \{y : y \subseteq x\},$$
$$\mathcal{F}(x) = \{y \subseteq x \mid \text{Finite}(y)\},$$
$$\bigcup x = \{z : y \in x, z \in y\},$$
$$\bigcap x = \{z \in \bigcup x \mid \forall y \in x\,(z \in y)\}.$$

⊣

Remark 2.4 Occasionally, we will relax the syntax of setformers, e.g., to define *domain*, *multi-image*, *back image*, and *composition* operators in the manner shown below, leaving it as understood how things could be written down in agreement with our official syntax and in analogy with the above Example 2.2 (see Exercise 2.9):

$$\mathbf{dom}\,(R) =_{\text{Def}} \{x : \langle x, y \rangle \in R\},$$
$$R\vec{\upharpoonright}D =_{\text{Def}} \{y : \langle x, y \rangle \in R \mid x \in D\},$$
$$R\overleftarrow{\upharpoonright}D =_{\text{Def}} \{x : \langle x, y \rangle \in R \mid y \in D\},$$
$$R \circ S =_{\text{Def}} \{\langle x, z \rangle : \langle x, y \rangle \in R, \langle y, z \rangle \in S\}.$$

Here and there, other deviations from the official syntax will appear without pedantic explanations; e.g., we expect that the reader will easily figure out by her-/himself that the setformer

$$\{\langle Y_i, Y_j \rangle : i, j \in \{1, \ldots, \kappa\}, u \in Y_i, w \in Y_j \mid \langle u, w \rangle \in E\}$$

stands for

$$\{\langle Y_i, Y_j \rangle : Y_i \in \{Y_1, \ldots, Y_\kappa\}, Y_j \in \{Y_1, \ldots, Y_\kappa\}, u \in Y_i, w \in Y_j \mid \langle u, w \rangle \in E\}.$$

⊣

Panel 2.5 Sets and computer science

The ease and conciseness with which sets can render complex mathematical or abstract objects have been the motivation behind the aim of representing information in a set-theoretic manner.

A new field, today called Computable Set Theory, emerged as a long-term research project initiated by Jacob T. Schwartz in the 1970s with the intention of cross-fertilizing Set Theory and Computer Science. This has led, on the one hand, to set-based programming languages such as SETL [104] or the more recent {log} [29] and CLP($\mathscr{SET}$) [31]. On the other hand, it has uncovered decidable fragments of set theory together with decidability algorithms implementable on a computer [19, 88].

One emblematic example is the **Multi-Level-Syllogistic with Singleton** fragment and its enaction into the proof-checker Referee/ÆtnaNova [72, 105]. Referee, which we will present in Chap. 5, assists its users in the development of computer-verified proofs of mathematical facts.

2.2.4 Transitive Sets

Yet another way of introducing new constructs, based on $\in$-*inductive* definitions, relies on the axiom of foundation. A basic example—of utmost importance for us— is the *transitive closure* of a set x:

$$\mathsf{trCl}\,(x) \; =_{\mathrm{Def}} \; x \cup \bigcup\nolimits_{y \in x} \mathsf{trCl}\,(y).$$

Thus, e.g., $\mathsf{trCl}\,(\emptyset) = \emptyset$; moreover, $x \subseteq \mathsf{trCl}\,(x)$ holds in general. When the converse inclusion $\mathsf{trCl}\,(x) \subseteq x$ also holds, all sets involved in a (bottom-up, set-theoretic) construction of x, belong to x. When x admits such a complete self-description, it is said to be *transitive*:

Definition 2.4 (Transitive set) A set x is *transitive* if $x = \mathsf{trCl}\,(x)$.

See also Example 3.1 for an example of these definitions.

We can capture with equal ease the notion of *hereditarily finite set* in terms of the transitive closure operation:

Definition 2.5 (Hereditarily finite set) A set x is *hereditarily finite* if its transitive closure is finite.

(We will return more accurately to this in Sect. 3.1 (see also Chap. 5, bottom of Fig. 5.2.)

Very peculiar hereditarily finite sets can be used to represent natural numbers either *à la* Zermelo (see Panel 2.4) or *à la* von Neumann (see what follows).

Among transitive sets, the von Neumann ordinals (cf. Fig. 2.3) are particularly important. They are used to represent all possible well orderings up to order isomorphism.[4]

```
0 =: 0,
1 =: (0),
2 =: (0, (0)),
3 =: (0, (0), (0, (0))),
    · ··    ·
ω =: (0, (0), (0, (0)), (0, (0), (0, (0))), .....),
ω + 1 =: (0, (0), (0, (0)), ... ., (0, (0), (0, (0)) .....)),
    ·    ·    ·.    ·
```

Fig. 2.3 Beginning of the sequence of ordinal numbers (From [68])

---

[4]Recall that a *well-ordering* is a dyadic relation $\lhd$ over a set S such that (1) $x \lhd x$ is false, for any x in S; (2) $x \lhd z$ follows from $x \lhd y$ and $y \lhd z$, for all x, y, z in S; and (3) for each nonnull subset T of S, there is an m in T such that $m \lhd t$ holds for every t in $T \setminus \{m\}$.

Definition 2.6 (von Neumann's ordinals) A transitive set well ordered by the membership relation $\in$ is a *(von Neumann) ordinal*.

We will return to this notion, as perfected by Raphael M. Robinson [100], in Fig. 3.1. For an alternative characterization of the *finite* von Neumann ordinals, which represent natural numbers, see Exercise 2.10.

Specially important, among ordinals, are the cardinal numbers:

Definition 2.7 (Cardinals and cardinality) A *bijection* is a set b satisfying the condition

$$b \subseteq \mathbf{dom}\,(b) \times \mathbf{img}\,(b) \wedge \forall z \exists y \exists x \left(b \upharpoonright^* \{z\} \subseteq \{y\} \wedge b \upharpoonright^{-1} \{z\} \subseteq \{x\}\right).$$

A *cardinal* is an ordinal c such that there is no bijection b satisfying

$$\mathbf{img}\,(b) \in c = \mathbf{dom}\,(b).$$

The *cardinality* of a set x, to be denoted $|x|$, is the cardinal c such that there is a bijection b satisfying

$$\mathbf{dom}\,(b) = c \wedge \mathbf{img}\,(b) = x.$$

Remark 2.5 To better understand this definition, think of $\in$ as of the comparator "less than" between ordinals and take into account that every element of an ordinal can be proved to be an ordinal (cf. Fig. 3.1).

In the light of this fact, the above definition says that a cardinal is an ordinal that cannot be put in one-one correspondence with any smaller ordinal.

Exercise 2.11 asks the reader to prove that every set x—possibly infinite—can be put in one-one correspondence with some ordinal o_x; in its light, the above definition implies that the cardinality of x—intended as its number of elements—is the smallest ordinal,

$$|x| = \bigcap \{o \in o_x \cup \{o_x\} \mid \text{there is a bijection } b \text{ with } \mathbf{dom}\,(b) = o \text{ and } \mathbf{img}\,(b) = x\},$$

which can be put in one-one correspondence with x.

2.2.5 Anti-Foundation

There are situations in which our basic axiomatic machinery is not powerful enough to enable the introduction of meaningful (and useful) set-theoretic objects. To profitably handle such situations, one or more axioms must be added. Consider, for example, the (abovementioned) set Ω, solution to the equation:

$$x = \{x\}. \tag{2.1}$$

Any solution to the above equation not only falsifies the axiom of foundation but challenges our very basic intuition of what a set is. We are now in a position to explain a bit more precisely on *why* this is the case. One root of the problem lies in the axiom of extensionality, stating that two sets are equal if and only if they have the same elements.[5] Foundation and extensionality can be jointly used—on *(hereditarily) finite* objects—to infer the existence of a mechanical equality check, as a consequence of the fact that any descending membership chain $x_1 \ni x_2 \ni \cdots$ must end. The existence of Ω—if we opt in favor of it—puts us in a position in which not only the foundation axiom must be dropped, but also extensionality must be reconsidered.

Unless one assumes $\in$ to be a well-founded relation, it is in fact convenient to replace the foundation axiom by Aczel's *anti-foundation* axiom, often called AFA [2]. Anti-foundation enforces that every system of set-theoretic equations of the form:

$$\begin{cases} x_1 = \{x_{1,1}, \ldots, x_{1,m_1}\}, \\ x_2 = \{x_{2,1}, \ldots, x_{2,m_1}\}, \\ \vdots \quad \vdots \qquad \ddots \\ x_n = \{x_{n,1}, \ldots, x_{n,m_n}\}, \end{cases}$$

with $x_{i,k}$ among $x_1, \ldots, x_n$, has *one and only one* set-theoretic solution. When adopted in its full extent, which permits n and the m_i's to be infinite (cf. Panel 2.6), anti-foundation subsumes extensionality; but this book will usually refer AFA to *finite* systems of set-theoretic equations. AFA legitimizes introduction of a constant Ω to be interpreted as the one and only self-singleton: $\Omega = \{\Omega\}$.

Remark 2.6 Various axiomatizations of Set Theory have been proposed where membership is not well founded; such variants differ in the criterion used to enforce set equality. Bisimilarity is the *coarsest* equivalence relation on graph nodes, and consequently, it induces the largest possible equality relation on (hyper-)sets.

2.3 Graph Theories

In this book a *graph*, namely, a pair $G = (V, E)$, will be a structure providing a finite interpretation for the language $\mathscr{L}_E$. Note that, by the very semantics of first-order logic,

$$V \neq \emptyset \text{ and } E \subseteq V \times V$$

[5]To be completely precise, the axiom spells out only one of the two implications. The reverse, however, follows from the standard basic assumptions made on the equality symbol.

> **Panel 2.6** Anti-foundation as a sentence
> The anti-foundation axiom was introduced by Aczel in [2]; Barwise and Moss
> [9, p. 5] indicate the paper by Forti and Honsell [40] as a precursor of Aczel's
> set theory.
>   How can one be fully formal in stating AFA? Expressing it as a first-order
> sentence is easier if we allow us to use the syntactic device of *setformers*.
> Aczel [2] does not propose a sentence for AFA, as we do here:
>
> $$\forall V \, \forall E \, \exists! \, \ell \, (\ell = \{\, \langle x, \{\eta : y \in V, \, \langle y, \eta \rangle \in \ell \mid \langle x, y \rangle \in E\}\rangle : x \in V\}).$$
>
> By expanding the quantifier $\exists! \, \ell$ according to its defining macro, we get
>
> $$\forall V \, \forall E \, \exists f \, \forall \ell \, (\ell = f \; \leftrightarrow \; \ell = \{\langle x, \{\eta : y \in V, \, \langle y, \eta \rangle \in \ell \mid \langle x, y \rangle \in E\}\rangle : x \in V\}),$$
>
> whose implication " $\rightarrow$ " corresponds to anti-foundation proper, whereas the
> implication of opposite orientation is sometimes called *hyperextensionality*.
>   The term "hyperset" was coined by [8]. In [2, 57], hyperextensionality is
> called "strong extensionality."

must hold. Unless indicated to the contrary, a graph is finite and the *edge* (or
adjacency or *arc*) relation is to be considered the one of a *directed* graph. *Un*directed
graphs can be captured, formally, by means of an axiom stating irreflexive symmetry
for the relation E:

$$\forall x, y [x \, E \, y \rightarrow (x \neq y \land y \, E \, x)];$$

this makes the basic *theory* of undirected graphs. Other significant graph theories
often result from *forbidding* specific subgraphs. For example, the theory of acyclic
graphs can be introduced by means of the following scheme which forbids, for any
positive integer n, (directed) cycles involving n nodes:

$$\forall x_1, \ldots, x_n \neg (x_1 \, E \, x_2 \land \ldots \land x_{n-1} \, E \, x_n \land x_n \, E \, x_1).$$

These same axioms, when referred to ($n \geq 3$ pairwise distinct nodes in) undirected
graphs, forbid undirected cycles, thus giving us (unrooted) *trees*. Likewise, if in
undirected graphs we forbid induced paths of length 2 by requiring

$$\forall x, y, z \neg [x \, E \, y \land y \, E \, z \land \neg (x \, E \, z)],$$

Panel 2.7 Graphs and set representations

The origins of graph theory can be traced back to the 1736 paper of Leonhard Euler on the Königsberg bridge problem. For many years, graph theory remained a subdiscipline of combinatorics. This ceased being the case in the last century, partly due to the tremendous number of applications of graphs that have been found. This has not been incidental, since, like sets, graphs are among the most natural mathematical structures. For this reason, it is difficult to give an immediate definition of a graph without falling into one of the two extremes: either a formal definition or just a delegation to a synonym— e.g., network, map, list of adjacencies. Graphs have been usually considered finite, and it is in fact their finite combinatorics that gives rise to deep and difficult problems. Many problems ask what properties are enjoyed by graphs as a consequence of some structural property. For example, the even degree of every vertex of a graph and its connectedness guarantee the existence of an Eulerian tour [35]; the lack of an odd cycle in a graph ensures that it is bipartite [53]; the lack of an odd-induced cycle of length at least five, and of the complement of such a cycle, ensures that a graph is perfect [23]; the fact that a graph has no induced subgraph isomorphic to the complete bipartite graph $K_{1,3}$ ensures that certain otherwise difficult computational problems are tractable, for example, the independent set problem [66, 103]. Graphs make no exception from most of mathematics, and their formal definition involves the concept of set: a graph is in fact a set of vertices paired with a set of ordered or unordered pairs of vertices. The axiomatic foundation of Set Theory becomes apparent when studying graphs endowed with infinitely many vertices.

we obtain the vertex-disjoint unions of complete graphs, that is, all graphs of the form $(V_1 \cup \cdots \cup V_m,\ E_1 \cup \cdots \cup E_n)$, where $E_i = \{\{x, y\} \mid x \in V_i,\ y \in V_i \setminus \{x\}\}$ holds for each i and $V_i \cap V_j = \emptyset$ holds when $i \neq j$.

A graph theory that will play an important role in this book is the theory of *claw-free* graphs: the sentence

$$\forall x, y, z, w[(y \mathbin{E} x \wedge z \mathbin{E} x \wedge w \mathbin{E} x \wedge w \neq y \wedge y \neq z \wedge z \neq w) \rightarrow (y \mathbin{E} z \vee y \mathbin{E} w \vee z \mathbin{E} w)],$$

characterizes, among undirected graphs, the ones which have no subgraphs isomorphic to the claw $K_{1,3}$ (drawn on p. 103).

After having thus hinted at some graph-theoretic terminology, in the ongoing, we will be much more thorough in this respect. We still refer the interested reader to [7, 113] for further graph theoretic notions.

Let $G = (V, E)$ be a graph. We will often use the notations $V(G) =_{\text{Def}} V$ and $E(G) =_{\text{Def}} E$. The elements of V will be called *nodes* or *vertices*. If the relation E holds between two nodes u and v, namely, $u \mathbin{E} v$, we will indicate this by the notation $\langle u, v \rangle \in E$, thinking of E as a set of ordered pairs. When dealing with an undirected graph, we can think of E as a set of doubletons and hence will write $\{u, v\} \in E$ if there is an edge between u and v. Before going further, it is useful to discuss (directed) graphs and undirected graphs separately.

2.3.1 Directed Graphs

In this book, a *graph* is a directed graph, unless the contrary is stated explicitly (as we will do in Part II). We now introduce some basic notions about graphs; see Example 2.4 for some examples of these.

Let $D = (V, E)$ be a graph. When $e = \langle u, v \rangle$ is an arc, namely, $e \in E$, we say that e *is from* u *to* v, that e is *outgoing from* u, and that e is *incoming to* v. When $e = \langle u, v \rangle$ is an arc, we also say that v is an *out-neighbor* of u and that u is an *in-neighbor* of v. Notice that we allow *self-loops* in a digraph, that is, arcs of the form $\langle v, v \rangle \in E(D)$, but forbid that two arcs both are from u to v (this is implicit in the equality criterion between ordered pairs: $\langle u, v \rangle = \langle u', v' \rangle \leftrightarrow u = u' \wedge v = v'$). A digraph without self-loops will be called *simple*.

Given a vertex $v \in V(D)$, the *out-neighborhood* of v is the set of its out-neighbors $N_D^+(v) =_{\text{Def}} \{u \in V(D) \mid \langle v, u \rangle \in E(D)\}$. The *in-neighborhood* of v is the set $N_D^-(v) =_{\text{Def}} \{u \in V(D) \mid \langle u, v \rangle \in E(D)\}$. If the *in-degree* of $v \in V(D)$, namely, $d_D^-(v) =_{\text{Def}} |N_D^-(v)|$, is 0, we say that v is a *source* in D, while v is a *sink* in D if its *out-degree* $d_D^+(v) =_{\text{Def}} |N_D^+(v)|$ is 0. We may skip the subscript D when this is clear from the context.

Containment relations and operations. If H is a graph with $V(H) \subseteq V(D)$ and $E(H) \subseteq E(D)$ we say that H is a *subgraph* of D. If $W \subseteq V(D)$, $D[W]$ is the graph $(W, \{\langle u, v \rangle \in E(D) \mid \{u, v\} \subseteq W\})$, called the subgraph of D induced by W. If $H = D[V(H)]$, then H is an *induced subgraph* of D. We write $D - W$ for the graph $D[V(G) \setminus W]$. When $W = \{v\}$, we simply write $D - v$, instead of $D - \{v\}$. If $F \subseteq V(D) \times V(D)$, we write $D + F$ for the graph $(V(D), E(D) \cup F)$; analogously, $D - F$ is the graph $(V(D), E(D) \setminus F)$.

Given graphs D_1 and D_2, we say that D_1 and D_2 are *isomorphic* if there exists a bijection $f : V(D_1) \rightarrow V(D_2)$ such that $\langle u, v \rangle \in E(D_1)$ if and only if $\langle f(u), f(v) \rangle \in E(D_2)$. In this case, f is called an *isomorphism* (between D_1 and D_2). An *automorphism* is an isomorphism between a graph and itself. It is well known that (under the operation of composition) the set of all automorphisms of a graph G forms a group, denoted here by $\mathfrak{aut}(G)$ and referred to as the *automorphism group* of G.

A *bijective labeling function* of a graph $D = (V, E)$ with labels Λ is a function $\lambda : V \rightarrow \Lambda$, where $|\Lambda| = |V|$. We will say that a *labeling* of D with labels Λ is the graph $(\Lambda, \{\langle \lambda(u), \lambda(v) \rangle \mid \langle u, v \rangle \in E\})$, where λ is a bijective labeling function of G with labels Λ. By bringing into play the automorphism group of D, it is possible to determine the number of *distinct* labelings with labels Λ of D (i.e., not having the same set of arcs). The proof of the following result can be found in, e.g., [10, Ch. 9]. See also Example 2.3.

Lemma 2.1 *Let D be a (directed) graph on n vertices. The number of distinct labelings of G with labels $\Lambda = \{1, \ldots, n\}$ is*

$$\frac{n!}{|\mathrm{aut}(D)|}.$$

Example 2.3 Let D be the following graph:

We can easily verify that D has exactly two automorphisms, namely, the identity and the one interchanging the two sinks. Thus, by Lemma 2.1, there are $3!/2 = 3$ distinct labeling of G with labels $\{1, 2, 3\}$:

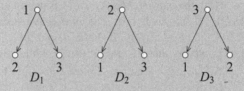

where $E(D_1) = \{\langle 1, 2\rangle, \langle 1, 3\rangle\}$, $E(D_2) = \{\langle 2, 1\rangle, \langle 2, 3\rangle\}$, $E(D_3) = \{\langle 3, 1\rangle, \langle 3, 2\rangle\}$. However, the graph H drawn below

has only the identity as automorphism; thus, it has $3!/1 = 6$ distinct labelings with labels $\{1, 2, 3\}$:

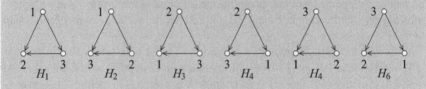

Connectivity. A *(directed) path* is a graph $P = (V, E)$ of the form

$$V = \{v_1, v_2, \ldots, v_k\}, \quad E = \{\langle v_1, v_2\rangle, \langle v_2, v_3\rangle, \ldots, \langle v_{k-1}, v_k\rangle\},$$

where $k \geqslant 1$ and the v_i's are *pairwise distinct*. We say that P *is from* vertex v_1 to v_k and that v_1 and v_k are its *end vertices*. P is also called an *s-t path*, where $s = v_1$ and $t = v_k$. We also say that v_k is *reachable* from v_1. The length of P is the number of its arcs, that is, $k - 1$. A path of length 0 is called *trivial*. We will often refer to a path by the natural order of its vertices and write $P = (v_1, v_2, \ldots, v_k)$. We say that a path P *is in* a graph D if P is a subgraph of D. We usually denote directed paths on k vertices with P_k.

A graph in which every vertex is reachable from any other vertex is said to be *strongly connected*.

Basic properties of graphs. If $P_k = (v_1, v_2, \ldots, v_k)$ is a directed path and $k \geqslant 3$, then the graph $C_k = P_k + \{\langle v_k, v_1\rangle\}$ is called a *(directed) cycle*. We will often designate a cycle by any one of the natural orders of its vertices and write, e.g., $C_k = (v_1, v_2, \ldots, v_k, v_1)$. A graph having no directed cycle as subgraph is called *acyclic*. Note that every acyclic graph has at least one sink and at least one source (Exercise 2.12). The acyclicity of a (finite) graph D can be equivalently characterized by requiring that $E(D)$ be a *well-founded* relation, i.e., that every nonnull set W of vertices of D has a vertex s such that $N^+(s) \cap W = \emptyset$ (Exercise 2.13).

Example 2.4 Let D be the *simple* graph drawn below

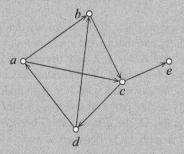

We have $N^-(b) = \{a, d\}$, $N^+(c) = \{d, e\}$. Vertex e is a *sink* because $N^+(e) = \emptyset$. Its *induced subgraph* $D[\{a, b, d\}]$ is *isomorphic* to its induced subgraph $D[\{c, d, b\}]$. $P = (a, b, c, e)$ is a *path* with *end vertices* a and e, while (a, b, c, d, b) is not a path. $C = (a, b, c, d, a)$ is a *cycle*, thus D is not acyclic. $D[\{a, b, c, d\}]$ is *strongly connected*.

2.3.2   Undirected Graphs

We continue with basic notions about undirected graphs; see Example 2.5 for some examples of these. Let $G = (V, E)$ be an undirected graph. If $e = \{u, v\}$ is an edge, u and v are called its *end vertices* (or *end points*) and they differ from one another; we say that e is *incident* to u and to v and that u and v are *adjacent*, or *neighbors*.

Given a vertex $v \in V(G)$, the *neighborhood* of v is the set of neighbors of v in G, $N_G(v) =_{\mathrm{Def}} \{u \in V(G) \mid \{v, u\} \in E(G)\}$. The *degree* of v is $d_G(v) =_{\mathrm{Def}} |N_G(v)|$. A vertex of degree 1 is called a *leaf*, one of degree 0 is said to be *isolated*. We may omit the subscript G when this is clear from the context.

Containment relations and operations. The notions of *subgraph, induced subgraph*, and the notations $G[W]$, $G - W$ (with $W \subseteq V(G)$), $G + F$, $G - F$ (with $F \subseteq E(G)$) are defined in analogy with the directed case. A *graph property* (or *graph class*) is a set of undirected graphs closed under isomorphisms. A graph class is *hereditary* if it is closed under taking induced subgraphs, or, equivalently, under deleting vertices.

Given undirected graphs H and G, we say that G is *H-free* if no induced subgraph of G is isomorphic to H. Graph H is also called a *forbidden induced subgraph*. The family of H-free undirected graphs is a hereditary graph class, for any undirected graph H. For example, look at the following small undirected graphs drawn on p. 103: the *net*, the *claw*, and $K_{2,3}$—the *net* is a *claw*-free graph, while $K_{2,3}$ is not claw-free.

Connectivity. A *path* is an undirected graph $P = (V, E)$ of the form

$$V = \{v_1, v_2, \ldots, v_k\}, \quad E = \{\{v_1, v_2\}, \{v_2, v_3\}, \ldots, \{v_{k-1}, v_k\}\},$$

where $k \geqslant 1$ and the v_i's are *pairwise distinct*. We say that P *connects* (or *joins*, or *is between*) vertices v_1 and v_k and that v_1 and v_k are its *end vertices*. The length of P is the number of its edges, that is, $k - 1$. A path of length 0 is called *trivial*. We will often designate a path by either one of the natural orders of its vertices and write, e.g., $P = (v_1, v_2, \ldots, v_k)$. We say that a path P *is in* an undirected graph G if P is a subgraph of G. We usually denote paths on k vertices with P_k.

Given an undirected graph G, the *distance* in G between vertices $u, v \in V(G)$ is the shortest length of a path in G between u and v; if no such path exists, then the distance is taken to be ∞. An undirected graph G is *connected* if there is a path between any two distinct vertices of G. In Chap. 5 (page 133) we will introduce an equivalent characterization of connected graphs, one not resorting to paths. A maximal connected subgraph of G is called a *connected component* (or just a *component*) of G. If C is a connected component of G, we will sometimes write C when we actually mean $V(C)$.

We say that $X \subseteq V(G)$ is a *cut set* of G if $G - X$ has more connected components than G. If $X = \{v\}$ is a cut set, then v is called a *cut vertex*. Exercise 2.17 asks the reader to show that any connected undirected graph has at least one vertex that is not a cut vertex.

Basic properties of undirected graphs. If $P_k = (v_1, v_2, \ldots, v_k)$ is a path and $k \geq 3$, then the undirected graph $C_k = P_k + \{\{v_k, v_1\}\}$ is called a *cycle*. We will often designate a cycle by any one of the natural orders of its vertices and write, e.g., $C_k = (v_1, v_2, \ldots, v_k, v_1)$. A connected undirected graph having no cycle as subgraph is called a *tree*. Any connected undirected graph G has a *spanning* tree, namely, a tree subgraph T such that G and T have the same vertices (see Exercises 2.15 and 2.16). A graph G such that $\{u, v\} \in E(G)$ holds for every distinct $u, v \in V(G)$ is said to be *complete*. An undirected complete graph on n vertices is denoted as K_n. If $C \subseteq V(G)$, then C is called a *clique* if $G[C]$ is complete; if on the opposite $G[C]$ has no edges, then C is an *independent* set.

A graph G is said to be *multipartite* if there is a partition of $V(G)$ whose blocks are independent sets. If this partition has k blocks, then the graph is called *k-partite*. A multipartite graph G is *complete* if any two vertices belonging to different blocks of the partition of $V(G)$ are adjacent. A 2-partite graph is called *bipartite*. A complete bipartite graph whose partition of the vertices has blocks of sizes n and m, respectively, is denoted as $K_{n,m}$.

A path P in a graph G is *Hamiltonian* if $V(P) = V(G)$. Analogously, a cycle C in G is *Hamiltonian* if $V(C) = V(G)$; if G has a Hamiltonian cycle, G is said to be *Hamiltonian*.

Given a graph G, a subset $M \subseteq E(G)$ is called a *matching* if no edges of M share an end vertex. A matching M is said to be *perfect* if every vertex of G is an end vertex of an edge of M.

Orientations. A (directed) graph D is an *orientation* of an undirected graph G if $V(D) = V(G)$ and for every $\{u, v\} \in E(G)$ it holds that exactly one of the arcs $\langle u, v \rangle$, $\langle v, u \rangle$ belong to $E(D)$. The undirected graph G is said to be the *undirected graph underlying D*. A (directed) graph is said to be *connected* (or *weakly connected*) if its underlying undirected graph is connected. An orientation of a complete undirected graph is said to be a *tournament*.

Example 2.5 Let G be the undirected graph drawn below

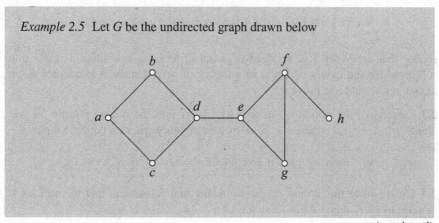

(continued)

Example 2.5 (continued)

We have $N(a) = \{b, c\}$. Vertex h is a *leaf* because it has *degree* 1. Look at the graph $K_{1,3}$, called *claw*, shown on page 103. The *induced subgraph* $G[\{d, b, c, e\}]$ of G is *isomorphic* to the claw, while $G[\{e, d, f, h\}]$ is not; hence, it is *claw-free*.

$P = (a, b, d, e)$ is a *path*, while (f, e, g, h) is not. $G[\{b, c, d, e, f\}]$ is a *tree*. G is a *connected* graph and d is a *cut vertex* of it. The set $\{e, f, g\}$ is a *clique* and the set $\{a, d, f\}$ is an *independent set*. $G[\{a, b, c, d\}]$ is *complete multipartite* (the vertex partition is $\{a, d\}, \{b, c\}$); it is in fact a *bipartite graph*. $G[V(G) \setminus \{c\}]$ has a *Hamiltonian path* $P = (a, b, d, e, g, f, h)$. G admits a *perfect matching*, for example, the set $M = \{\{a, b, \}, \{c, d\}, \{e, g\}, \{f, h\}\}$.

$G[\{d, e, f, g\}]$ is isomorphic to the undirected graph *underlying* $D[\{e, c, b, d\}]$, where D is the graph shown in Example 2.4. Furthermore, $D[\{a, b, c\}]$ is an *acyclic tournament*.

Exercises

2.1 Argue that the following prenex formulations of the extensionality and foundation axioms (with leading universal quantifiers omitted) are logically equivalent to the ones seen in Sect. 2.2:

$$\exists e \big((e \in x \; \leftrightarrow \; e \in y) \; \rightarrow \; x = y \big), \qquad \exists m \, \forall y \in x \big(y \notin m \; \wedge \; m \in x \big).$$

2.2 Would the following *transitive embedding* sentence,

$$\forall x \, \exists t \, \big[x \in t \; \wedge \; \forall v \, \forall w \, \big((v \in t \wedge w \in v) \; \rightarrow \; w \in t \big) \big],$$

stating that every set x is a subset of a transitive set t, constitute a reasonable substitute for the axiom of union in a theory of sets? Unless it is adopted as an axiom, how could one prove it?

2.3 Generalize the notion of ordered pair $\langle x, y \rangle$ into the one of *n-tuple* (or just "*tuple*") $\langle x_1, \ldots, x_n \rangle$, where $n \in \mathbb{N}$, so that the equality criterion between tuples is

$$\langle x_1, \ldots, x_n \rangle = \langle y_1, \ldots, y_m \rangle \text{ if and only if } n = m \wedge x_1 = y_1 \wedge \cdots \wedge x_n = y_n.$$

2.4 Derive from the axioms of extensionality and foundation that the null set $\emptyset$ belongs to every nonnull transitive set.

2.5 Argue, without making use of the axiom of foundation, that the set i of which the sentence

$$\forall x \exists i \left(x \in i \ \wedge \ \forall y \in i \exists u \in i \forall z (z \in u \ \leftrightarrow \ z = y) \right)$$

asserts the existence must be infinite when the set x to which that i is related is not a singleton.

2.6 By adopting the abbreviations

$$V \subseteq T \ \leftrightarrow_{\text{Def}} \ \forall w \in V \ w \in T \quad \text{and} \quad V \in\in X \ \leftrightarrow_{\text{Def}} \ \exists w \in X \ V \in w,$$

derive from the axioms of pairing, transitive embedding (see Exercise 2.2), power set, and comprehension the claims

$$\forall x \exists p \forall v (v \in p \ \leftrightarrow v \subseteq x),$$
$$\forall x \exists u \forall v (v \in u \ \leftrightarrow v \in\in x),$$
$$\forall x \exists t \left(x \subseteq t \wedge (\forall v \in t \ v \subseteq t) \wedge \forall t' \left((x \subseteq t' \wedge \forall v \in t' \ v \subseteq t') \ \rightarrow \ t \subseteq t' \right) \right),$$

stating the existence of the *powerset*, *unionset*, and *transitive-closure* set of any x.

2.7 Explain why the following formulation of the axiom of choice is equivalent to the one informally stated in Remark 2.2:

$$\forall x \exists c \forall w \exists! v \in c \forall u \in x (v \in u \ \leftrightarrow \ w \in u).$$

2.8 Discuss how to carve an equivalent of the theory of natural numbers with successor inside a set theory whose axioms make it possible to specify, by means of a formula $\zeta(y)$ the property: y is a set of the form $\{\cdots\{\emptyset\}\cdots\}$. This can be done much as in Panel 2.4, but you should avoid relying on the infinity axiom.

Formulate $\zeta(y)$ for a theory endowed with these axioms: extensionality, foundation, pairing, transitive embedding (see Exercise 2.2), and comprehension.

2.9 Rewrite the setformers that appear in abridged form in Remark 2.4 so as to make them compliant with the syntax of setformers introduced on p. 43.

2.10 (Alternative characterization of the von Neumann natural numbers) Denote by $\bar{n}$ the the $n + 1$-st *finite* von Neumann ordinal, which encodes the natural number n. In particular, $\bar{0} = \emptyset$. (These $\bar{n}$'s can be called *von Neumann's numerals*.)
  Prove by induction on n that $\bar{n} = \{\bar{0}, \bar{1}, \ldots, \overline{n-1}\}$.

2.11 (*) Prove that for every set x, there exists a bijection $\flat$ such that $\mathbf{dom}\,(\flat)$ is an ordinal and $\mathbf{img}\,(\flat) = x$.
  Show that when x is finite the domain of such a bijection is uniquely determined, but that the situation is radically different when x is infinite.

2.12 Prove that if G is a graph devoid of cycles, then G has at least one sink and one source.

2.13 Show that a finite graph G is devoid of cycles if and only if every nonnull set W of vertices of G has a vertex s such that $N^+(s) \cap W = \emptyset$.

2.14 Show that any undirected graph admits an acyclic orientation.

2.15 Prove that any connected undirected graph has a spanning tree.

2.16 Let G be a connected undirected graph with n vertices and m edges. Describe an $O(n + m)$-time algorithm computing a spanning tree of G.

2.17 Let $G = (V, E)$ be a connected undirected graph. Show that there exists at least one vertex $v \in V$ such that v is not a cut vertex of G. Describe two proofs of this fact, using the following strategies:

1. Consider a longest path P in G. What can you say about the endpoints of P?
2. Use the fact that G, being connected, has a spanning tree T (recall Exercise 2.15). What can you say about the leaves of T?

Can you strengthen your proofs to show that G has at least two vertices that are not cut vertices of G?

Chapter 3
Sets, Graphs, and Set Universes

For a fixed axiomatic system of Set Theory, either the full variety of models of the axioms or a deductive system yielding theorems—or both—often is to be taken into account. In certain situations, in fact, one inclines toward a view more firmly geared to semantics and in others toward a more syntactically oriented view. The simple combinatorial character of the arguments discussed in this book allows us to opt for a simplified semantic approach. Most often, in fact, our work will regard the specific model of *hereditarily finite sets*. Technically speaking, this is a common (initial) part of all models of our (set) theories, and the sporadic cases in which a richer collection of sets must be taken into account will be duly signaled along the presentation.

This chapter complements the preceding one with a short discussion on set *universes* and by revisiting a bijection between *hereditarily finite* sets and natural numbers, first defined by Wilhelm Friedrich Ackermann in 1937 and which will play an important role in the rest of this book. The detailed inner structure of a hereditarily finite set will be a *pointed membership graph* (see below); as a matter of fact, the structures which are isomorphic to pointed membership graphs will become our privileged representation of hereditarily finite sets. After introducing this important representation, this chapter will begin to illustrate its usefulness through the use of "*diags*," specific diagrams which are degraded versions of the pointed membership graphs, for solving special cases of the satisfiability problem referred to set-theoretic formulae. As will be seen in later chapters, already those graphs which are isomorphic to membership graphs whence the special vertex designated as "point" has been removed can serve to analyze combinatorial problems regarding sets.

The collection HF of all hereditarily finite sets is usually constructed in stages, through an infinite sequence of enlargements. One technique for performing this construction results from a simplification of von Neumann's classical *cumulative hierarchy*, which was meant to serve as a universe of sets modeling the entire theory ZFC (cf. Panels 3.1 and 3.2). Von Neumann's construction, which assigns a *rank*,

© Springer International Publishing AG 2017
E.G. Omodeo et al., *On Sets and Graphs*, DOI 10.1007/978-3-319-54981-1_3

possibly transfinite, to every set, as well as its scaled-down version tailored for HF, where ranks remain finite, will be analyzed in this chapter.[1]

Non-well-founded variants of HF, called $HF^{1/2}$ and HF^1, will also be introduced, but it would be hard and perhaps less natural to construct these universes as hierarchies; much more straightforwardly, they can be described in terms of bisimulations and *bisimilarity*, notions which will also be presented in this chapter.

The shift from HF to $HF^{1/2}$—from sets to "*hypersets*," which is a popular name for the aggregates akin to sets that form a non-well-founded universe—is best understood if one represents sets / hypersets by particular graphs, dubbed *pointed membership graphs*, in the way discussed next.

Of any given set / hyperset s, consider the graph (V_s, E_s) whose

- vertices form the smallest set / hyperset V_s which has s as an element and for which $y \in V_s$ holds whenever $y \in x \in V_s$;
- arcs are all pairs $\langle x, y \rangle$ such that $x, y \in V_s$ and $x \ni y$.

Then consider an arbitrary triple (V', E', s') where (V', E') is a graph, $s' \in V'$, and there is a graph isomorphism between (V_s, E_s) and (V', E') in which s' corresponds to s. A key observation, surfacing over and over again in this book, is that the triple (V', E', s') conveys all relevant information about the original s.

Let us call *pointed graph* a triple (V, E, s_*) where $s_* \in V$ and (V, E) is a graph all of whose vertices are reachable from s_*. A pointed graph may or may not qualify to *represent* (to within isomorphism, of course) a set or a hyperset in the manner just described, and, in fact, a problem addressed in this chapter is

0. elicit the structural properties which make a pointed graph a genuine representation of a (hyper)set.

Panel 3.1   Sets, classes, and axiomatic systems

To be is to be the value of a bound variable.   (W. V. O. Quine, 1948)

We often specify a property that some sets enjoy(e.g., finitude, being hereditarily finite or being an ordinal) either by means of a formula involving previously specified properties or—when we refer to a well-founded set universe—through a recursive characterization. Respective examples are:

$$\text{Finite}\,(F) \;\leftrightarrow_{\text{Def}}\; \big(\forall g \in (\mathcal{P}(\mathcal{P}(F)) \setminus \{\emptyset\})\big)\, \exists m\big(g \cap \mathcal{P}(m) = \{m\}\big),$$
$$\text{HF}\,(H) \;\leftrightarrow_{\text{Def}}\; \text{Finite}\,(H) \wedge (\forall k \in H)\big(\text{HF}(k)\big).$$

(continued)

---

[1]A "slower" construction of the hereditarily finite sets, which relies on the *adjunction* operator $x \overset{\text{with}}{\longmapsto} x \cup \{x\}$ instead of on the operator $x \mapsto \mathcal{P}(x)$ is analyzed in [52] and [5].

Panel 3.1 (continued)
Antinomies of the past have brought to light cases in which there can be no *set* encompassing a specified property. For example, neither a set $\mathscr{U}$ to which all sets belong nor a set On of all ordinals can exist; notwithstanding, we can define the properties "being a set" and "being an ordinal" as follows:

$$\mathscr{U}(X) \leftrightarrow_{\text{Def}} X = X,$$
$$\text{On}(R) \leftrightarrow_{\text{Def}} (\forall x \in R)(x \subseteq R) \wedge (\forall x \in R)(\forall y \in R)(x \in y \vee y \in x \vee x = y).$$

Since, after all, such properties cut off "collections" of some sort from the universe of sets, they are often called *classes*; one also feels authorized to write, e.g., $r \in$ On instead of On(r). A class which is not a set, for example, $\mathscr{U}$ or On, is called a *proper* class. It often depends on the strength of the axioms whether a class is proper or not, e.g., HF is a set if infinity, comprehension, and replacement are postulated, whereas HF coincides with $\mathscr{U}$ if one postulates, *à la* Tarski [109], that only finite sets exist.

The distinction between proper classes and sets smoothly extends to functions: the function sending every set to its ordinal rank (see Fig. 3.1) is a proper class of ordered pairs, but its restriction to HF, whose multi-image is $\mathbb{N}$, may well be a set (under axioms enforcing that HF be a set).

———————

Two main types of set theories differ in how they treat proper classes. For some theories, proper classes simply *do not belong* to the universe of discourse. The best known such theory is Zermelo-Fraenkel, ZF, whose axioms are essentially the ones introduced in this book, with foundation and infinity axioms, without AFA, and with the axiom of choice often left aside (when this axiom does enter the game, the acronym of the theory becomes ZFC). For other theories, such as the von Neumann-Gödel-Bernays NGB, classes are the main entities of discourse, while sets are those classes which are entitled to *belong to other classes*. Unlike ZF, which needs infinitely many axioms, finitely many first-order sentences suffice to axiomatize NGB; hence, an automated deduction system knowledgeable about sets is more easily instructed for NGB than for ZF (cf. [13, 97], and [11]).

Panel 3.2 The cumulative hierarchy of sets and other set universes
Axiomatic set theory had already evolved (by the essential contributions of Zermelo [114], Fraenkel, and Skolem) into today's version ZF, when von Neumann modeled it by his renowned *cumulative hierarchy* [69, 70]. Having been conceived downstream, after decades-long investigations on the foundations of mathematics, that structure perhaps does not fully deserve

(continued)

Panel 3.2 (continued)

the relevance of *intended model*: suffice it to say that ZF admits, besides that model (having as its own domain of support a proper class[a] and encompassing sets whose cardinality exceeds the countable), also models of countable cardinality.[b] Also, Gödel later proposed a model alternative to the one due to von Neumann, the universe of *constructibles* [44], which comes to coincide with the former only when a specific axiom so caters.

Anyway, the cumulative hierarchy is, for anybody who undertakes an advanced study on sets, a valuable conceptual tool of some preliminary exposure to which—if only at an intuitive, but nonetheless rigorous, level— will pay off when one arrives at the axiomatic formalization, traditionally based on first-order predicate calculus.

One of the benefits that can ensue from placing the semantics before the formal-logical description is the gradualness made possible by such an approach. Preparatory to the cumulative hierarchy in which, as proposed by von Neumann, one level corresponds to every ordinal (even to each *transfinite* ordinal), one can consider similar but less demanding hierarchies. For example, the one of the *hereditarily finite* sets may be taken as a reference model for an axiomatic theory close in spirit to the one due to Zermelo but focused exclusively on *finite* sets [109]; other hierarchies, the so-called *superstructures*, play an important rôle in setting up the ground for nonstandard analysis [26]. The study of these scaled-down versions of the von Neumann's hierarchy brings to light algorithmic manipulations which make sense insofar as one deals with relatively simple sets only, but which bear a lot of significance for those whose interests are more deeply oriented toward computer science than toward foundational issues.[c]

Various hierarchies similar to the one due to von Neumann held a historically crucial rôle in the investigations on axiomatic theories antithetic to ZF. Models stemming from this more speculative framework—namely, a "cumulative proto-hierarchy" proposed by Fraenkel in the far 1922 [41] and the abovementioned class of Gödel's constructible sets—aided in the study on the independence of the axiom of choice and of the continuous hypothesis. Moreover, as this book examines on a small-scale instance (referring to *finite hypersets* only), it is easy to obtain models of theories of non-well-founded sets [2, 9] from standard hierarchies.

---

[a]See Panel 3.1.

[b]This constitutes the so-called *Skolem paradox*.

[c]The importance of algorithms for the manipulation of nested sets is witnessed by the availability of sets (with their correlated associative *maps*) in several programming languages: SETL [27, 28], Maple, Python, etc.

Three other problems, tightly interrelated with (0), will be tackled:

1. Can we, given a pointed graph (V, E, s_*), split V into disjoint nonnull blocks $Y_1, \ldots, Y_\kappa$, shrink the graph accordingly (i.e., reorganize its structure consistently with the chosen split[2]), and thus obtain the representation of a (hyper)set?
2. Assuming that the answer to the preceding question is "yes," find a "least-effort" way of splitting V as wanted; this is to say, make sure that no partition of V coarser than the proposed $\{Y_1, \ldots, Y_\kappa\}$ yields a graph representing a (hyper)set.
3. How can we retrieve, from a graph representing a (hyper)set, the represented s?

As the reader may expect, by satisfactorily solving problems (1) and (2), one would get a cheap solution to (0) as well; in order to recognize whether or not a pointed graph represents a (hyper)set, one could in fact proceed in this manner:

- check that the answer to (1), for the given (V, E, s_*), is affirmative;
- find the coarsest partition of the set V of vertices that satisfies requirement (2);
- check that the partition found consists of only singletons (so that the shrunk graph does not differ from the given one).

In essence this is the way we will solve (0) for hypersets. The abovementioned notion of bisimilarity will, in fact, be the equivalence relation associated with the sought coarsest partition, and hence it must coincide with the identity relation over the vertices of a pointed graph that represents a hyperset.

For ordinary (i.e., well-founded) sets, this chapter will instead offer a direct characterization of the representing graphs. We will identify a new property of graphs, *extensionality*. In order that a pointed graph represents a set, we will see, it must enjoy acyclicity (see Chap. 2) and extensionality.

Problem (2) can be coped with, in the well-founded case, by means of a technique which Andrzej Mostowski devised for much more imposing systems than our graphs. We will exploit his technique to "*decorate*" every pointed graph (V, E, s_*) representing a set: concretely, we will determine an injection $v \mapsto \mathfrak{m}\, v$ from V into HF which satisfies the biimplication $\langle u, w \rangle \in E \leftrightarrow \mathfrak{m}\, u \ni \mathfrak{m}\, w$ and the general rule

$$\mathfrak{m}\, u = \{\mathfrak{m}\, w : w \in V \mid \langle u, w \rangle \in E\} \text{ for all } u \in V;$$

the set represented by the pointed graph will then be $\mathfrak{m}\, s_*$. The case of hypersets is even simpler: the very anti-foundation axiom enforces that there is a unique way of decorating a pointed graph representing a hyperset.

---

[2]In order that this reorganization be doable, we will require, for each pair Y_i, Y_j of blocks, that either no arc $\langle u, w \rangle \in E$ with $u \in Y_i$ and $w \in Y_j$ exists or to each $u \in Y_i$ there corresponds some $w \in Y_j$ such that $\langle u, w \rangle \in E$; the shrinking operation will then yield the pointed graph $(\{Y_1, \ldots, Y_\kappa\}, \{\langle Y_i, Y_j \rangle : i, j \in \{1, \ldots, \kappa\}, u \in Y_i, w \in Y_j \mid \langle u, w \rangle \in E\}, Y_*)$ with $s_* \in Y_* \in \{Y_1, \ldots, Y_\kappa\}$.

Mostowski's way of decorating a graph should be regarded, rather than as a mono-use technique, as paradigm of a general method for assigning (hyper)sets injectively to the vertices of a graph. Injectivity seldom comes for free; often it must be enforced either by preliminary manipulations of the graph or by a decorating rule enriched with "gadgets" (cf. Panel 3.4). Decorations find applications in solvable cases of the decision problem for set theory, as we explain next. A set-theoretic formula φ is given, whose syntax is subject to heavy restraints; two questions must be answered:

Satisfiability: Do there exist assignments $x_1 \mapsto \pmb{x}_1, \ldots, x_n \mapsto \pmb{x}_n$ of (hyper)sets to the distinct free variables $x_1, \ldots, x_n$ of φ under which φ becomes true?
Satisfaction: When such assignments exist, can we concretely exhibit one?

Instances of these two problems are addressed in the last section of this chapter, which treats two fragments of our first-order language about sets. In both cases, analysis of the input formula φ leads either to its rejection (when it turns out that φ cannot be satisfied) or to the selection of a graph specially contrived to diagram a collection of satisfying assignments. On the one hand, such graph retains enough information about the assignments it represents as to enable verification that such assignments make φ true. On the other hand, proving the correctness of the proposed decision algorithm calls for our ability to decorate the vertices of the graph with sets, some of which will be the sought values $\pmb{x}_i$.

3.1 Hereditarily Finite Sets and Hypersets

Any model of our axiomatic set theories allows—regardless of the actual axiomatic choices—for the definition of collections built from the null set $\emptyset$ by means of the power-set operator. The initial part of any such "universe" is most important for our purposes in this book: it is the collection of the entities known as *hereditarily finite sets*

$$\mathsf{HF} = \bigcup_{n \in \mathbb{N}} \mathsf{HF}_n,$$

whose *levels* HF_n are inductively defined as follows:

$$\mathsf{HF}_0 \;\; =_{\mathrm{Def}} \;\; \emptyset;$$
$$\mathsf{HF}_{m+1} =_{\mathrm{Def}} \;\; \mathcal{P}(\mathsf{HF}_m), \text{ for every } m \in \mathbb{N}.$$

Example 3.1 We arrange in two sequences, below, the definitions of

numbers   and   **levels:**

$0 =_{\text{Def}} \emptyset,$ $\mathsf{HF}_0 =_{\text{Def}} \emptyset\,;$

$1 =_{\text{Def}} \{0\},$ $\mathsf{HF}_1 =_{\text{Def}} \mathcal{P}(\mathsf{HF}_0)\,;$

$2 =_{\text{Def}} 1 \cup \{1\},$ $\mathsf{HF}_2 =_{\text{Def}} \mathcal{P}(\mathsf{HF}_1)\,;$

$3 =_{\text{Def}} 2 \cup \{2\},$ $\mathsf{HF}_3 =_{\text{Def}} \mathcal{P}(\mathsf{HF}_2)\,;$

$\cdots$ $\cdots$

Expanding these we can effectively determine the levels as follows:

$$\mathsf{HF}_1 = \{0\} = 1 = \{\mathsf{HF}_0\},$$
$$\mathsf{HF}_2 = \{0, 1\} = 2 = \{0,\ \mathsf{HF}_1\},$$
$$\mathsf{HF}_3 = \{0, 1, \{1\}, 2\} = \{0, 1, \{1\}, \mathsf{HF}_2\},$$
$$\mathsf{HF}_4 = \Big\{0, 1, \{1\}, 2, \{\{1\}\}, \{2\}, \{0, \{1\}\}, \{0, 2\}, \{1, \{1\}\}, \{1, 2\},$$
$$\{0, 1, \{1\}\}, 3, \{\{1\}, 2\}, \{0, \{1\}, 2\}, \{1, \{1\}, 2\}, \mathsf{HF}_3\Big\},$$

etc.

Note that (natural) numbers, intended as above, are just peculiar sets. Likewise, all levels HF_n, as well as the elements of each of these levels, are sets.

The transitive closure of the set $x = \{\{\emptyset\}\} \in \mathsf{HF}_3$ is $\text{trCl}\,(x) = \{\emptyset, \{\emptyset\}\} \neq x$; therefore x is not a transitive set. HF_3 is itself a hereditarily finite set (e.g., $\mathsf{HF}_3 \in \mathsf{HF}_i$ for all $i \geq 4$); moreover $\text{trCl}\,(\mathsf{HF}_3) = \mathsf{HF}_3$; hence HF_3 is transitive. Each HF_n, like its subscript n, is a transitive hereditarily finite set.

In the above characterization of HF, we have made use of the operator $\mathcal{P}$, whose introduction is legitimized by the power-set axiom, but one can avoid exploiting such a demanding axiom and characterize HF in terms of more elementary operators such as, e.g., the k-element collection operators (with $k \in \mathbb{N}$) or the *with* operator mentioned at p. 41 and at p. 41, respectively.

As a matter of fact, under the foundation axiom, the sets in HF might get assigned as values x_i to the unknowns x_i by solutions of systems of equations of the form

$$\begin{cases} x_0 = \{x_{0,1}, \dots, x_{0,k_0}\}, \\ x_1 = \{x_{1,1}, \dots, x_{1,k_1}\}, \\ \ \vdots \ \ \vdots \qquad \ddots \\ x_n = \{x_{n,1}, \dots, x_{n,k_n}\}, \end{cases}$$

where the variables $x_0, x_1, \dots, x_n$ appearing as left-hand sides are pairwise distinct and each doubly indexed x appearing in a right-hand side is drawn from among $x_0, x_1, \dots, x_n$. Note that we are now resorting exclusively to k-element collection operators.

Under the foundation axiom, a system of the above form admits a solution if and only if it satisfies a certain condition, as the reader is invited to prove as Exercise 3.2. On the other hand, if AFA is postulated instead of foundation, such a system will unconditionally admit one and only one solution (cf. p. 48). This indicates that the systems of equations of the said form capture a collection of aggregates strictly larger than HF; we can still say that such aggregates are *hereditarily finite*, but membership is no longer well founded over them. Throughout this book, after [8], we denote that larger collection as $HF^{1/2}$ and sometimes call it the collection of *rational hyper*sets.

Remark 3.1 The universe $HF^{1/2}$ is not the only universe which makes sense to regard it as the collection of hereditarily finite non-well-founded sets. A more advanced possibility is to consider as hereditarily finite hypersets also solutions to *infinite* systems of set-equations, just as long as the right-hand sides of such equations are finite. The resulting universe is then denoted as HF^1 and consists of those finite hypersets such that all elements in their transitive closures are, in their turn, finite hypersets.

As a convenient adjustment of notation, let us denote HF also as HF^0. The distinctive features of HF^0, $HF^{1/2}$, and HF^1 will clearly emerge from later sections; particular prerequisites are the notions of extensional and hyperextensional graphs, which will show up in Sect. 3.4.

It will in fact turn out, after that point, that the sets in HF^0 correspond to finite extensional acyclic graphs and the hypersets in $HF^{1/2}$ to finite hyperextensional graphs; as for the entities in HF^1, they will be representable by hyperextensional graphs possibly endowed with *infinitely many vertices*, in which every vertex has nevertheless a finite out-degree. The chain $HF^0 \subsetneq HF^{1/2} \subsetneq HF^1$ of inclusions will be obvious.   ⊣

3.2   The Full-Fledged Cumulative Hierarchy

> *Thus, in our system, all objects are sets. We do not postulate the existence of any more primitive objects. To guide his intuition, the reader should think of our universe as all sets which can be built up by successive collecting processes, starting from the empty set.*
>   (Paul Joseph Cohen, [24, p. 50])

The full-fledged set universe is *von Neumann's cumulative hierarchy*

$$\mathscr{V} = \bigcup_{i \in On} \mathscr{V}_i,$$

whose *levels* $\mathscr{V}_n$ are now inductively defined on the entire class On of *ordinals* (cf. Panel 3.1, Figs. 3.1 and 3.2), using—this time in an essential way—the power-set axiom:

$$\mathscr{V}_n =_{Def} \bigcup_{i \in n} \mathcal{P}(\mathscr{V}_i).$$

$$\begin{aligned}
\mathrm{On}(R) &\leftrightarrow_{\mathrm{Def}} \forall x \in R\,(x \subseteq R)\;\&\\
&\qquad\qquad \forall x \in R\,\forall y \in R\,(x \in y \vee y \in x \vee x = y)\\
x^+ &=_{\mathrm{Def}} x \cup \{x\}\\
\mathrm{rank}(x) &=_{\mathrm{Def}} \bigcup\{\mathrm{rank}(y)^+ : y \in x\}
\end{aligned}$$

$$\begin{aligned}
\mathrm{On}(O_1)\;\&\;\mathrm{On}(O_2) &\rightarrow O_1 \cap O_2 \in \{O_1, O_2\}\\
\mathrm{On}(O) &\rightarrow \mathrm{On}(O^+)\\
\mathrm{On}(O) &\rightarrow \big(Y \in O \leftrightarrow (\mathrm{On}(Y)\;\&\;Y \subsetneq O)\big)\\
\mathrm{On}(\bigcup\{o \in S\,|\,\mathrm{On}(o)\})&\\
\mathrm{On}(O)\;\&\;Y \in O &\rightarrow Y^+ \subseteq O\\
\mathrm{On}(O) &\rightarrow O = \bigcup\{y^+ : y \in O\}\\
\mathrm{On}(O_1)\;\&\;\mathrm{On}(O_2) &\rightarrow O_1 \in O_2 \vee O_2 \in O_1 \vee O_1 = O_2\\
\mathrm{On}(X)\;\&\;X = M^+ &\rightarrow M = \bigcup X\;\&\;\mathrm{On}(M)\\
\mathrm{rank}(X) = /\,0 &\leftrightarrow X = /\,0\\
\mathrm{rank}(\bigcup X) &= \bigcup\{\mathrm{rank}(y) : y \in X\}\\
\mathrm{rank}(X \cup Y) &= \mathrm{rank}(X) \cup \mathrm{rank}(Y)\\
X \subseteq Y &\rightarrow \mathrm{rank}(X) \subseteq \mathrm{rank}(Y)\\
\mathrm{Finite}(F)\;\&\;\{t \in F\,|\,\neg\mathrm{Finite}(\mathrm{rank}(t))\} = /\,0 &\rightarrow \mathrm{Finite}(\mathrm{rank}(F))\\
\mathrm{On}(R) &\leftrightarrow R = \mathrm{rank}(R)\\
X \in Y &\rightarrow \mathrm{rank}(X) \in \mathrm{rank}(Y)\;\&\;\mathrm{rank}(X) \subseteq \mathrm{rank}(Y)\\
\mathrm{On}(X)\;\&\;\mathrm{Finite}(X)\;\&\;X \neq \emptyset\;\&\;V = \bigcup X &\rightarrow X = V^+\;\&\;\mathrm{On}(V)\\
\mathrm{rank}(\{X, Y\}) &= \mathrm{rank}(X \cup Y)^+\\
\mathrm{On}(R)\;\&\;\mathrm{Finite}(R) &\rightarrow \exists h\,\big(\mathrm{Finite}(h)\;\&\\
&\qquad \forall y\,(\mathrm{rank}(y) \subseteq R \rightarrow y \in h)\big)\\
\mathrm{Finite}(\mathrm{rank}(X)) &\leftrightarrow \mathrm{Finite}(X)\;\&\;\forall y \in X\big(\mathrm{Finite}(\mathrm{rank}(y))\big)
\end{aligned}$$

Fig. 3.1   Laws on ordinals and on the successor and rank operations, as defined under foundation (see Exercise 3.5)

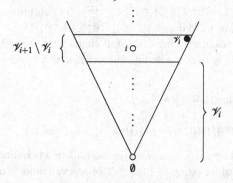

Fig. 3.2   Cumulative hierarchy founded on the null set. Each $\mathscr{V}_i$ belongs to the next level $\mathscr{V}_{i+1}$ and includes as subsets, besides having them as elements, all $\mathscr{V}_j$'s subscripted by j's lower than i. The inclusion of each $\mathscr{V}_i$ in $\mathscr{V}_{i+1}$ is strict; in fact, each ordinal number i belongs to the *layer* $\mathscr{V}_{i+1} \setminus \mathscr{V}_i$. Starting with $\mathscr{V}_3 \setminus \mathscr{V}_2$, to which the set $\{1\}$ belongs, each layer has also elements which are not numbers

This definition is in accordance, up to the first infinite ordinal, with the above-seen construction of HF: obiously $\mathscr{V}_0 = \emptyset$ holds; moreover, one gets $\mathscr{V}_{m+1} = \mathcal{P}(\mathscr{V}_m)$; see Exercise 3.3. (By the way, the latter equality holds even beyond natural numbers, if one takes the successor $m + 1$ of every ordinal m to be $m \cup \{m\}$ by definition.)

Our discussion will normally revolve around either HF or $\mathsf{HF}^{1/2}$. In other words, the typical entity of our universe of discourse will be a set whose transitive closure is finite. In the rare cases when we must allow infinite sets to enter the scene, the axiom of infinity will be tacitly assumed, along with the power-set axiom if needed.

Albeit we will not take any special care to specify the weakest axiomatic system which our reasonings presuppose, let us point out once more that almost all sets in this book (save for Chap. 8) are finite and, definitely, all of the arguments we will be using are finitary.

Since sets enter the universe in successive stages of its construction, for each set x, we can consider the first level at which x becomes available. The index of that level is called *rank* of x:

Definition 3.1 (Rank) The *rank* of a set x, to be denoted $\mathsf{rank}(x)$, is the minimum n such that $x \in \mathscr{V}_{n+1}$ (so that, in particular, $\mathsf{rank}(\emptyset) = 0$).

The simple Exercise 3.4 asks the reader to show that $\mathsf{rank}(x)$ also equals the minimum n such that $x \subseteq \mathscr{V}_n$; see Example 3.2 below.

The following alternative characterization of rank can be given and will be used extensively in this book, in the majority of cases:

$$\mathsf{rank}(x) = \max_{y \in x} \big(\mathsf{rank}(y) + 1\big). \qquad (3.1)$$

Example 3.2 Let $x = \{\emptyset, \{\{\emptyset\}\}\} \in \mathsf{HF}_4 \setminus \mathsf{HF}_3$. By definition, the rank of x is 3. We also have that $x \subseteq \mathsf{HF}_3 = \{\emptyset, \{\emptyset\}, \{\{\emptyset\}\}, \{\emptyset, \{\emptyset\}\}\}$.

Since x is a hereditarily finite set, the characterization (3.1) of the rank definition shows how to compute this number also in a bottom-up fashion:

- $\mathsf{rank}(\emptyset) = 0$,
- $\mathsf{rank}(\{\emptyset\}) = \max(0 + 1) = 1$,
- $\mathsf{rank}(\{\{\emptyset\}\}) = \max(1 + 1) = 2$,
- and thus $\mathsf{rank}(\{\emptyset, \{\{\emptyset\}\}\}) = \max(0 + 1, 2 + 1) = 3$.

In more graphical terms, one can also see the rank of a hereditarily finite set as the deepest level of a nesting of $\{\cdot, \ldots, \cdot\}$. In an even more graphical manner, in Sect. 3.4, we will see that if we interpret the inverse $\ni$ of membership as edges of a graph, then the rank also equals the length of a longest path from the vertex representing the set to the vertex representing $\emptyset$.

Set universes are rich, and among their elements, as already announced, transitive sets play an important role. The reason why is that a large class of dyadic relations can be represented by transitive sets. This result stems from a technique introduced by Mostowski in 1949 [67], summarized below; for a proof, see [50].

Lemma 3.1 (Mostowski's collapsing lemma)   *Let E be a well-founded relation on a class V such that, indicating by $E\!\restriction^{\!\curvearrowright}\{x\}$ the class of all y's in V for which $\langle x, y \rangle$ belongs to E,*

- *$E\!\restriction^{\!\curvearrowright}\{x\}$ is a set for each x in V;*
- *$E\!\restriction^{\!\curvearrowright}\{x\} \neq E\!\restriction^{\!\curvearrowright}\{x'\}$ holds when x, x' are distinct elements in V.*

Then there exist a transitive class M and an isomorphism $\mathfrak{m}$ between (V, E) and $(M, \in)$. The transitive class M and the isomorphism $\mathfrak{m}$ are unique.

The function $\mathfrak{m}$ mentioned in the above lemma is often dubbed a *decoration* of G when $G = (V, E)$ is a graph. Mostowski's collapsing lemma hence tells us, in particular, that any graph in which there are neither cycles nor distinct vertices endowed with the same out-neighborhood can be uniquely decorated (see Example 3.3). As we will discuss in Sect. 3.4, AFA can be viewed as an axiom (scheme) enforcing the validity of this consequence of Mostowski's lemma on a collection of graphs, dubbed "*hyperextensional*" graphs, where cycles are not forbidden.

Example 3.3   As annotated in the illustration below,

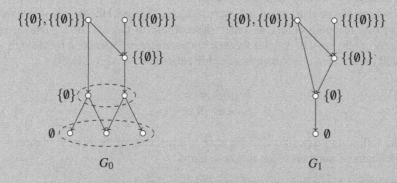

G_0 G_1

there is only one assignment $u \mapsto \mathfrak{m}u$ of sets to the vertices u that satisfy the condition $\mathfrak{m}u = \{\mathfrak{m}v : \langle u, v \rangle \text{ is an arc}\}$ for each vertex u of $G_0 + G_1$ (this is evident, because there are no cycles; hence, $\mathfrak{m}$ can be determined by proceeding bottom-up from the sinks to the sources). No two nodes have the same out-neighborhood in G_1; hence, the restriction of $\mathfrak{m}$ to G_1 is its injective decoration *à la* Mostowski. On the contrary, G_0 has three sinks, to each of which $\mathfrak{m}$ must assign the value $\emptyset$; another "collision" then arises, and hence $\mathfrak{m}$ must assign the same value, $\{\emptyset\}$, to two nodes. In a definite sense, either one of G_0 and G_1 portrays the set $\mathrm{trCl}(\{\,\{\{\emptyset\}, \{\{\emptyset\}\}\},\ \{\{\{\emptyset\}\}\}\,\})$, but G_1 so does more economically.

Remark 3.2 The statement of Mostowski's collapsing lemma carefully uses the term "class" (see Panel 3.1) when referring to the domain V of the function $\mathfrak{m}$, as well as to the multi-image M of $\mathfrak{m}$. This is done in order to state the result in full generality. We will exploit Lemma 3.1 only in situations in which V and M are sets.

3.3 The Ackermann Encoding of Hereditarily Finite Sets

The following elegant encoding of elements of HF was first proposed in [1]:

> **Definition 3.2 (The Ackermann encoding)** For every $x \in$ HF, the *Ackermann number* of x is recursively defined to be
>
> $$\mathbb{N}_A(x) = \Sigma_{y \in x}\, 2^{\mathbb{N}_A(y)}.$$

$\mathbb{N}_A(\cdot)$ is a bijection between HF and $\mathbb{N}$ and embodies our intuition that (well-founded) hereditarily finite sets and natural numbers are closely related. In terms of the binary representation of natural numbers, the membership relation becomes so readable through this encoding: the set x whose Ackermann number is n has for elements those y's whose Ackermann numbers correspond to 1's in the binary representation of n. See Example 3.4 below.

It is easy to see that, as already set forth, $\mathbb{N}_A(\cdot)$ maps HF onto $\mathbb{N}$ injectively. Hence, this mapping enacts an order (isomorphic to the one of $\mathbb{N}$) among the elements of HF: we will call this the *Ackermann order*, will denote it by the sign $\prec$, and will denote by h_i the i-th element of HF relative to $\prec$. Accordingly:

$$\mathbb{N}_A(h_i) = i,$$
$$h_i \prec h_j \text{ iff } i < j.$$

The following proposition can be read as a restating of the bitwise comparison between natural numbers in set-theoretic terms.

Proposition 3.1 *Given $h_i, h_j \in$ HF, the following holds:*

$$h_i \prec h_j \text{ iff } \max_{\prec}(h_i \setminus h_j) < \max_{\prec}(h_j \setminus h_i).$$

Here, conventionally (since the null set has no maximum), we treat $\max_{\prec} \emptyset$ as if it were an entity strictly smaller than any $h \in$ HF.

Proof Since $h_i \prec h_j$ amounts precisely to $i < j$, it suffices to compare the base-2 representation of i with the one of j and to apply the definitions of the Ackermann encoding and the Ackermann order. $\dashv$

Example 3.4 The sets whose Ackermann numbers are the first natural numbers are:

- $\mathbb{N}_A(\emptyset) = 0$,
- $\mathbb{N}_A(\{\emptyset\}) = 2^{\mathbb{N}_A(\emptyset)} = 2^0 = 1$,
- $\mathbb{N}_A(\{\{\emptyset\}\}) = 2^{\mathbb{N}_A(\{\emptyset\})} = 2^1 = 2$,
- $\mathbb{N}_A(\{\emptyset, \{\emptyset\}\}) = 2^{\mathbb{N}_A(\emptyset)} + 2^{\mathbb{N}_A(\{\emptyset\})} = 2^0 + 2^1 = 3$,
- $\mathbb{N}_A(\{\{\{\emptyset\}\}\}) = 2^{\mathbb{N}_A(\{\{\emptyset\}\})} = 2^2 = 4$.

Given n, we can also proceed in a top-down manner in finding the set x such that $\mathbb{N}_A(x) = n$, by recursively inspecting the binary representation of n. For example, if $n = 17$, then $17 = (10001)_2$, which means that $x = \{y, z\}$ with $\mathbb{N}_A(y) = 4$ and $\mathbb{N}_A(z) = 0$. From the above calculations, we already know y and z, but we iterate this binary expansion one more time. We have $\mathbb{N}_A(y) = 4 = (100)_2$; thus, $y = \{v\}$ with $\mathbb{N}_A(v) = 2$; $v = \{\{\emptyset\}\}$; thus, we finally get $x = \{\{\{\{\emptyset\}\}\}, \emptyset\}$.

Let $x' = \{\{\{\{\emptyset\}\}\}\}$, namely, $x' = x \setminus \{\emptyset\}$. We have $\mathbb{N}_A(x') = 16$ and thus $x' \prec x$. Equivalently, by Proposition 3.1, we also obtain $x' \prec x$ since

- $\max_{\prec} \emptyset$ is by convention strictly smaller than any $h \in \mathsf{HF}$, and
- $\max_{\prec}(x \setminus x') = \max_{\prec}\{\emptyset\} = \emptyset$.

Every one of the levels HF_k has finitely many elements—by precise assessment of this number of elements, one discovers that it is (if $k > 1$) the hyperexponential amount

$$2^{{\cdot}^{{\cdot}^{\cdot^2}}} \Big\} \; k-1 \text{ times} \; .$$

Denoting by $\beth(k)$ the function

$$\beth(k) =_{\mathrm{Def}} \begin{cases} 0 & \text{if } k = 0, \\ 2^{\beth(k-1)} & \text{otherwise,} \end{cases}$$

that results from iterating exponentiation, namely, the function $m \mapsto 2^m$, k times, we can characterize the hereditarily finite sets of a given rank r to be the ones whose Ackermann numbers fall in the interval $[\beth(r), \beth(r + 1) - 1]$; specifically:

Lemma 3.2 *For all* $h \in \mathsf{HF}$, *it holds that*

$$\mathbb{N}_A(h) \in [\beth(\mathrm{rank}(h)), \beth(\mathrm{rank}(h) + 1) - 1].$$

The (plain inductive) proof of this lemma is left to the reader as Exercise 3.8.

3.4 Sets and Hypersets as Membership Graphs

By considering $\in$ as an edge relationship, what graphs do we get? We will now address this issue. The following definition should be restrained to hereditarily finite sets; for, as a general rule, the graphs treated herein are finite:

> **Definition 3.3 (Membership graphs)**   The *membership graph* of a set x is the graph
>
> $$\left(\mathsf{trCl}\,(x),\ \{\langle u, v\rangle :\ u, v \in \mathsf{trCl}\,(x) \mid u \ni v\}\right).$$
>
> The *pointed membership graph* of a set x is the structure
>
> $$\left(\mathsf{trCl}\,(\{x\}),\ \{\langle u, v\rangle :\ u, v \in \mathsf{trCl}\,(\{x\}) \mid u \ni v\},\ x\right),$$
>
> namely, the membership graph of $\{x\}$ with x as a designated vertex.
>   (More generally, a graph is called *pointed*[a] if one of its vertices has been designated from which all vertices of the graph are reachable.)
>
> _____
>
> [a]In the literature, our pointed graphs are sometimes called *pointed accessible* graphs.

(Here and there, we may use the locution "membership *di*graph" in order to stress that these are directed graphs.)

Remark 3.3 To simplify the characterizations just given of the two kinds of membership graphs, it may be helpful to remark the obvious identities

$$\{\langle u, v\rangle : u \in \mathsf{trCl}\,(x),\quad v \in \mathsf{trCl}\,(x)\quad \mid u \ni v\} = \{\langle u, v\rangle : u \in \mathsf{trCl}\,(x),\quad v \in u\},$$
$$\{\langle u, v\rangle : u \in \mathsf{trCl}\,(\{x\}), v \in \mathsf{trCl}\,(\{x\})\ \mid u \ni v\} = \{\langle u, v\rangle : u \in \mathsf{trCl}\,(\{x\}), v \in u\}$$

between setformers. Thanks to the transitivity of $\mathsf{trCl}\,(x)$ and of $\mathsf{trCl}\,(\{x\})$, in fact, the missing iterator on the right of each equality ($v \in \mathsf{trCl}\,(x)$ and $v \in \mathsf{trCl}\,(\{x\})$, respectively) follows from the clause $v \in u$ serving as an iterator in the right-hand sides, which is rewritten as the restraining condition $u \ni v$ on the left. $\dashv$

Clearly, in a membership graph, two vertices cannot have the same out-neighborhood; for this would contradict the axiom of extensionality. On the other hand, distinct sets x, x' may well have $\mathsf{trCl}\,(x) = \mathsf{trCl}\,(x')$ (Fig. 3.3 provides an example). Membership graphs have no cycles as far as "ordinary" sets are concerned—that is to say, in a universe of sets which models foundation, e.g., in HF—then the graph of $\mathsf{trCl}\,(\{x\})$ has only one source, namely, x, and hence solely x candidates to be its point; in this case, $x = x'$ will follow from $\mathsf{trCl}\,(\{x\}) = \mathsf{trCl}\,(\{x'\})$.

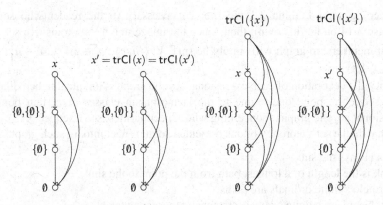

Fig. 3.3 For $x = \{\emptyset, \{\emptyset, \{\emptyset\}\}\}$ and $x' = x \cup \{\{\emptyset\}\}$, $\mathsf{trCl}(x) = \mathsf{trCl}(x')$ holds

A membership graph is a pleasantly concrete but redundant object: to better grasp its nature, we must allow its set $\mathsf{trCl}(x)$ of vertices to be superseded by an equinumerous set V.

Before moving on to "extraordinary" sets whose universe systematically violates foundation, we want to capture the structural properties of those graphs which are *isomorphic* to membership graphs in the well-founded case. Of course, one such property is acyclicity, which we recall here, thus taking the opportunity to also introduce the notion of rank of a vertex[3]:

Definition 3.4 (Acyclic graph and rank)   We say that a graph $G = (V, E)$ is *acyclic* if $\langle v_k, v_1 \rangle \notin E$ holds for any path $(v_1, \ldots, v_k)$ contained in G.

We call *rank* of a vertex u of an acyclic graph G the maximum value r for which G contains a path $(u_0, u_1, \ldots, u_r)$ with $u_0 = u$.

As hinted above, we also need a natural graph-theoretic counterpart of the set-theoretic notion of extensionality:

Definition 3.5 (Extensional graph)   We say that a graph $G = (V, E)$ is *extensional* if for all distinct $u, v \in V$, it holds that

$$N^+(u) \neq N^+(v). \tag{3.2}$$

If $u, v \in V$ are such that $u \neq v$ and $N^+(u) = N^+(v)$, we say that u and v *collide*.

To get an isomorphism between any given extensional acyclic graph $G = (V, E)$ and a membership graph, it suffices to consider the decoration $\mathfrak{m}$ of G *à la Mostowski*

---

[3]Determining the ranks of all vertices of an acyclic graph is a task of low computational complexity, cf. Exercise 3.13.

(the very general Lemma 3.1 will not be necessary to the reader who solves Exercise 3.14 on her/his own); then, since the image $m \restriction V$ is a transitive set, the sought membership graph will simply be $(m \restriction V, \{\langle u, v \rangle : u \in m \restriction V, v \in u\})$ (cf. Remark 3.3).

Since the decoration of each extensional acyclic graph is unique, the hereditarily finite sets correspond one-to-one to the isomorphism classes of certain pointed extensional acyclic graphs (concerning whose peculiarities, see Exercise 3.15).

What do the set-theoretic notions presented so far reflect into, in such graphs?

- $\emptyset$ is always the sink;
- rank is the length of a longest path from the point to the sink;
- Zermelo's finite ordinals are paths;
- von Neumann's finite ordinals are acyclic tournaments.

As we will see in Chap. 6, the Ackermann encoding has its analogue for the vertices of an extensional acyclic graph.

Let us now lift the focus of our discussion to *hypersets*, namely, to non-well-founded sets complying with the anti-foundation axiom AFA. In this case, the graph-theoretic view is even more insightful and rewarding than for ordinary sets. Actually, Peter Aczel himself popularized this view in his influential book [2], where pointed graphs related to the ones we are proposing are called *pictures of sets*. Among all pictures of a set, Aczel considers its *canonical picture*, which is precisely our pointed membership graph—whose decoration, being the identity function, is injective—and its *canonical tree picture*, where lack of redundancy is traded for elegance of representation. Aczel's tree picture is infinite in the case of a hereditarily finite hyperset *proper*; this is why we prefer to disallow it in general. Thus, for us, the presence of a cycle in a pointed graph will flag that the graph represents—if anything—a proper hyperset. We also want that our hyperset-representing graphs have an injective decoration, and, precisely in this sense, we will insist on the lack of redundancy.

No changes to the definitions (see Definition 3.3) of membership graphs and pointed membership graphs are needed to adjust them to hypersets; those definitions presuppose, though, the transitive closure operator $\mathrm{trCl}(\cdot)$, and we cannot retain the $\in$-recursive characterization of that operator given in Sect. 2.2.4. It will be the case anyhow, for every hyperset s, that an $\subseteq$-minimal superset t of s exists such that $y \in t$ follows from $y \in x \in t$ (cf. Exercises 2.2 and 2.6); it shall then be understood that $\mathrm{trCl}(s) = t$ holds for this t (so that s itself is a transitive hyperset when $t = s$).

Example 3.5 Consider the system of set-equations displayed below in ordinary notation (left) and as a graph (right):

$$
\begin{cases}
x_0 = \{x_0, x_1, x_8\} \\
x_1 = \{x_2, x_5\} \\
x_2 = \{x_3\} \\
x_3 = \{x_4\} \\
x_4 = \{\} \\
x_5 = \{x_4, x_6\} \\
x_6 = \{x_7\} \\
x_7 = \{\} \\
x_8 = \{x_8\}
\end{cases}
$$

According to AFA, the <u>a</u>nti-<u>f</u>oundation <u>a</u>xiom introduced at the end of Sect. 2.2.5, this system admits one and only one solution $x_0 \mapsto x_0, \ldots, x_8 \mapsto x_8$.

We can easily read off either representation of this system the values x_1 through x_7, which are the ordinary sets $x_4 = x_7 = \emptyset$, $x_6 = x_3 = \{\emptyset\}$, $x_2 = \{\{\emptyset\}\}$, $x_5 = \{\emptyset, \{\emptyset\}\}$ and $x_1 = \big\{\{\{\emptyset\}\}, \{\emptyset, \{\emptyset\}\}\big\}$.

As for the values of x_8 and x_0, the guarantee that hypersets Ω and x_0 exist which satisfy the respective equations $x_8 = \{x_8\}$ and $x_0 = \big\{x_0, \Omega, \big\{\{\{\emptyset\}\}, \{\emptyset, \{\emptyset\}\}\big\}\big\}$ comes directly from AFA. It is obvious that $x_0 \neq x_8$ are distinct, because x_0 has an element which is not a hyperset proper.

The only vertex of the above graph that we can pick as its point is x_0 (this in fact is the only vertex from which all other vertices can be reached); however, the fact that multiple vertices get the same value in the system of equations described by this pointed graph indicates that the graph does not irredundantly represent x_0; an acceptable representation of x_0 will result from coalescing vertex x_7 with vertex x_4 and vertex x_6 with vertex x_3.

In conclusion, the pointed membership graph of x_0 will be isomorphic to the following one, whose point is shown in black:

(continued)

Example 3.5 (continued)

Exercise 3.16 invites the reader to discuss a system of equations similar to the one examined in this example.

AFA has a graph-theoretic counterpart in the notion of *bisimilarity* that, much as in the case of sets, will turn out to subsume the notion of extensionality. We begin by introducing the notion of bisimulation among nodes of the same graph.

Definition 3.6 (Bisimulation)   A *bisimulation* on a given graph $G = (V, E)$ is a relation $B \subseteq V \times V$ such that for all $u_0, u_1 \in V$ for which $\langle u_0, u_1 \rangle \in B$ holds, the following two conditions hold as well (see Fig. 3.4):

- $\forall v_1[\langle u_1, v_1 \rangle \in E \rightarrow \exists v_0(\langle u_0, v_0 \rangle \in E \wedge \langle v_0, v_1 \rangle \in B)]$;
- $\forall v_0[\langle u_0, v_0 \rangle \in E \rightarrow \exists v_1(\langle u_1, v_1 \rangle \in E \wedge \langle v_0, v_1 \rangle \in B)]$.

In words, B is a bisimulation if and only if $\langle u_0, u_1 \rangle \in B$ implies that to every E-out-neighbor v_1 of u_1, there corresponds at least one E-out-neighbor v_0 of u_0 such that $\langle v_0, v_1 \rangle \in B$ holds and, symmetrically, to every E-out-neighbor v_0 of u_0, there corresponds at least one E-out-neighbor v_1 of u_1 such that $\langle v_0, v_1 \rangle \in B$ holds. A trivial example of bisimulation is $\{\langle x, x \rangle : x \in V\}$.

We leave the proof of the following lemma as Exercise 3.17 to the reader:

Lemma 3.3 (Union of bisimulations)   *Consider a graph $G = (V, E)$ and a set $\mathscr{B}$ of bisimulations on G. The union $\bigcup \mathscr{B}$ is, in its turn, a bisimulation on G; moreover, if $\mathscr{B}$ is the set of* all *bisimulations on G, then $\bigcup \mathscr{B}$ is an equivalence relation over V.*

Based on the observation just made, we now introduce the notion of *bisimilarity*:

Definition 3.7 (Bisimilarity)   The largest bisimulation on a given graph G is called *bisimilarity* on G; it will be denoted by $\equiv_G$ (or just by $\equiv$).

Bisimilarity holds between nodes u, v when they are in relation B for *some* bisimulation.

Definition 3.8 (Hyperextensionality)   A graph G is said to be *hyperextensional* if $\equiv_G$ is the identity relation. See also Example 3.6.

Fig. 3.4 Diagram
illustrating the properties of a
bisimulation B on E

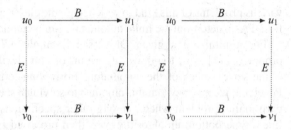

Any hyperextensional graph is also extensional—this fact is proved in
Lemma 3.4 below. Hyperextensionality is a natural adaptation of the notion of
extensionality to the case of cyclic graphs. The adaptation is based on a parsimony
principle: *whatever can be considered equal is equal*. Hyperextensionality is the
graph-theoretic analogue of AFA, which stems from it when the edge relation is the
converse $\ni$ of membership.

Example 3.6 The three graphs drawn below are hyperextensional:

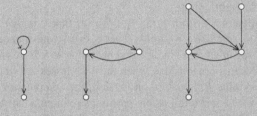

however, the following three graphs are not hyperextensional. We draw with
dashed ellipses the bisimilarity classes on these graphs.

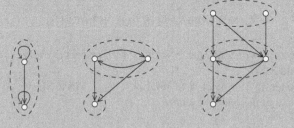

We invite the reader to check that these three bisimilarity relations indeed
satisfy the two conditions from Definition 3.6.

Remark 3.4 The question whether a graph is extensional or hyperextensional can
be answered by techniques that fall under "partition refinement as an algorithmic
paradigm" (cf. [82]). Under the heading *stable partitioning*, Panel 3.3 presents

various problems of this kind to which, e.g., the classical state minimization problem for DFAs ('deterministic finite automata') can be reduced easily.

For minimizing a given DFA (see Example 3.7 below), John E. Hopcroft proposed in [48] a top-down algorithm of complexity $O(|\mathcal{Q}| \log |\mathcal{Q}|)$, where $\mathcal{Q}$ is the set of states of the automaton. Borrowing terminology and notation from Panel 3.3, we may say that this amounts to solving a specialized version of the stable partitioning problem where, among other specificities, $\mathfrak{R}$ consists of functions. A linear-time bottom-up algorithm was then proposed in [83] for the case when $\mathfrak{R}$ consists of a single function[4] (this can hence be used for minimizing a DFA whose alphabet is singleton). Then Robert Paige and Robert E. Tarjan, in [82] (cf. also [51]), combined the key point "process the smaller half" of Hopcroft's strategy with novel ideas to design an algorithm, running in $O(|R| \log |S|)$ time and $O(|R| + |S|)$ space, for the stable partitioning problem with $\mathfrak{R} = \{R\}$. This hence is an upper bound for the complexity of computing bisimilarity on a graph $G = (S, R)$, in general, but when the input graph is acyclic, the problem can be solved by an $O(|R|)$ algorithm [30], deep-rooted in the Ackermann order of the well-founded hereditarily finite sets.

Panel 3.3 Stable partitioning

Let π be a partition of S, with $\pi = S/\sim_\pi$. We say that π is *stable*, relative to an $R \subseteq S \times S$, iff $\sim_\pi \circ R \subseteq R \circ \sim_\pi$ holds. That is, every equivalence class of $\sim_\pi$ can be "safely" regarded as a single element, as far as R is concerned. From this point of view, it is obvious that when the $\sim_\pi$-classes are singletons, π is stable for any possible R. Speaking in general, the computational challenge is to produce the *coarsest* of all stable partitions refining a partition given as input.

For a fixed $R \subseteq S \times S$, stability can be defined locally, with respect to a subset Q of the domain S, as follows:

$$\forall p \in \pi \quad \emptyset \in \left\{ p \cap (R^{\triangleleft}Q), \, p \setminus (R^{\triangleleft}Q) \right\};$$

the *stability* of π will hence mean stability with respect to every $\sim_\pi$-class (see Figs. 3.5 and 3.6).

More generally, for a set $\mathfrak{R}$ of relations, $\mathfrak{R} \subseteq \mathcal{P}(S \times S)$, we say that π is stable relative to the entire $\mathfrak{R}$ when

(continued)

---

[4]In this special case, where $\mathfrak{R} = \{f\}$ and f is a function from the entire S into S, stability amounts to the requirement that $f(b) \in q$ must follow from $\{a, b\} \subseteq p$ and $f(a) \in q$, with p, q blocks. This partitioning problem is treated at length in [3, pp. 157–162], which gives a top-down algorithm for its solution whose running time is $O(|S| \log |S|)$.

Panel 3.3 (continued)

$$\forall R \in \mathfrak{R} \; \forall q \in \pi \; \forall p \in \pi \; \big(p \cap (R \hspace{-0.3em}\restriction\hspace{-0.3em} q) \neq \emptyset \;\rightarrow\; p \subseteq R \hspace{-0.3em}\restriction\hspace{-0.3em} q\big).$$

The *stable partitioning problem*, in its strongest formulation, is the problem of determining the partition of S that

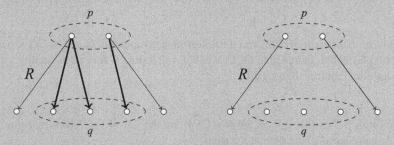

Fig. 3.5 *R*-stability of a partition π with respect to one of its blocks, q. (Every arrow here represents an ordered pair in R, while ovals represent blocks in π.) Within any block p of π, either each element has one or more *R*-out-neighbors in q (*left*) or no element has any such neighbor (*right*)

- is finer than a given initial partition $\pi^\star$ of S,
- is stable with respect to a given $\mathfrak{R} \subseteq \mathcal{P}(S \times S)$,
- is the coarsest of all partitions that satisfy the preceding two conditions.

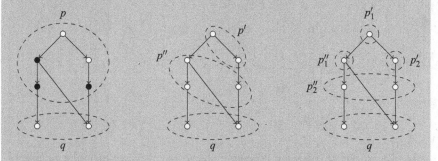

Fig. 3.6 *R*-instability of a partition π with respect to one of its blocks, q. (Arrows represent ordered pairs in R, while *ovals* represent blocks in π.) Some elements of p (drawn in *black*) have *R*-out-neighbors in q whereas some others do not, causing instability. We can refine π by splitting (*center*) p into two blocks p' and p'' that behave "the same" with respect to q. In this situation, q is called a *splitter*. Notice that the resulting partition is still not stable with respect to either one of the new blocks p' and p''. The coarsest stable partition of π is drawn on *the right*

(continued)

Panel 3.3 (continued)

A number of sophisticated algorithms are available today to solve this problem either in full generality or in restricted forms. For example, the minimization problem for deterministic finite automata can be reduced to the case in which $\Re$ consists of a number of *functions* equal to the cardinality of the alphabet.

A variety of problems can easily be reduced to stable partitioning; see [43].

Hence, in the light of [82] and [30], testing a given graph $G = (V, E)$ for being hyperextensional requires time $O(|E| \log |V|)$, and testing G for extensionality, when G is acyclic, requires time $O(|E|)$. ⊣

Example 3.7 Either one of the two DFAs on the left of the following figure can be reduced to the one on the right.

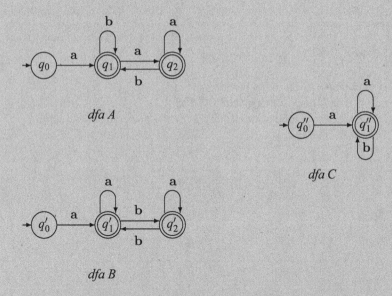

dfa A

dfa C

dfa B

In fact, q_1 and q_2 both are accepting states, and they are bisimilar relative to the transitions labeled **a**, as well as relative to the transitions labeled **b**; likewise, q'_1 and q'_2 are bisimilar. Since q_1 and q_2 behave the same, they can be merged into a single state. The same can be said of q'_1 and q'_2.

Note that any extensional acyclic graph has exactly one sink, and from every vertex, there is a directed path to it. The following lemma shows that also hyperextensional simple graphs, besides also being extensional, enjoy the same two properties.

Lemma 3.4 *Let G be a hyperextensional simple digraph. The following hold:*

(i) G is extensional;
(ii) G has a unique sink;
(iii) there is a directed path from every $v \in V$ to the sink of G.

Proof To see that *(i)* holds, it suffices to observe that if distinct $u, v \in V$ have $N^+(u) = N^+(v)$, then the equivalence relation that puts u and v in the same class and keeps every other vertex in a singleton class, that is, the equivalence relation induced by the partition $\{\{u, v\}\} \cup \{\{w\} : w \in V \setminus \{u, v\}\}$, is a nontrivial bisimulation on G.

For *(ii)*, note that G is extensional, by *(i)*, and thus it has at most one sink. If it does not have any sink, then G must have at least two vertices because it is simple and does not admit self-loops. Hence, the equivalence relation that puts all vertices of G in the same equivalence class is a nontrivial bisimulation over G. This contradicts hyperextensionality of G.

To show *(iii)*, take, for a contradiction, a vertex $v \in V$ such that there is no directed path from v to a sink of G. Let C be the set of all vertices u such that there is a directed path from v to u. By assumption on v, each vertex in C has at least one out-neighbor, and all such out-neighbors are in C. Since $N^+(v) \neq \emptyset$, we have that $|C| \geqslant 2$. The equivalence relation induced by the partition $\{C\} \cup \{\{w\} : w \in V \setminus C\}$ is a nontrivial bisimulation over G, contradicting the hyperextensionality of G.   ⊣

The following lemma shows that hyperextensional graphs indeed generalize extensional acyclic graphs.

Lemma 3.5 *If G is an extensional acyclic graph, then G is hyperextensional.*

Proof Suppose that G is an extensional acyclic graph that admits a nontrivial bisimulation B, and let u_0, v_0 be distinct vertices of G such that $u_0 B v_0$. By the extensionality of G, $N^+(u_0) \neq N^+(v_0)$. Therefore, we may assume w.l.o.g. that there exists an $u_1 \in N^+(u_0) \setminus N^+(v_0)$. Since $u_0 B v_0$, there exists an $v_1 \in N^+(v_0)$—so that plainly $u_1 \neq v_1$—such that $u_1 B v_1$. By continuing this process for long enough, since the number of vertices of G is finite, we will sooner or later encounter a vertex u_i, or v_i, already visited. This contradicts the acyclicity of G.   ⊣

In the following two lemmas, we state the triviality of the automorphism group of an extensional acyclic graph and of a hyperextensional graph, i.e., that the only automorphism for such a graph $G = (V, E)$ is the identity function $id_G : V \to V$ such that

$$id_G(v) = v, \text{ for all } v \in V .$$

Such graphs are also called *rigid*.

In Exercise 3.19, the reader is invited to prove Lemma 3.7. Lemma 3.6 then immediately follows from it, since thanks to Lemma 3.5 we know that any extensional acyclic graph is also hyperextensional. In Exercise 3.20, the reader is invited to give a direct proof of it, without resorting to Lemma 3.5.

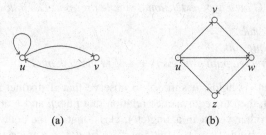

(a)   (b)

Fig. 3.7 Both graphs above, being strongly connected, have all vertices bisimilar to one another. However, they both have the identity as unique automorphism: on the left, in fact, vertices u and v have a different number of out-neighbors; similarly, on the right, vertex u is the only one with 2 out-neighbors, vertex w is the only one with 2 in-neighbors, and vertex v is the only one that has w as its only out-neighbor

Lemma 3.6 will be used in Chap. 6 when counting transitive well-founded sets of cardinality n. Thanks to Lemma 2.1, it implies that the number of distinct labelings of an extensional acyclic graph on n vertices, with n distinct labels, is $n!$.

Lemma 3.6 *Any extensional acyclic graph is rigid.*

Lemma 3.7 *Any hyperextensional graph is rigid.*

Next, we show that the converse of Lemma 3.6 holds. Therefore, on acyclic graphs, extensionality and rigidity are equivalent.

Lemma 3.8 *If G is a rigid acyclic graph, then G is extensional.*

Proof Let $G = (V, E)$ and r be the length of the longest path in G: we prove our claim by induction on r.

In case $r = 0$, since G is acyclic and rigid, V must be a singleton; hence, trivially, G is extensional.

In case $r > 0$, obtain the graph $G' = (V', E')$ from G by deleting all vertices whose rank is r. We claim that G' is rigid. In fact, if there was a nonidentity automorphism of G', this could be extended to a nonidentity automorphism of G simply by sending every $v \in V \setminus V'$ to itself. Moreover, clearly, G' is acyclic, and therefore the inductive hypothesis can be applied to conclude that G' is extensional.

At this point, if G were not extensional, there would exist two distinct nodes $u, v \in V \setminus V'$ such that $N^+(u) = N^+(v)$. Hence, the function $f : V \to V$ defined as $f(u) = v, f(v) = u$, and $f(x) = x$ for $x \in V \setminus \{u, v\}$, would be a nonidentity automorphism of G, contradicting our hypothesis.   ⊣

Remark 3.5 Notice however that the converse of Lemma 3.7 does not hold, namely, there are rigid graphs G that are not hyperextensional. See Fig. 3.7 for two examples.

3.5 Graphs for Set-Formulae

In this section, we give a basic example of algorithmic interplay between sets and graphs. In particular, we discuss very basic set-satisfiability problems and the role graphs play in their solution. This will be done in two steps: first, in Sect. 3.5.1, we will deal with set-satisfiability problems for unquantified formulae, and then, in Sect. 3.5.2, we will generalize our techniques to the case when one universally quantified variable can appear in the formula submitted to the satisfiability test.

Panel 3.4 Labeled graphs

A natural question arises when discussing the usages of graphs to abstractly model binary relations: what about *labels*? More precisely, what about the possibility of labeling either arcs or vertices?

Lemmas 3.6 and 3.7 tell us that extensional acyclic graphs and hyperextensional graphs admit only *one* way of labeling their nodes. The "label" of a vertex v in a (hyper)extensional graph is essentially given by its out-neighborhood set $N^+(v)$.

The question then becomes: can we use membership graphs in those situations in which a standard directed labeled graph would be used? The answer to this question is *yes*. In [30] the reader can find a way—by no means unique—to turn any labeled graph (with labels on either nodes, edges, or both) into a membership graph. The technique consists in replacing labels by "gadgets" that essentially are unique sets introduced as elements.

3.5.1 Ground-Level Satisfaction

We will now discuss how to solve, by means of graph algorithms, the following two problems about sets.

Problem 3.1 (Set-satisfiability of literal conjunctions) Given a set $\mathscr{C}$ of literals of the forms $x = y$, $x \neq y$, $y \in x$, $y \notin x$, $x = \emptyset$, where x, y stand for variables ranging over sets, establish whether the conjunction of the literals in $\mathscr{C}$ can be made true through an assignment $x \mapsto x$ of sets to its variables.

In case $\mathscr{C}$ is satisfiable, we are also interested in a satisfying assignment:

Problem 3.2 (Set-satisfaction of literal conjunctions) Check whether Problem 3.1 admits a positive answer; if so, also produce a satisfying set assignment.

We have restricted these two problems to *conjunctions* of literals so that they are solvable in polynomial time, as we will see next. In fact, allowing any

propositional combination of set literals leads to an NP-complete problem; for
example, Exercise 3.22 asks the reader to model the satisfiability problem for
formulae of propositional logic with a propositional combination of set literals.
Note also that any propositional combination of equality and membership literals
can be reduced to a finite (though not necessarily polynomial) number of instances
of Problem 3.1 by syntactic manipulations: see Example 3.9 on page 85.

At present, the given formula $\mathscr{C}$ involves no quantified variables. In a generalization of the problem which we will discuss in Sect. 3.5.2, a universally quantified
formula will enter into play, and unrestrained use of propositional connectives will
be admitted.

As a first step for solving Problem 3.1, we want to figure out which variables
are equal to one another. This subproblem can be stated as the one of finding the
connected components of an undirected graph $G_=$ whose vertices are the variables
occurring in $\mathscr{C}$ plus $\emptyset$ and in which each literal $x = y$ in $\mathscr{C}$ is represented as the
edge $\{x, y\}$. For every vertex x of $G_=$, we denote by C_x the connected component to
which x belongs. Having the connected components of $G_=$ allows us to propagate
the equalities in $\mathscr{C}$, by observing that two variables x and y belong to the same
component if and only if they must be equal. Hence, if there is an inequality $x \neq y$
in $\mathscr{C}$ such that $C_x = C_y$, we end reporting failure (i.e., unsatisfiability of $\mathscr{C}$).

It remains to handle the membership literals in $\mathscr{C}$, which we do by constructing
a graph $G_\in$ whose vertices are the connected components of $G_=$ and where each
literal $x \in y$ in $\mathscr{C}$ is represented as the arc $\langle C_y, C_x \rangle$. At this point, by virtue of the
equalities already found, for every arc $\langle C_y, C_x \rangle$ of $G_\in$ and any pair x', y' with $x' \in C_x$
and $y' \in C_y$, we know that $x' \in y'$ follows from $\mathscr{C}$. Hence, if there is a literal $x \notin y$
in $\mathscr{C}$ such that $\langle C_y, C_x \rangle$ is an arc of $G_\in$, we end reporting failure. Likewise, we fail if
there is an outgoing arc from $C_\emptyset$.

A less manifest source of unsatisfiability may lie in the presence of a cycle in
$G_\in$; for such a cycle would represent a constraint which well-founded sets cannot
meet. If a cycle is present, we end reporting failure.

Unless we have ended reporting failure, we can now report that $\mathscr{C}$ is satisfiable,
without yet producing any representation of a satisfying assignment. See also
Example 3.8.

Example 3.8 Consider the following formula as input for Problem 3.1:

$$\mathscr{C} \ : \ x = y \ \& \ y = z \ \& \ y \notin z \ \& \ z \neq v \ \& \ x \in u \ \& \ v \in u \ \& \ v = \emptyset.$$

The undirected graph $G_=$ and its connected components are:

(continued)

Example 3.8 (continued)
The only inequality literal in $\mathscr{C}$ is $z \neq v$, and z and v belong to different connected components of $G_=$. Therefore, we continue by constructing $G_\in$:

$$C_x = C_y = C_z \qquad\qquad\qquad C_v = C_\emptyset$$

The only negated membership literal in $\mathscr{C}$ is $y \notin z$, and $\langle C_z, C_y \rangle$ is not an arc of $G_\in$. Finally, since there are no arcs outgoing from $C_\emptyset$ and $G_\in$ is acyclic, we conclude that $\mathscr{C}$ is satisfiable.

In what precedes, we have obtained an algorithm for solving Problem 3.1 based on two simple graph algorithms. This can be summarized as follows.

Theorem 3.1 *Problem 3.1 is solvable in time $O(n + m)$, where n and m are, respectively, the number of variables and of literals in $\mathscr{C}$.*

Proof The undirected graph $G_=$ has at most n vertices and m edges, and its construction can be done in time $O(n + m)$. Computing the connected components of $G_=$ is doable in time $O(n + m)$ by a graph visit, and for each vertex v of $G_=$, we store the index of the connected component to which v belongs. For each inequality $x \neq y$ in $\mathscr{C}$, we check whether $C_x = C_y$ by checking in constant time whether the indices of the connected components associated with x and with y are the same. Likewise, for every literal $x \notin y$ in $\mathscr{C}$, we check in constant time whether $\langle C_y, C_x \rangle$ is an arc of $G_\in$. The graph $G_\in$ has at most n vertices and m arcs, and testing whether it is acyclic is doable in time $O(n + m)$ by another graph visit. ⊣

Example 3.9 Problem 3.1 can be used to check for validity formulae involving propositional connectives other than & , e.g., the following one:

$$x \in y \ \& \ (z \in x \vee z \in y) \ \rightarrow \ y \notin z. \qquad (3.3)$$

A propositional analysis splits the negation of (3.3) into the disjunction of the following two formulae:

$$x \in y \ \& \ z \in x \ \& \ y \in z \qquad (3.4)$$

$$x \in y \ \& \ z \in y \ \& \ y \in z \qquad (3.5)$$

(continued)

Example 3.9 (continued)

both of which are instances of Problem 3.1. Our algorithm for solving Problem 3.1 shows that neither of these two conjunctions is satisfiable, because the graph $G_\in$ corresponding to each of them contains a cycle. We therefore conclude that the original formula (3.3) is valid.

Notice that we cannot guarantee that an arbitrary formula is reducible to a polynomial number of instances of Problem 3.1. One sees, in fact, that establishing whether or not a propositional combination of set literals is satisfiable is NP-complete (Exercise 3.22).

To convert the satisfiability response for Problem 3.1—when positive—into an assignment satisfying $\mathscr{C}$ as required by Problem 3.2, we must decorate each vertex C of $G_\in$ with a set X_C. Since between the variables belonging to different vertices of $G_\in$ there may be some inequality literals in $\mathscr{C}$, we conservatively choose the set-decoration of the vertices of $G_\in$ to be injective. Similarly, since between the variables belonging to different vertices of $G_\in$ there may be some nonmembership literals in $\mathscr{C}$, we will ensure that $X_C \in X_{C'}$ holds only when $\langle C', C \rangle$ is an arc of $G_\in$.

To achieve the former goal, to each X_C such that $C \neq C_\emptyset$, we can assign a unique element s_C chosen such that it is distinct from all the sets that will decorate the vertices of $G_\in$. For example, each s_C can be chosen so that it has cardinality at least as large as the number of vertices of $G_\in$. In order to achieve the second goal, we set $X_{C_\emptyset} = \emptyset$, and then scan the vertices of $G_\in$ in topological order, and for each vertex C of $G_\in$ we set:

$$X_C = \{X_{C'} : \langle C, C' \rangle \text{ is an arc of } G_\in\} \cup \{s_C\}.$$

As one readily sees, the assignment $x \mapsto X_{C_x}$ (closely akin to the Mostowski decoration related to Lemma 3.1) is a satisfying set assignment for $\mathscr{C}$. Observe that for our purposes, it is immaterial what particular values the sets s_C take (Exercise 3.23 suggests another strategy for choosing these s_C). Hence, the graph $G_\in$ would, by itself, be an acceptable description of the assignment satisfying $\mathscr{C}$.

Example 3.10 Resume with the formula from Example 3.8 as input for Problem 3.2 and the acyclic graph $G_\in$ built for it. One way to assign the sets s_C and the resulting set-decoration of the vertices of $G_\in$ is:

(continued)

Example 3.10 (continued)

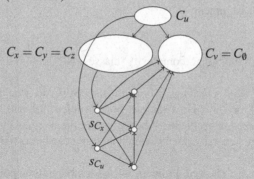

where thus $X_{C_\emptyset} = \emptyset, X_{C_x} = \{s_{C_x}\}, X_{C_u} = \{\{s_{C_x}\}, \emptyset, s_{C_u}\}$.

Remark 3.6 The construction just carried out illustrates a use of the extensional acyclic graphs that we have abstracted from membership graphs in Sect. 3.4.

If, in the same setting of Problem 3.1, we are to decide whether there is an assignment of *hyper*sets to the variables rendering $\mathscr{C}$ true, only a slight change to the solution method proposed above is needed: it suffices to withdraw the check that the graph $G_\in$ is acyclic. Much as before, when the satisfiability response is affirmative, we can exploit $G_\in$ as a guideline for constructing the assignment $x \mapsto X_{C_x}$. In particular, one must again expand the sets s_C so as to ensure the injectivity of the assignment $C \mapsto X_C$; to do this, one can resort—to mention one among many possible—to the technique suggested above, but insisting, to stay on the safe side, that the cardinality of each s_C *exceeds* the number of vertices of $G_\in$.

3.5.2 Quantified-Level Satisfaction

We will now tackle the satisfaction problem for formulae of the form $\forall y\, \mu$:

Problem 3.3 (Set-satisfiability of formulae with a single prefixed universal quantifier) Let φ be a formula of the form $\forall y\, \mu$, where y occurs in μ and μ stands for a propositional combination of literals of the forms $x = v$, $x \in v$, $x \neq v$, and $x \notin v$, with x, v variables. Establish whether or not the variables other than y, inside μ, can be replaced by sets so that φ becomes true.

Much as before, along with a positive answer to Problem 3.3, we are also interested in getting an explanation:

Problem 3.4 (Set-satisfaction of formulae with a single universal quantifier)
Check whether Problem 3.3 admits a positive answer; if so, also produce a
satisfying set assignment.

Panel 3.5 Computational complexity classes NP and #P; hardness and
completeness

A problem is called a *decision problem* if the answer to it on any input is of
the form "yes" or "no." For example, both Problem 3.1 and the 3CNFSAT
problem are decision problems. When studying the complexity of a decision
problem, we are generally interested in how fast it can be solved, as a function
of its input size.

As Theorem 3.1 showed, Problem 3.1 is solvable in time polynomial in its
input size. However, there is no known algorithm that can solve the 3CNFSAT
problem in polynomial time. Whether such a procedure exists or not is a major
open question in computer science. This is even more so, because a large
number of problems are equivalent to or at least as difficult as 3CNFSAT.
For example, Exercise 3.22 asks the reader to show that this is the case with
Problem 3.3.

To make such correspondences more formal, we need to introduce a few
more concepts. The class NP of decision problems contains exactly those
problems that, for any "yes" input I, admit a "yes certificate" testifying, after
a polynomial-time computation, that I is indeed a "yes" input. For example, a
certificate for the 3CNFSAT problem can be the assignment of truth values to
the propositional variables and thus 3CNFSAT $\in$ NP.

Given two decision problems Π and Π', we say that there is a *polynomial-
time reduction* from Π' to Π if there exists a procedure ρ that transforms any
input I for Π' to an input $\rho(I)$ for Π such that:

(i) $\rho(I)$ runs in time polynomial in the size of I, and
(ii) I is a "yes" input for Π' if and only if $\rho(i)$ is a "yes" input for Π.

A problem Π is said to be:

- *NP-hard* if for all problems $\Pi' \in$ NP, there is a polynomial-time reduction
  from Π' to Π. Thus, Π is as hard, up to a polynomial factor, as any
  problem in NP. 3CNFSAT is in fact an NP-hard problem.
- *NP-complete* if Π is NP-hard and $\Pi \in$ NP. 3CNFSAT $\in$ NP; thus, it is also
  NP-complete. In Sect. 4.5, we will study two other NP-complete problems
  related to sets and graphs.

For every decision problem Π in NP, we can consider its *counting version*
#Π asking for the *number* #(I) of solutions to an input I. Clearly, if I is a
"no" input, then #$(I) = 0$, and if I is a "yes" input, then #$(I) > 0$. For

(continued)

Panel 3.5 (continued)
example, #3CNFSAT asks for the number of satisfying truth assignments to the variables of the input formula. The class made up of such counting problems is called #P. One can analogously define the notions of hardness and completeness for the class #P, by additionally requiring that the polynomial-time reductions also "keep track" of the number of solutions. #3CNFSAT is a #P-complete problem. We will show that the two problems from Sect. 4.5 are also #P-complete.

As suggested already, Problem 3.1 can be seen as the degraded case of the problem being discussed now, in which μ can only be a conjunction of literals and y does not occur in μ (save hidden in the use of $\emptyset$—this constant could in fact be superseded by a variable $x_\emptyset$ subject to the condition $\forall y\, y \notin x_\emptyset$). For this reason, the reduction called for in Exercise 3.22 also shows that Problem 3.3 is NP-hard. Addressing the satisfiability issue for the more general scheme $\forall y_0 \cdots \forall y_{m+1}\, \mu$ would pose an even more challenging problem, as will emerge from Chap. 8.

Notice also that the input formulae for Problem 3.3 are complex enough to express the Boolean set operators $\bullet \setminus \bullet$, $\bullet \cap \bullet$, $\bullet \cup \bullet$, $\bullet \triangle \bullet$. For example, a union literal $a = b \cup c$ can be written as

$$\forall y\, (y \in a \leftrightarrow (y \in b \vee y \in c)),$$

while a symmetrical-difference literal $a = b \triangle c$ can be written as

$$\forall y\, (y \notin a \leftrightarrow (y \in b \leftrightarrow y \in c)).$$

Example 3.11 Using only *Boolean terms* of either one of the forms $x = y \cap z$, $x = y \cup z$, or the constant $\emptyset$, we can easily encode 3CNFSAT—the satisfiability problem for propositional logic formulae in conjunctive normal form, where each clause is a disjunction of exactly 3 literals. Consider a 3CNF formula:

$$\varphi = (L_{1,1} \vee L_{1,2} \vee L_{1,3}) \& \cdots \& (L_{m,1} \vee L_{m,2} \vee L_{m,3}),$$

with $P_1, \ldots, P_k$ all of its propositional variables. We define the formula φ^{set} to which φ will reduce, associating with each literal P_i, $\neg P_i$ appearing in φ a distinct set-theoretic variable denoted X_{P_i}, $X_{\neg P_i}$, respectively. Let φ^{set} be:

$$(\&_{i=1}^{k} X_{P_i} \cap X_{\neg P_i} = \emptyset) \, \&$$
$$(X_{L_{1,1}} \cup X_{L_{1,2}} \cup X_{L_{1,3}}) \cap \cdots \cap (X_{L_{m,1}} \cup X_{L_{m,2}} \cup X_{L_{m,3}}) \neq \emptyset.$$

(continued)

Example 3.11 (continued)

It is immediate to see that φ^{set} can be brought in $\forall y\mu$, where μ is in 3CNF-form.

As a matter of fact, the following, slightly different, encoding can be brought in $\forall y\mu$, where μ is now in 2CNF-form:

$$(\&_{i=1}^{k} X_{P_i} \cap X_{\neg P_i} = \emptyset \ \& \ \emptyset \in X_{P_i} \cap \{X_{P_i}\} \ \& \ \emptyset \in X_{\neg P_i} \cap \{X_{\neg P_i}\}) \ \&$$
$$(\dot{X}_{L_{1,1}} \neq X_{L_{1,2}} \vee X_{L_{1,1}} \neq X_{L_{1,3}}) \ \& \ \cdots \ \& \ (X_{L_{m,1}} \neq X_{L_{m,2}} \vee X_{L_{m,1}} \neq X_{L_{m,3}}).$$

Full details and more information on reductions and their complexity can be found in [73] and following citations therein.

See Example 3.12 below for another problem about sets reducible to Problem 3.3.

Example 3.12 Consider the *set unification* problem of establishing whether or not a formula

$$\&_{i=1}^{k} x_i = \{x_{i,1}, \ldots, x_{i,m_i}\}.$$

is satisfiable by sets. On the one hand, such a formula is an instance of Problem 3.3, since it can be written as

$$(\forall y) \left(\&_{i=1}^{k} \left(y \in x_i \leftrightarrow \bigvee_{j=1}^{m_i} y = x_{i,j} \right) \right).$$

On the other hand, the satisfiability of propositional formulae can also be modeled as a set unification problem, which constitutes an alternative to Exercise 3.22 for showing that Problem 3.3 is NP-hard. For example, consider the following propositional formula

$$(q_1 \vee \neg q_2 \vee q_3) \ \& \ (q_1 \vee q_2 \vee \neg q_3) \ \& \ (\neg q_1 \vee q_2 \vee \neg q_3).$$

The corresponding input to the set unification problem can be constructed as

$$F = \emptyset \ \& \ T = \{F\} \quad \&$$
$$\{p_1, n_1\} = \{F, T\} \qquad \& \ \{p_2, n_2\} = \{F, T\} \qquad \& \ \{p_3, n_3\} = \{F, T\} \quad \&$$
$$\{F, T\} = \{F, p_1, n_2, p_3\} \ \& \ \{F, T\} = \{F, p_1, p_2, n_3\} \ \& \ \{F, T\} = \{F, n_1, p_2, n_3\}.$$

Exercise 3.24 invites the reader to formalize this reduction.

Generally, in the case at hand, no relation over the unquantified variables of $\forall y \mu$ qualifies as the natural counterpart of equality, like the connected graph components did in the preceding section; however, there are finitely many ways of partitioning the set of those variables into disjoint blocks. Trying one partition amounts to assuming that each block collects variables which we choose to map to the same value, whereas variables belonging to distinct blocks are meant to take different values. Accordingly, each partition enables us to simplify the given formula so that only one variable from each block remains after the simplification, all occurrences of "=" and "$\neq$" disappear from the formula, and we can insist on satisfying the simplified formula *injectively*—namely, through an assignment of distinct values to its unquantified variables (see Panel 3.6). It should be clear that the formula given at the outset is satisfiable if and only if at least one of the formulae resulting from the choice of a partition and from the appropriate subsequent simplification is injectively satisfiable. We will, henceforth, assume w.l.o.g. that $\forall y \mu$ does not involve equality and aim at satisfying it injectively; moreover we will work under the tacit assumption that the quantified variable is subject to one or more constraints of the form $y \in x$, x unquantified, and to all inequalities $y \neq x$ with x distinct from y.

Panel 3.6 Bounding the universally quantified variable of $\forall y \mu$

Trivial simplifications applicable to any set-theoretic formula φ include the replacement of each equality of the form $v = v$ by `true`: then, inside φ, each inequality $v \neq v$ becomes `false` and the constants `true`, `false` can propagate through the propositional connectives (e.g., "ψ & `true`" reduces to ψ and "`false` $\rightarrow \psi$" to `true`). When acyclicity of $\in$ is postulated, trivial simplifications will alike cause all literals $v \in v$, $v \notin v$ to vanish from φ.

Before testing φ for satisfiability, if x, v are unquantified (hence existential) distinct variables appearing in φ, we can analyze separately φ_v^x and the opposite alternative $x \neq v$ & φ, where φ_v^x results from all trivial simplifications triggered by the replacement of x by v all over φ. In order to satisfy φ, we must be able to satisfy one of the alternatives: hence, working with this idea systematically, through case-based analysis, we can reduce set-satisfiability to the *injective* set-satisfiability problem for formulae where no literal of either form $x = v$, $x \neq v$ involves two unquantified variables.

Suppose, next, that φ is of the form $\forall y \mu$ with μ devoid of quantifiers. The above remarks entail that there is no loss of generality in requiring that y occurs in each equality literal of φ and in seeking to satisfy φ with different set-values for the variables $x_1, \ldots, x_n$ other than y in φ. We can rewrite φ as the conjunction of $\&_{i=1}^{n} \mu_{x_i}^y$ with

0. $\forall y \left(\left(\&_{i=1}^{n}(y \neq x_i \ \& \ y \notin x_i) \ \& \ y \in y \right) \rightarrow \mu \right)$,

---

(continued)

Panel 3.6 (continued)

1. $\forall y \left(\left(\&_{i=1}^n (y \neq x_i \ \& \ y \notin x_i) \ \& \ y \notin y \right) \rightarrow \mu \right)$, and with
2. $\forall y \left(\left(\left(\&_{i=1}^n y \neq x_i \right) \ \& \ \bigvee_{i=1}^n y \in x_i \right) \rightarrow \mu \right)$,

whence we can discard 0., inasmuch as true, when $\in$-acyclicity is postulated. More generally, the conjunction of $\&_{i=1}^n \mu_{x_i}^y$ with 0. and 1. is translatable into multiple instances of Problem 3.1 so that a negative answer to any of those will engender a negative answer to our present problem, whereas uniformly positive answers will leave us with the injective satisfiability of 2. as the only pending problem.

To see more clearly how to handle 1., assume momentarily $\in$-acyclicity and define for disjoint sets $\Xi, \bar{\Xi}$ such that $\Xi \cup \bar{\Xi} = \{x_1, \ldots, x_n\}$:

$$\gamma(\Xi) \ \leftrightarrow_{\mathrm{Def}} \ \&_{i=1}^n (y \neq x_i \ \& \ y \notin x_i) \ \& \ \&_{x \in \Xi} x \in y \ \& \ \&_{x \in \bar{\Xi}} x \notin y.$$

Since each implication $\left(\&_{i=1}^{n-1} \&_{j=i+1}^n x_i \neq x_j \right) \rightarrow \exists y \, \gamma(\Xi)$ is plainly provable, a sound treatment of 1. consists in checking the injective satisfiability of all ground-level formulae $\gamma(\Xi)_{x_0}^y \rightarrow \mu_{x_0}^y$ with $\Xi \subseteq \{x_1, \ldots, x_n\}$. One can handle 0. in analogy with 1. in non-well-founded set theory, which admits $\in$-cycles.

A key idea, in the ongoing, is that every entity in the universe of sets can be coarsely classed as follows, relative to any one-one assignment $x \mapsto \boldsymbol{x}$ of values to the variables other than y appearing in μ:

X-sets: these form the multi-image $\mathscr{X}$ of the mapping $x \mapsto \boldsymbol{x}$;

Y-sets: these are the elements of $\bigcup \mathscr{X} \setminus \mathscr{X}$; they differ from all X-sets but can nevertheless be relevant for the determination of the truth value of the formula $\forall y \mu$—particularly so because any Y-set belonging to the symmetrical difference $\boldsymbol{x} \bigtriangleup \boldsymbol{x}'$ of two X-sets witnesses that $\boldsymbol{x} \neq \boldsymbol{x}'$;

Z-sets: these lie outside $\bigcup \mathscr{X} \cup \mathscr{X}$ and play no significant role in the determination of the value of $\forall y \mu$.

There are finitely many X-sets, i.e., $\mathscr{X}$ is finite; however, it is possible that infinitely many Y-sets exist. A second key idea is that there are a finite number of Y-sets which "behave differently" and hence deserve being regarded as distinct entities in the evaluation of $\forall y \mu$ under the assignment $x \mapsto \boldsymbol{x}$. To be specific, we call *behavior* of a Y-set y the triple $y^-, y^+, \mathring{y}$ consisting of the place, $y^- = \{x \in \mathscr{X} \mid y \in \boldsymbol{x}\}$, occupied by y; the subset, $y^+ = \{\boldsymbol{x} \in \mathscr{X} \mid \boldsymbol{x} \in y\}$, formed by the X-sets belonging to y; and the truth value $\mathring{y} = (y \in y)$ (which is false when y is an ordinary set, but which may also be true when y lies in the non-well-founded universe).

A third key idea is that we can adequately represent X-sets, as well as Y-sets exhibiting different behaviors, as vertices of a graph whose arcs reflect the membership relation to the extent necessary to support the evaluation of $\forall y \, \mu$; we can insist that no two vertices representing X-sets can have the same out-neighbors, but it would be pointless to impose an analogous requirement on the vertices which represent Y-sets, because—due to its syntactic limitations—our formula cannot directly describe relationships involving two or more Y-sets. In looking for a graph modeling $\forall y \, \mu$, search can be limited to a finite inventory, which we can blindly span until a graph—if any—is found which evaluates $\forall y \, \mu$ to true.

After these preparatory considerations, we are now ready to introduce the *diag* notion, which adapts the idea which brought the graph $G_\in$ into play in Sect. 3.5.1 to the present, more demanding context.

Definition 3.9 (Diag) A *diag* is a graph

$$\Delta = (X \cup Y, \, E)$$

such that

1. $N^+(x) \neq N^+(x')$ for any pair x, x' of distinct vertices in X;
2. $X \cap Y = \emptyset$;
3. $X \cap N^-(y) \neq \emptyset$ for each vertex y in Y;
4. for $X_0 \subseteq X$ and $X_1 \subseteq X$, there can be at most two vertices y, y' in Y—one with a self-loop, the other one devoid of it—whose sets of in-neighbors and out-neighbors in X are X_0 and X_1, respectively.

When X consists of the variables distinct from y occurring in the given μ, we can *evaluate* the formula $\forall y \, \mu$ in Δ under the identity mapping $x \mapsto x$ with x in X; to do this, momentarily think of $X \cup Y$ (whose cardinality cannot exceed $|X| + 2^{2|X|}$, by Conditions 3 and 4.) as being the universe of discourse and of $\in$ as designating the arc relation E. As noted above, arcs—if any—between distinct vertices y, y' in Y nohow affect the outcome of this evaluation; hence, it would make sense (though at the price of a little pedantry) to augment the definition of a diag with the two inclusions

5. $N^\pm(y) \subseteq X \cup \{y\}$ for each vertex y in Y.

As explained above, one can exploit the diags whose X consists of the unquantified variables of $\forall y \, \mu$ to figure out how to injectively satisfy μ or to assess that this is an impossible task. Clearly enough, condition 1. is there to ensure that distinct variables cannot receive the same value; also, when working under the $\in$-acyclicity postulate, one will only take acyclic diags into account. The correctness of the proposed method then rests on two facts:

soundness: a diag in which $\forall y \, \mu$ evaluates to true can be obtained from any injective assignment satisfying this formula;

completeness: an injective assignment satisfying $\forall y \, \mu$ can be obtained from any diag evaluating this formula to true.

Both of these are checked straightforwardly, and we limit ourselves to a few indications on how to carry out their validation.

3.5.2.1 Soundness

Given the injective assignment $x \mapsto x$ of domain X, we put $\mathscr{X} = \{x : x \in X\}$ and then begin constructing the graph $(X \cup Y, E)$ by taking

$$Y = \left\{ y \subseteq X \mid y \neq \emptyset \ \& \ \left(\bigcap_{x \in y} x \right) \nsubseteq \left(\mathscr{X} \cup \bigcup_{x \in \mathscr{X} \setminus y} x \right) \right\}.$$

Intuitively, each y in Y represents a region of the Venn diagram associated with $\mathscr{X}$ which is populated by at least one set not belonging to $\mathscr{X}$; thus, for each vertex y in Y, we can pick a set y in $\left(\bigcap_{x \in y} x \right) \setminus \left(\mathscr{X} \cup \bigcup_{x \in \mathscr{X} \setminus y} x \right)$ and define the neighbors of y guided by the selected y: $N^+(y) \setminus \{y\} = \{x \in X \mid x \in y\}$, $N^-(y) \setminus \{y\} = y$, and y has a self-loop if and only if $y \in y$. For each pair x, x' belonging to $\mathscr{X}$, we put $\langle x, x' \rangle$ in E if and only if $x' \in x$. From the assumption that the given assignment satisfies $\forall y\, \mu$, it plainly follows that the graph so constructed is a diag evaluating $\forall y\, \mu$ to true.

3.5.2.2 Completeness

Conversely, out of a given diag $\Delta = (X \cup Y, E)$—which must be acyclic if we are working under the $\in$-acyclicity postulate—we obtain an injective set-valued assignment $v \mapsto v$ of domain $X \cup Y$ as follows. Regard the elements of X as variables $x_1, \ldots, x_{n'}$ and the ones of Y as additional variables $x_{n'+1}, \ldots, x_{n'+n''}$, where x_i and x_j are distinct when $i \neq j$; also regard each arc $\langle x_i, x_j \rangle \in E$ such that either $i = j$ or $i \leq n'$ or $j \leq n'$ as a membership literal $x_j \in x_i$: with our assignment, we will enforce that $x_j \in x_i$ holds for these literals, whereas $x_j \notin x_i$ will be enforced for any other pair $\langle x_i, x_j \rangle$ with $i, j \in \{1, \ldots, n' + n''\}$. To achieve this, we will proceed much in the same manner in which we have converted a satisfiability response to Problem 3.1 into an assignment as required by Problem 3.2. For brevity, here we discuss only the case when Δ is acyclic: while scanning the vertices x_i of Δ in topological order, put

$$x_i = \begin{cases} \{x_j : \langle x_i, x_j \rangle \in E\} & \text{when } i \leq n', \\ \{x_j : \langle x_i, x_j \rangle \in E \mid j = i \vee j \leq n'\} \cup \{s_i\} & \text{when } i > n', \end{cases}$$

for each vertex, choosing the sets s_i so that none of them equals any vertex image x resulting from this construction.

From the assumption that $\forall y\, \mu$ gets evaluated to true by the given Δ, it plainly follows that the assignment so constructed satisfies $\forall y\, \mu$.

Exercises

3.1 Is the function that sends every set x to the singleton $\{x\}$ a set? Is the class of all finite sets a proper class? Is the class of all (finite) graphs a set?

3.2 In a set theory postulating foundation and extensionality and enough to legitimize the use of the k-element collection operators $\{S_1, \ldots, S_k\}$ with $k \in \mathbb{N}$, consider a system

$$\bigwedge_{i=0}^{n} x_i = \{x_{i,1}, \ldots, x_{i,k_i}\}$$

of set-equations. Suppose that, herein, each doubly subscripted variable $x_{i,j}$ is drawn from among $x_0, x_1, \ldots, x_n$ and that these left-hand side variables are pairwise distinct. Prove that the following is a necessary and sufficient condition for solvability of the shown system of equations: There is a permutation $0 \mapsto \pi_0, 1 \mapsto \pi_1, \ldots, n \mapsto \pi_n$ of the subscripts such that every variable occurring in the right-hand side of each equation $x_{\pi_i} = \{x_{\pi_i,1}, \ldots, x_{\pi_i,k_{\pi_i}}\}$ is one of $x_{\pi_0}, \ldots, x_{\pi_{i-1}}$.

3.3 Prove that for every pair i, j of ordinals, the following biimplications are true:

$$i \subsetneqq j \leftrightarrow i \in j, \qquad \mathscr{V}_i \subsetneqq \mathscr{V}_j \leftrightarrow i \in j.$$

Moreover, prove that every $\mathscr{V}_i$ is a transitive set.

3.4 Show that $\mathsf{rank}(x)$ equals the minimum n such that $x \subseteq \mathscr{V}_n$ for any set x.

3.5 (*) Prove all laws listed in Fig. 3.1, regarding ordinals and the successor and rank operations.

3.6 Prove that in a theory of sets with foundation, the recursive property

$$\mathit{Is\_\mathscr{V}}(L) \quad \leftrightarrow_{\mathrm{Def}} \quad L = \bigcup \{\mathcal{P}(\ell) : \ell \in L \mid \mathit{Is\_\mathscr{V}}(\ell)\}$$

rightly characterizes the levels of von Neumann's cumulative hierarchy, namely, that $\mathit{Is\_\mathscr{V}}(L)$ is logically equivalent to the existence of an ordinal o such that $L = \mathscr{V}_o$.

3.7 Prove that $\mathsf{rank}(h_i) < \mathsf{rank}(h_j)$ implies $i < j$, for every $h_i, h_j \in \mathsf{HF}$.

3.8 Prove Lemma 3.2.

3.9 Explain how to assign to every graph G a natural number $\ddot{\imath}(G)$ so that for any two graphs $G_0 = (V_0, E_0)$, $G_1 = (V_1, E_1)$ whose sets of vertices are initial segments $V_0 = \{0, \ldots, n_0\}$ and $V_1 = \{0, \ldots, n_1\}$ of $\mathbb{N}$,
the equality $\ddot{\imath}(G_0) = \ddot{\imath}(G_1)$ holds if and only if G_0 and G_1 are isomorphic.

3.10 Prove that $|\mathscr{V}_n| = \beth(n)$ holds for each n in $\mathbb{N}$.

3.11 Put $\overline{\mathscr{V}}_n = \mathscr{V}_n \setminus \mathscr{V}_{n-1}$, for each $n > 0$ in $\mathbb{N}$. Prove that $|\overline{\mathscr{V}}_n| = \left(2^{|\overline{\mathscr{V}}_{n-1}|} - 1\right) |\mathscr{V}_{n-1}|$.

3.12 Let G be an extensional acyclic graph, and let R be the set of vertices of maximum rank in G. Show that $G - R$ is an extensional acyclic graph.

3.13 Let G be an acyclic graph with m arcs. Describe an algorithm that computes the rank of every vertex of G in total time $O(m)$.

3.14 Consider an acyclic graph $G = (V, E)$. Specify how to determine a function m from V into sets so that the equality $\mathsf{m}u = \{\mathsf{m}v : \langle u, v \rangle \in E\}$ holds for every $u \in V$. Prove that (1) there is only one such function m; (2) the image $\mathsf{m} \upharpoonright V$ is a transitive set; (3) each value $\mathsf{m}u$ is hereditarily finite, actually $\mathsf{m}u$ (as a set) and u (as a vertex) have the same rank; and (4) m is injective if and only if G is extensional.

3.15 Prove that the hereditarily finite sets are in one-one correspondence with the isomorphism classes of those pointed extensional acyclic graphs (V, E, s) which enjoy the two properties: $\{s\} \times (V \setminus \{s\}) \subseteq E$ and, for some $x \in V$, all maximum-length paths $(v_0, v_1, \ldots, v_r)$ of (V, E) have $v_1 = x$. Is the second of these truly needed?

3.16 In the system of equations of Example 3.5, replace the last equation by the following: $x_8 = \{x_0, x_5, x_8\}$; then consider the values x_0, x_8 in the solution to this new system. Draw the pointed membership graphs of x_0 and x_8; do these differ from one another? At a more fundamental level, does the inequality $x_0 \neq x_8$ hold?

3.17 Prove Lemma 3.3.

3.18 Prove that in every strongly connected graph, all nodes are bisimilar to one another.

3.19 Prove Lemma 3.7.

3.20 Prove Lemma 3.6 without resorting to Lemma 3.7.

3.21 A special case of stable partitioning is the *Venn partitioning problem*: determine, given a set $\mathscr{A}$ of sets, the coarsest of all partitions of $S = \bigcup \mathscr{A}$ that are stable with respect to $\mathfrak{R} = \{a \times S : a \in \mathscr{A}\}$.
  Give a slick characterization of the solution to the Venn partitioning problem.

3.22 Show that testing for satisfiability an arbitrary propositional combination of literals of the form $x = y, x \neq y, y \in x, y \notin x, x = \emptyset$ is an NP-hard problem (recall Panel 3.5), by reducing from the propositional satisfiability problem. *Hint:* Model every propositional letter as a set s satisfying the constraint $F \in s \leftrightarrow T \notin s$, where F and T are two set variables encoding false and true, respectively.

3.23 Consider the problem on p. 86 of choosing the sets s_C so that each vertex C of $G_\in$ is decorated by a set different from all sets s_C. Show that the wanted diversities can be enforced by choosing sets s_C all of the same rank r, larger than or equal to the number of vertices of $G_\in$.

3.24 Provide a formal description of the technique, illustrated in Example 3.12, for reducing the propositional satisfiability problem to the set unification problem.

3.25 Show that in the soundness proof (p. 94) for the diag-based test for the injective satisfiability of $\forall y\,\mu$, one can manage to limit the size of $|Y|$ so that $|Y| < |X|$. *Hint:* Observe that when $x_1, \ldots, x_n$ are distinct sets, a set d of cardinality $|d| < n$ exists such that $(x_i \triangle x_j) \cap d \neq \emptyset$ for $0 < i < j \leqslant n$.

3.26 Extend to the case of hypersets the completeness proof (p. 94) for the diag-based test for the injective satisfiability of $\forall y\,\mu$.

3.27 Establish that the formula

$$\left(u = t \setminus \{c\} \wedge s = t \cup \{x, v\} \wedge \left(z = v \vee (z = c \wedge z \in s) \right) \right) \rightarrow$$
$$u \cup \{x, z\} \in \{s \setminus \{c\}, s \setminus \{v\}, s\}$$

is valid by first restating its negation in the form $\forall y\,\mu$ (with μ devoid of quantifiers and of any set-theoretic operator) and then checking that the latter is unsatisfable.

Part II
Graphs as Sets

Chapter 4
The Undirected Structure Underlying Sets

As we have seen in Sect. 3.4, each set S is described in full by its pointed membership graph. Likewise, each transitive set is described in full by its membership graph. By "forgetting" the orientation of the edges of a membership graph, one obtains the *un*directed graph

$$G_s = \left(\text{trCl}\,(s),\ \left\{\{x,\, y\} : x \in \text{trCl}\,(s),\ y \in x\right\}\right).$$

This chapter studies the undirected graph which a well-founded set S induces in this manner when its transitive closure is finite. Since our main objects in this chapter are undirected graphs, we will use here the term *graph* to denote a undirected graph and *digraph* to denote a directed graph.

After embodying the main characteristics of G_s in the definition of *set graph*, we will find out additional features (e.g., connectedness) entailed by those, thereby revealing salient properties of a generic finite set $\text{trCl}\,(s)$.

Set graphs comprise two well-known classes of graphs. As shown in this chapter,

- every graph which has a Hamiltonian path is a set graph;
- claw-free graphs form the largest hereditary class of graphs every connected member of which is a set graph.

In Chap. 5, by taking into account that connected claw-free graphs are set graphs, we will obtain some hidden facts about them with relative ease.

We then continue by presenting two graph transformations under which the class of set graphs is closed: substitution of a vertex by a set graph and removal of a cut vertex accompanied by addition of certain edges between its neighbors.

Then this chapter discusses the algorithmic complexity of establishing whether or not a given undirected graph G is a set graph. This amounts to establishing whether an orientation can be imposed on the edges of G in a way that reflects the properties of membership over ordinary sets: well-foundedness and extensionality. To wit, neither cycles nor distinct vertices endowed with the same set of out-neighbors are permitted in the graph resulting from the orientation. The said problem turns out

© Springer International Publishing AG 2017

E.G. Omodeo et al., *On Sets and Graphs*, DOI 10.1007/978-3-319-54981-1_4

to be NP-complete, much like the analogous problem arising for hypersets: Does a given G admit a *hyper*extensional orientation?

In this chapter we will often play with some particular graphs, which are portrayed below (Fig. 4.1).

In the table below we gather the definitions of some graph classes that will appear in this chapter.

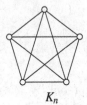

K_n

The complete graph K_n on n vertices. Between any two vertices of K_n there is an edge.

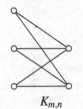

$K_{m,n}$

The complete bipartite graph $K_{m,n}$ with partite sets of respective sizes m and n. Between any two vertices in different partite sets there is an edge, and there are no other edges.

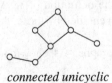

connected unicyclic graph

A connected graph G having a unique cycle.

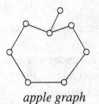

apple graph

An apple graph, that is, a graph made up of one cycle of length at least 4 plus one edge pendent from one of its vertices.

block graph

A graph in which every bi-connected component forms a complete graph.

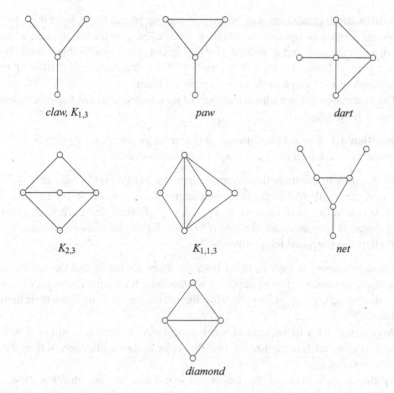

claw, $K_{1,3}$ *paw*   *dart*

$K_{2,3}$ $K_{1,1,3}$ *net*

diamond

Fig. 4.1 Some special graphs

4.1 Set Graphs

In this section we undertake the study of the undirected graphs underlying sets, which we call set graphs. Set graphs were introduced in [64] and later studied also in [63] and [65]; most of the results in this chapter come from the three articles just referenced. Here, by "graph" we mean an undirected graph, and by "digraph" we mean a directed graph. When necessary, we will stress the fact that we are dealing with undirected graphs.

 Definition 4.1 (Set graph) A graph is called a *set graph* if it admits an extensional acyclic orientation (an *EAO*, for short).

Recall from Sect. 3.4 that two vertices of directed graph are said to *collide* if they have the same out-neighborhood. Every set graph must be connected, because every acyclic orientation of a disconnected graph G has at least two sinks, and these collide. As we will see below, any set graph must satisfy two other basic constraints, both referring to the cardinality of some appropriately chosen sets of its vertices. Both conditions will be used in various places throughout this chapter. Moreover, from a computational point of view, these constraints, applied to given sets of vertices, make up a method of filtering out those graphs that definitely are not set graphs. However, we will see in Sect. 4.5.1 that deciding whether or not a given graph is a set graph is an NP-complete problem.

The first constraint, which we call the *cut set condition*, generalizes the connectedness of set graphs.

Proposition 4.1 (Cut set condition) *If G is a set graph, then for every $X \subseteq V(G)$, the graph $G - X$ has at most $2^{|X|}$ connected components.*

Proof Arguing by contradiction, suppose there are an EAO D of G and an $X \subseteq V(G)$ such that the number of connected components $Y_1, \ldots, Y_t$ of $G - X$ exceeds $2^{|X|}$. Since D is acyclic, there exist $y_1 \in Y_1, \ldots, y_t \in Y_t$ such that each y_i is a sink in $D[Y_i]$. Since D is extensional, the sets $N^+(y_i) \subseteq X$ must be pairwise distinct, which contradicts the supposed inequality $t > 2^{|X|}$. ⊣

The reader is urged to bear in mind both the graph-theoretic and the set-theoretic view of our arguments. For example, in set-theoretic terms, the above proof simply states that, as sets, $y_1, \ldots, y_t$ are pairwise distinct subsets of X, and thus their number is at most $2^{|X|}$.

Proposition 4.1 will be seen at work repeatedly in Sect. 4.3, where it will be applied to a cut set X of cardinality one. We refer to this application of it as the *cut vertex condition*.

By direct application of the cut vertex condition, we see that the claw (see page 103 and Example 4.1) is not a set graph. Removal of its vertex of degree three results in a graph with three connected components. A similar, but more general, necessary condition is given in Exercise 4.6. We continue with another condition that set graphs must satisfy.

Proposition 4.2 (Same neighbors condition) *If G is a set graph, then for every $X \subseteq V(G)$, the set $Y = \{y \in V(G) \mid N(y) = X\}$ has cardinality at most $|X| + 1$.*

Proof Let X, Y be subsets of $V(G)$ as in the claim. Observe first that in any acyclic orientation of G, either $N^+(y_1) \subseteq N^+(y_2)$ or $N^+(y_2) \subseteq N^+(y_1)$ must hold for every pair $y_1, y_2 \in Y$. If there were an exception to this, we would find an $x_1 \in N^+(y_1) \setminus N^+(y_2)$ and an $x_2 \in N^+(y_2) \setminus N^+(y_1)$; but then we would have arcs $\langle x_1, y_2 \rangle$ and $\langle x_2, y_1 \rangle$, hence the cycle $(y_1, x_1, y_2, x_2, y_1)$, a contradiction.

If G admits an EAO D, from the above observation and from the obvious inclusion $N^+(y) \subseteq X$ holding for each $y \in Y$, we get $|Y| \leq |X| + 1$. ⊣

Also Proposition 4.2 entails that the claw $K_{1,3}$ is not a set graph, since its three vertices of degree one have the same neighborhood of cardinality one. Proposition 4.2 implies that also the complete bipartite graph $K_{2,4}$ (see page 103) is not a set graph. See Exercise 4.7 for a full characterization of the complete bipartite graphs that are set graphs.

Example 4.1 Consider the claw $K_{1,3}$ shown below:

Removal of vertex a disrupts this graph into three connected components: $\{b\}$, $\{c\}$, and $\{d\}$. Also, we have $N(b) = N(c) = N(d) = \{a\}$. These two remarks show that the claw fails to satisfy the "cut vertex" condition of Proposition 4.1 and the "same neighbors" condition of Proposition 4.2.

From either remark, it follows that the claw is not a set graph; notice, incidentally, that this does not yield that all set graphs are claw-free: for example, the graph at the beginning of Example 4.2 contains claws, but it is a set graph.

4.2 Graph Classes Included in the Class of Set Graphs

In this section we show that the class of set graphs contains two well-known classes of (undirected) graphs, namely, graphs with a Hamiltonian path and connected claw-free graphs [66, 103]. Among other properties, claw-free graphs generalize the well-known class of line graphs, namely, those graphs which one obtains from a graph G by replacing every edge by a vertex and connecting two resulting vertices if the corresponding edges have a vertex in common. Exercise 4.10 asks the reader to show this fact.

When a digraph D is acyclic and there is a directed path from x to y, then x and y cannot collide in D. Consequently, recalling that a *Hamiltonian path* in an undirected graph is one that meets every vertex exactly once:

Theorem 4.1 *If G has a Hamiltonian path, then G is a set graph.*

Proof Let $(x_1, x_2, \ldots, x_n)$, with $n = |V(G)|$, be a Hamiltonian path in G. To obtain an EAO D of G, orient every edge $\{x_i, x_j\}$ of G as $\langle x_j, x_i \rangle$ where $j > i$. Clearly, D is acyclic; moreover, since we have a path in D from x_j to x_i for each pair i, j such that $1 \leqslant i < j \leqslant n$, D is also extensional. ⊣

Example 4.2 Consider the graph below in which we highlight a Hamiltonian path, and label its vertices as in the proof of Theorem 4.1:

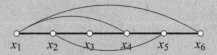

An EAO of it constructed as in the proof of Theorem 4.1 is

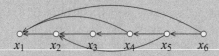

Set graphs that have no Hamiltonian paths exist: one such is the so-called *net* shown below (left), which is a set graph because it admits an EAO (right).

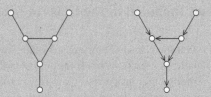

Exercise 4.4 asks the reader to generalize the net into a class of set graphs devoid of Hamiltonian paths.

We now prove the main result in this section, namely, that all connected claw-free graphs are set graphs. We will however prove a more general variant of this result, which works for $K_{1,r+2}$-free graphs, for any fixed $r \geqslant 1$, in which a suitably weakened variant of extensionality holds.

Definition 4.2 (*r*-extensionality) Given a digraph D and $A \subseteq V(D)$, we say that A is an *r-collision* of D if $|A| \geqslant r$ and for any $u, v \in A$ we have $N^+(u) = N^+(v)$. We say that D is *r-extensional* if no $(r + 1)$-collision exists.

Note that extensionality is the same as 1-extensionality. Moreover, if A is a collision in an acyclic digraph D, then A is an independent set in the undirected graph underlying D.

Theorem 4.2 *For any* $r \geq 1$, *every connected* $K_{1,r+2}$-*free graph* G *admits an* r-*extensional acyclic orientation with a unique sink. Such an orientation of* G *can be found in linear time in the size of* G.

Proof The proof goes by induction, the case $|V(G)| = 1$ being plain. Let $v \in V(G)$ be a vertex that is not a cut vertex of G (such a vertex always exists, see Exercise 2.17). Since $G - v$ is connected and $K_{1,r+2}$-free, let D' be an r-extensional acyclic orientation with a unique sink of $G - v$ as resulting from the inductive hypothesis. We consider two cases:

Case 1. If the sink of D' belongs to $N_G(v)$, then construct the orientation D of G with $V(D) = V(D') \cup \{v\}$ and $E(D) = E(D') \cup \{\langle x, v \rangle \ : \ x \in N_G(v)\}$. Clearly D is acyclic and has v as unique sink. Since any collision of D is also a collision of D', r-extensionality of D' implies r-extensionality of D.

Case 2. See Fig. 4.2. If the sink of D' does not belong to $N_G(v)$, then construct the orientation D of G with $V(D) = V(D')$ and $E(D) = E(D') \cup \{\langle v, x \rangle \ : \ x \in N_G(v)\}$. Orientation D is acyclic and has a unique sink. To see that it is also r-extensional, observe that any collision of D of size greater than r must contain v. Let A be such a collision. Consider the acyclic subdigraph $D[N_G(v)]$ and let w be a sink of it. Since the sink of D does not belong to $N_G(v)$, vertex w is not the sink of D, and thus it has an out-neighbor, say z. Observe that there is no arc between the vertices in A and z, since an arc directed in D from z toward an element of A would create a cycle in D, while an arc directed from a vertex in A toward z would contradict the fact that w is a sink of $D[N_G(v)]$. Since A is an independent set of size greater than r, the set $A \cup \{z, w\}$ contains an induced subgraph isomorphic to $K_{1,r+2}$, which contradicts the $K_{1,r+2}$-freeness of G.

The algorithm suggested by this proof is summarized as Algorithm 4.1 and has complexity $O(|V(G)| + |E(G)|)$. For choosing a vertex of G that is not a cut vertex, in such a way that the overall complexity stays linear, in Algorithm 4.1 we also maintain a spanning tree T of G, together with the set of leaves of T. (Recall that a spanning tree of a connected graph G can be computed in time $O(|V(G)| + |E(G)|)$; see also Exercise 2.16.)   ⊣

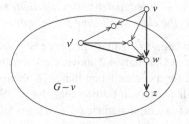

Fig. 4.2 Illustration of the Case 2 of the proof of Theorem 4.2. We assume that v collides with v'; in the notation in the proof, this reads $\{v, v'\} \subseteq A$. The set $\{w, v, v', z\}$ induces a claw in G

Algorithm 4.1: Finding an r-extensional acyclic orientation of connected $K_{1,r+2}$-free graphs

---

Input: A connected $K_{1,r+2}$-free graph G, a spanning tree T of G, whose set of leaves is L.
Output: An r-extensional acyclic orientation of G and its unique sink.
orient(G, T, L)
      if $|V(G)| = 1$ **then**
         let $V(G) = \{v\}$;
         return (G, v).
      end
      let v be arbitrary vertex in L;
      $L := L \setminus \{v\}$;
      remove v from T, and if its parent u in T becomes a leaf, add u to L;
      $(D, s) := $ orient$(G - v, T, L)$;
      $V := V(D) \cup \{v\}$;
      if $s \in N_G(v)$ **then**
         $E := E(D) \cup \{\langle x, v \rangle : x \in N_G(v)\}$;
         return $((V, E), v)$;
      else
         $E := E(D) \cup \{\langle v, x \rangle : x \in N_G(v)\}$;
         return $((V, E), s)$.
      end
end

Since any acyclic orientation of a disconnected graph has at least as many sinks as connected components, we have the following corollary:

Corollary 4.1 *For every $r \geqslant 1$, a $K_{1,r+2}$-free graph G admits an r-extensional acyclic orientation if and only if G has at most r connected components.*

As a particular case, when $r = 1$, we get the result about connected claw-free graphs claimed in at the beginning of this section and in the Introduction:

Corollary 4.2 *If G is a connected claw-free graph, then G is a set graph.*

Since the claw is not a set graph, Corollary 4.2 entails that

Corollary 4.3 *The largest hereditary graph class (i.e., closed undertaking induced subgraphs, or, equivalently, closed under vertex deletion) every connected member of which is a set graph is the class of all claw-free graphs.*

Proof Let $\mathscr{C}$ denote the largest hereditary graph class every connected member of which is a set graph. From Corollary 4.2 above, we know that claw-free graphs are included in $\mathscr{C}$. Suppose for a contradiction that in $\mathscr{C}$ there is a graph containing a claw K. From $\mathscr{C}$ being hereditary, it follows that $K \in \mathscr{C}$. Since K is connected, K must be a set graph by an assumption made on $\mathscr{C}$. This is false, though, as observed in Example 4.1. $\dashv$

4.3   Characterizations of Set Graphs in Some Graph Classes

As we will see in Sect. 4.5.1, it is an NP-complete problem to decide whether a given graph is a set graph. For this reason, it is interesting to obtain characterizations of set graphs in some graph classes and, possibly, polynomial time algorithms for recognizing set graphs in such classes. More specifically, given a class $\mathscr{C}$ of graphs, we will characterize the graphs from $\mathscr{C}$ that are set graphs.

Our choice of the graph class $\mathscr{C}$ will be guided by the necessary cut vertex condition from Proposition 4.1 and by the two sufficient conditions from Theorem 4.1 and Corollary 4.2 (having a Hamiltonian path and being connected and claw-free, respectively).

In Sect. 4.3.1 we observe that thanks to the cut vertex condition, recognizing set graphs is trivial in the class on trees, that is, graphs without (undirected) cycles. Thus, we will study the problem when $\mathscr{C}$ is the class of connected graphs with a unique cycle. In Sect. 4.3.2 we study a class of graphs where being a set graph is equivalent to having a Hamiltonian path. In Sect. 4.3.3 we study a class of graphs where being a set graph is equivalent to being connected and claw-free and to being connected and satisfying the cut vertex condition; this result will be strengthened in Sect. 4.3.4, by identifying the largest hereditary class of graphs where these three latter conditions are equivalent.

4.3.1   Unicyclic Graphs

Observe that when a set graph G has no cycles, that is, when G is a tree, Proposition 4.1 implies that G can have no vertex of degree 3 or more. This implies that a tree T is a set graph if and only if T is a path. Our next results characterizes set graphs in the class of connected graphs with a unique (undirected) cycle, also called *unicyclic* graphs.

Given a graph G, a cycle C in G and a sub-tree T of G, we say that T is *pendent* from C if $V(C) \cap V(T) = \{r\}$, where r is a vertex of degree 1 in T; r will be called the *articulation* vertex of T. Our next definition captures the desired characterization (see also Fig. 4.3).

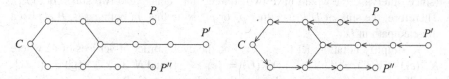

Fig. 4.3   A jellyfish graph (*left*) and an extensional acyclic orientation of it (*right*)

Definition 4.3 (Jellyfish graph)   A connected unicyclic graph G, having a cycle C, is said to be a *jellyfish graph* if there exist (possibly trivial) paths P, P', P'' in G, pendent from C and having respective articulation vertices r, r', r'', such that $G = C \cup P \cup P' \cup P''$, $r \neq r''$ and $d(r, r') = d(r', r'') = 1$.

The following lemma will be used in the proof of Theorem 4.3.

Lemma 4.1   *If D is an extensional acyclic digraph, then for every $X \subseteq V(D)$, there is at most one component S of the undirected graph underlying $D - X$ such that*

$$(\forall x \in X)(\forall s \in S)\big(\langle s, x \rangle \notin E(D) \big).$$

Proof Assuming that there were two such components S and S' of $D - X$, the sinks of $D[S]$ and $D[S']$ would also be sinks in D, contrary to the extensionality of D.   ⊣

We are now ready to prove the equivalence between set graphs and jellyfish graphs in the class of unicyclic graphs.

Theorem 4.3   *A unicyclic graph is a set graph if and only if it is a jellyfish graph.*

Proof We first prove the forward implication. Let D be an EAO whose underlying graph G is connected and unicyclic. Denoting by C the cycle of G, let us examine the orientation of the arcs between vertices of C. If there were at least two sinks s and s' in $D[V(C)]$, we would get a contradiction from Lemma 4.1 by taking as cut set $V(C) \setminus \{s, s'\}$. Conversely, if there were two sources in $D[V(C)]$, this would imply the existence of at least two sinks in $D[V(C)]$. Hence there is exactly one source t and one sink s in $D[V(C)]$. We claim that $s \in N^+(t)$. If not, then consider the two vertices s_1 and s_2 of C such that $s \in N^+(s_1) \cap N^+(s_2)$. Since D is extensional and neither s_1 nor s_2 is a source in $D[C]$, there exists, w.l.o.g., an $s_3 \in N^+(s_1) \setminus (N^+(s_2) \cup V(C))$. Taking $V(C) \setminus \{s\}$ as cut set in Lemma 4.1, the components of $G - (V(C) \setminus \{s\})$ containing s, and s_3, respectively, produce the desired contradiction.

Consider now $x \in V(G)$ with degree at least three. Since D is extensional, we get $x \in V(C)$, implying that all vertices in $V(G) \setminus V(C)$ lie on paths pendent from C. Moreover, by Proposition 4.1, two such paths cannot have the same articulation vertex.

Let $P = (p_1, \ldots, p_k)$ be a path in G pendent from C, having p_1 as its articulation vertex. If there were two sinks p_i and p_j, $i \neq j$, in $D[P]$, then we would get a contradiction from Lemma 4.1 by taking $P \setminus \{p_i, p_j\}$ as cut set. If the sink of $D[P]$ is p_t, where $1 < t < k$, then we obtain again a contradiction, as $N^+(p_{t-1}) = N^+(p_{t+1})$ (since otherwise we would have two sources, and hence also two sinks, in $D[P]$). Therefore, the sink of $D[P]$ is either p_1 or p_k. Note that, in both cases, P is also a directed path in D.

To simplify notation, let $(c_1, \ldots, c_\ell)$ be the cyclic order of vertices of C, where $N^+(c_1) \cap V(C) = \emptyset$, $N^+(c_\ell) \cap V(C) = \{c_1, c_{\ell-1}\}$, and $N^+(c_t) \cap V(C) = \{c_{t-1}\}$, for all $1 < t < \ell$. Consider also a path pendent from C, $P = (p_1, \ldots, p_k)$, with articulation vertex p_1. If p_k is the sink in $D[P]$, then $p_1 = c_1$, else we can apply Lemma 4.1 by taking $V(C) \setminus \{c_1\}$ as cut set. If p_1 is the sink of $D[P]$, then $p_1 \notin \{c_1, \ldots, c_{\ell-2}\}$. Indeed, if p_1 were some c_t, $1 \leqslant t \leqslant \ell - 2$, then we would have a collision between p_2 and c_{t+1}.

To sum up, we have that either $p_1 = c_1$, in which case p_k is the sink in $D[P]$, or that $p_1 \in \{c_{\ell-1}, c_\ell\}$, in which case p_1 is the sink in $D[P]$. In all of these cases, G is a jellyfish graph.

For the reverse implication, note that each jellyfish graph admits an EAO, which can be constructed as in Fig. 4.3 and along the same lines as in the forward implication. ⊣

Exercise 4.12 asks the reader for a linear-time algorithm for checking whether a given graph is a jellyfish graph, and Exercise 4.13 asks the reader for a linear-time algorithm for constructing an EAO of a jellyfish graph.

4.3.2 Complete Multipartite Graphs

Our next proposition generalizes the result requested in Exercise 4.7, by showing that in complete multipartite graphs having a Hamiltonian path is equivalent to being a set graph.

Proposition 4.3 *If G is a complete multipartite graph, then G is a set graph if and only if G has a Hamiltonian path.*

Proof The reverse implication follows from Theorem 4.1.

For the forward implication, we show by induction the following stronger claim: if D is an EAO of G having a vertex s as sink, then there exists a Hamiltonian path in G ending in s. When $|V(G)| = 1$, the claim is obvious. Thus, let G be a complete multipartite graph such that $|V(G)| \geqslant 2$ and suppose that D is an EAO of G having s as sink.

First, we claim that $D - s$ is an EAO of $G - s$. Assume, for a contradiction, that in $D - s$ there is a collision between x and y. Thus, x and y are not adjacent in G, and hence they belong to the same block of the partition of G. Additionally, we have that the symmetric difference of $N^+(x)$ and $N^+(y)$ in D consists solely of s, so say that $s \in N^+(x) \setminus N^+(y)$. Since G is multipartite, there must be an edge in G between s and y, which hence is oriented as $\langle s, y \rangle$ in D. This contradicts the fact that s is the sink of D, and we reach the desired contradiction.

Second, in D, the sink of $D - s$ is an in-neighbor of s.

Therefore, we can apply the inductive hypothesis to the complete multipartite graph $G - s$, having the EAO $D - s$, to obtain a Hamiltonian path in $G - s$ ending in a neighbor of s in G. By appending s to it, we obtain a Hamiltonian path in G. ⊣

Recall that a graph is called P_4-*free* if it does not have a path on four vertices as induced subgraph. Since the Hamiltonian path problem can be solved in linear time on P_4-free graphs [56], and complete multipartite graphs are P_4-free (see Exercise 4.15), we obtain the following result:

Theorem 4.4 *Set graphs can be recognized in linear time on the class of complete multipartite graphs.*

4.3.3   Block Graphs

By Corollary 4.2, a sufficient condition for a graph G to be a set graph is that G is connected and claw-free. On the other hand, a necessary condition for it to be a set graph is the cut vertex condition (Proposition 4.1): "for every vertex v of G, the graph $G - v$ has at most two connected components."

A class of graphs in which either one of the above conditions is both necessary and sufficient for a graph to be a set graph is the class of *block graphs*, that is, graphs in which every *biconnected component*, that is, every maximal connected subgraph without cut vertices, is a complete graph.[1] (Exercise 4.16 asks the reader to prove an alternative characterization of block graphs, one in terms of forbidden induced subgraphs, cf. [6, 49].)

Proposition 4.4 *Let G be a block graph. Then, the following conditions are equivalent:*

(1) G is a set graph.
(2) G is connected and satisfies the cut vertex condition.
(3) G is connected and claw-free.

Proof The conditions (1) $\Rightarrow$ (2) and (3) $\Rightarrow$ (1) follow from Proposition 4.1 and Corollary 4.2, respectively.

(2) $\Rightarrow$ (3): Let G be a connected block graph satisfying the cut vertex condition. Suppose for a contradiction that G contains a claw K induced by the vertex set $\{a, b, c, d\}$, where a is the vertex of degree 3 in K. Since G satisfies the cut vertex condition, we may assume w.l.o.g. that b and c belong to the same connected component of $G - a$. This implies that there exists a b-c path avoiding a, which, together with the path (b, a, c) forms a cycle. This implies that b and c are contained in some maximal connected subgraph of G without cut vertices. Since b and c are not adjacent, this contradicts the assumption that every such subgraph of G is complete.   ⊣

Exercise 4.17 asks the reader to use Proposition 4.4 to write an efficient algorithm for deciding whether a given block graph is a set graph.

---

[1] Notice that here we do not require block graphs to be connected, as opposed to e.g., [15].

4.3.4   A Generalization of Claw-Free Graphs and Block Graphs

The equivalences from Proposition 4.4 motivate the following questions about the largest hereditary classes of graphs where they hold:

1. What is the largest hereditary class of graphs $\mathcal{G}_1$ such that for every $G \in \mathcal{G}_1$ and every induced subgraph H of G, H is a set graph if and only if H is connected and satisfies the cut vertex condition?
2. What is the largest hereditary class of graphs $\mathcal{G}_2$ such that for every $G \in \mathcal{G}_2$, and every induced subgraph H of G, H is a set graph if and only if H is connected and claw-free?

Before characterizing the graph classes $\mathcal{G}_1$ and $\mathcal{G}_2$, we observe that both $\mathcal{G}_1$ and $\mathcal{G}_2$ generalize block graphs and claw-free graphs.

Remark 4.1 The class of block graphs is included in $\mathcal{G}_1 \cap \mathcal{G}_2$. Indeed, every induced subgraph H of a block graph G is a block graph. This follows either directly from the definition of block graphs or from their characterization in terms of forbidden induced subgraphs from Exercise 4.16. Proposition 4.4 applied to H then proves this observation.

Remark 4.2 The class of claw-free graphs is included in $\mathcal{G}_1 \cap \mathcal{G}_2$. Indeed, every induced subgraph H of a claw-free graph G is claw-free. If H is a set graph, then it must be connected and satisfy the cut vertex condition. Likewise, if H is connected, then it is a set graph by Corollary 4.2.

Theorem 4.5 below, apart from characterizing the classes $\mathcal{G}_1$ and $\mathcal{G}_2$, also shows that $\mathcal{G}_1 = \mathcal{G}_2$. To this aim, we need to introduce one more graph class. An *apple* of order $k \geqslant 4$ is the graph obtained from a cycle with k vertices by adding to it a new vertex and connecting it to precisely one vertex of the cycle (recall page 102). Thus, a graph G is called *apple-free* if it does not contain any apple as an induced subgraph [71] (see also [14, 16, 59]).

We also refer the reader to the graphs $K_{2,3}$, dart, and $K_{1,1,3}$, drawn in Fig. 4.1.

Theorem 4.5 *For every graph G, the following conditions are equivalent:*

(1) For every induced subgraph H of G, it holds that H satisfies the cut vertex condition if and only if H is claw-free.
(2) For every induced subgraph H of G, it holds that H is a set graph if and only if H is connected and claw-free.
(3) Graph G is (apple, $K_{2,3}$, dart, $K_{1,1,3}$)-free.

Proof The implication (1) $\Rightarrow$ (2) follows from Proposition 4.1 and Corollary 4.2.

The implication (2) $\Rightarrow$ (3) is straightforward since the apples, the $K_{2,3}$, the dart and the $K_{1,1,3}$ are graphs that are set graphs (notice that they contain a Hamiltonian path, so Theorem 4.1 applies), but they are not claw-free.

It remains to show (3) $\Rightarrow$ (1). Let G be an (apple, $K_{2,3}$, dart, $K_{1,1,3}$)-free graph and let H be an induced subgraph of G. Clearly, if H is claw-free, then H satisfies the cut vertex condition. Suppose now that H satisfies the cut vertex condition, and suppose for a contradiction that H contains a claw K induced by the vertex set $\{a, b, c, d\}$, where a is the vertex of degree 3 in K. Let $k \geqslant 2$ be the minimum distance in $H - a$ between two leaves of K. Note that k is finite since the graph $H - a$ has at most two connected components. We may assume, without loss of generality, that $P = (b, v_1, \ldots, v_{k-1}, c)$ is a path of length k connecting b and c in $H - a$. By the minimality of P, vertex d is not on P. However, d has a neighbor on P since otherwise G would contain either an induced apple (if a does not dominate P) or a dart (otherwise). Let v_j be a neighbor of d on P. Then, by the choice of P, we have that the length of the path $(d, v_j, v_{j-1}, \ldots, v_1, b)$ is at least k, and also the length of the path $(d, v_j, v_{j+1}, \ldots, v_{k-1}, c)$ is at least k. Consequently $j + 1 \geqslant k$ and $k - j + 1 \geqslant k$, which implies $j = 1$ and $k = 2$. However, now we see that G contains either an induced $K_{1,1,3}$ (if a is adjacent to v_1) or an induced $K_{2,3}$ (otherwise), a contradiction. $\dashv$

In the following corollary, we restate the above result to better match the two questions posed at the beginning of this section.

Corollary 4.4 *Let G be a (apple, $K_{2,3}$, dart, $K_{1,1,3}$)-free graph. Then, the following conditions are equivalent:*

(1) G is a set graph.
(2) G is connected and satisfies the cut vertex condition.
(3) G is connected and claw-free.

This further elucidates the connection between claw-free graphs and set graphs and provides another class of graphs in which recognizing set graphs—a problem that is NP-complete for general graphs, as we will see in Sect. 4.5.1—is solvable in polynomial time.

In Exercises 4.19 and 4.20 we explore a characterization of (apple, $K_{2,3}$, dart, $K_{1,1,3}$)-free graph in terms of claw-free graphs.

4.4 Set Graph Transformations

In this section we show that set graphs are closed under two graph transformations. The first one is the *substitution* of a single vertex of a set graph with another set graph and allows creating larger set graphs. The second one is the *suppression* of a cut vertex and allows creating smaller set graphs from non-biconnected set graphs.

> **Definition 4.4 (Vertex substitution)** If $G_1 = (V_1, E_1)$, $G_2 = (V_2, E_2)$ are graphs with $V_1 \cap V_2 = \emptyset$ and $x \in V_1$, the *substitution* $H = G_1(x \to G_2)$ of G_2 for x in G_1 is defined to be the graph obtained by deleting x from G_1 and joining each vertex of G_2 to each neighbor of x in G_1 (Fig. 4.4).

Proposition 4.5 *Set graphs are closed under vertex substitution with a set graph.*

Fig. 4.4 Illustration of the vertex substitution transformation (Definition 4.4)

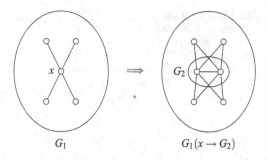

G_1 $G_1(x \to G_2)$

Proof Let $H = G_1(x \to G_2)$, where G_1 and G_2 are set graphs and x is a vertex of G_1. Let D_i be an EAO of G_i, for $i = 1, 2$. An EAO D of H is obtained as follows:

- All edges lying completely within $V(G_1)\backslash\{x\}$ get oriented as in D_1.
- All edges lying completely within $V(G_2)$ get oriented as in D_2.
- For every $v \in V(G_2)$ and every $w \in N_{D_1}^+(x)$, the edge $\{v, w\}$ gets oriented as $\langle v, w \rangle$.
- For every $w \in V(G_1)$ such that $x \in N_{D_1}^+(w)$ and every $v \in V(G_2)$, the edge $\{v, w\}$ gets oriented as $\langle w, v \rangle$.

Clearly, D is acyclic. Suppose that there is a collision in D between vertices u and v (so that $\langle u, v \rangle, \langle v, u \rangle \notin E(D)$). Then we cannot have $\{u, v\} \subseteq V(G_2)$, as this would contradict the extensionality of D_2. Similarly, because of the extensionality of D_1, we cannot have $\{u, v\} \cap V(G_2) = \emptyset$. Therefore, $|\{u, v\} \cap V(G_2)| = 1$, say $u \in V(G_2)$ and $v \notin V(G_2)$. But now, since $\langle u, v \rangle, \langle v, u \rangle \notin E(D)$, it follows from the construction that $N_D^+(v) \cap V(D_2) = \emptyset$; thus $N_D^+(v) = N_{D_1}^+(v) \subseteq V(D_1)$, and hence $N_D^+(u) = N_{D_1}^+(x)$. This collision between x and v in D_1 contradicts its extensionality. ⊣

As corollary of the above transformation, we get that set graphs are closed under the following standard graph-theoretic operations.

Corollary 4.5 • *If G is a set graph, then the graph obtained from G by introducing a new vertex v adjacent to all vertices of G is also a set graph. Vertex v is called an* universal *vertex.*
• *If G is a set graph and $v \in V(G)$, then the graph obtained from G by introducing a new vertex v' adjacent precisely to v and to every neighbor of v is also a set graph. Vertex v' is called a* true twin *of v.*

Exercise 4.22 asks the reader to prove Corollary 4.5. Implications converse to the ones of Corollary 4.5 do not hold in general: e.g., the set graph in Fig. 4.5 can be obtained from the claw $\{\alpha_1, \alpha_0, \alpha_2, \alpha_4\}$ by the addition of α_3 either as an universal vertex or as a true twin of α_1.

We will now introduce an operation that *suppresses* a cut vertex (see also Fig. 4.6).

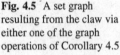 **Fig. 4.5** A set graph
resulting from the claw via
either one of the graph
operations of Corollary 4.5

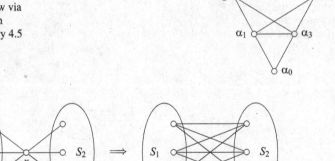

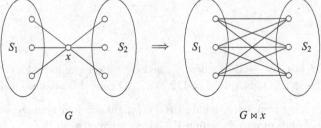

G $G \bowtie x$

Fig. 4.6 Illustration of the cut vertex suppression transformation (Definition 4.5)

> **Definition 4.5 (Cut vertex suppression)**   Let G be a graph and let x be a cut
> vertex of G such that $G - x$ has two connected components S_1 and S_2. We say
> that $G \bowtie x$ is the graph (V, E) where
>
> - $V = V(G) \setminus \{x\}$,
> - $E = (E(G) \setminus \{\{x, s\} : s \in N_G(x)\}) \cup \{\{s_1, s_2\} : s_1 \in S_1 \cap N_G(x) \wedge s_2 \in S_2 \cap N_G(x)\}$.

If G is a set graph and $x \in V(G)$ is a cut vertex of G, by Proposition 4.1, $G - \{x\}$
has precisely two components; therefore, we can consider the graph $G \bowtie x$. We
start with a preliminary lemma about cut vertices of set graphs.

Lemma 4.2   *If D is extensional and acyclic, then for every $x \in V(D)$, there is at
most one component S of $D - x$ such that $N^+(x) \cap S \neq \emptyset$.*

Proof Arguing by contradiction, suppose there are two such components S and S'
of $D - x$, so that $s \in S$, $s' \in S'$, and $\langle x, s \rangle$, $\langle x, s' \rangle$; then denote by t and t' the sinks of
$D[S]$ and $D[S']$, respectively. Since D is extensional and $\{x\}$ is a cut set, one of these
two local sinks, say t, has x as out-neighbor. However, this contradicts the acyclicity
of D, since we obtain the cycle starting with t, x, s and then following the path from
s to t in $D[S]$.   ⊣

Proposition 4.6   *If G is a set graph having a cut vertex x then $G \bowtie x$ is a set graph.*

Proof Let G be a set graph having a cut vertex x which gives rise to components S_1
and S_2 of $G - x$. Denoting by D an EAO of G, observe first that x is not the sink
of D, since otherwise the sinks of $D[S_1]$ and $D[S_2]$ would collide. From Lemma 4.2,
we thus get that x has out-neighbors in precisely one of S_1 and S_2, say S_1. Moreover,
since x is a cut vertex, there are vertices in S_2 having x as out-neighbor.

Obtain the EAO D' of $G \bowtie x$ in the following way:

- orient all edges lying completely within one of S_1, S_2 as in D,
- orient each edge $\{s_1, s_2\}$ with $s_1 \in S_1$ and $s_2 \in S_2$ as $\langle s_1, s_2 \rangle$ if and only if $\langle s_1, x \rangle \in E(D)$.

To see that D' is acyclic, note that any possible cycle of D' must contain arcs $\langle s_1, s_2 \rangle$ and $\langle s_2', s_1' \rangle$, with $s_1, s_1' \in S_1$, $s_1 \neq s_1'$, and $s_2, s_2' \in S_2$. Moreover, choose such vertices s_1, s_1', s_2, s_2' so that this cycle continues from s_1' to s_1 using only vertices of S_1. However, this would produce a cycle in D on the vertices s_1, x, s_1', followed by the vertices on the directed path from s_1' to s_1 in S_1 belonging to the assumed cycle of D'.

To see that D' is also extensional, argue by contradiction and suppose that there is a collision in D' between s and s' belonging to S_1. Since D is extensional, there must exist a vertex $z \in V(D)$ such that, w.l.o.g., $z \in N_D^+(s) \setminus N_D^+(s')$. If $z \neq x$, then $z \in S_1$ and hence also $z \in N_{D'}^+(s) \setminus N_{D'}^+(s')$. On the other hand, if $z = x$ then, according to the construction, s receives as out-neighbor an element of S_2, which is not the case for s'. If there is a collision between two vertices of S_2, the argument is identical.

We now have to consider a collision between $s \in S_1$ and $s' \in S_2$, and here we must analyze a few cases. If $N_{D'}^+(s) \cap S_1 \neq \emptyset$ and $N_{D'}^+(s) \cap S_2 \neq \emptyset$, then $\langle s, s' \rangle \in E(D')$, contradicting the fact that s and s' collide. Suppose now that $N_{D'}^+(s) \subseteq S_1$, implying that $N_{D'}^+(s') = N_D^+(x)$ and hence that there is a collision in D between x and s. The last case that needs to be considered is when $N_{D'}^+(s) \subseteq S_2$, which implies that s is a sink in $D'[S_1]$. However, this cannot be true, else s would also be a sink in $D[S_1]$ and would hence collide with the sink of $D[S_2]$ in D. $\dashv$

4.5 Computational Complexity

In this section we show that the problem of deciding whether a given graph is a set graph is NP-complete (Sect. 4.5.1). We refer the reader to Panel 3.5 for an informal discussion on what an NP-complete problem is. In Sect. 4.5.2 we show that the analogous problem corresponding to hypersets, that is, given a graph G, does G admit a hyperextensional orientation, is also NP-complete. We mainly follow the exposition from [63].

4.5.1 Recognizing Set Graphs

In this section, we prove that the following problem is NP-complete.

Problem 4.1 (Set graph recognition)   Given a graph G, decide whether G is a set graph.

Moreover, we will show that also its counting version, namely, "Given a graph G, return the number of EAOs of G," is intractable, in the sense that it belongs to the complexity class of #P-complete problems (recall again Panel 3.5).

The NP-complete Hamiltonian path problem [42] is the following one: "Is there, in a given graph G, a *Hamiltonian path*, i.e., a path meeting every vertex exactly once?" In order to obtain the NP-completeness of Problem 4.1 (and the #P-completeness of its counting version), we will show that the following variant of the Hamiltonian path problem can be reduced to Problem 4.1, and that this reduction preserves a correspondence between the solution sets of the two problems.

Problem 4.2 (Hamiltonian path on graphs with two leaves)   Given a graph G with exactly two leaves, decide whether G has a Hamiltonian path.

Exercise 4.23 asks the reader to show that also Problem 4.2 is NP-complete, and moreover, since the Hamiltonian path problem is #P-complete (see e.g., [84, Ch.18], [32]), that it is also #P-complete.

Given a graph $G = (V, E)$, denote by $S(G)$ the *subdivision graph of G*, that is, the bipartite graph obtained by applying a single subdivision to each edge of G. Stated formally, $S(G) = (V \cup X, F)$, where

- $X = \{x^e : e \in E\}$,
- $F = \{\{u, x^{uv}\} : \{u, v\} \in E\}$.

A vertex of X is called an *edge vertex*. If $P = (v_1, v_2, \ldots, v_n)$ is a Hamiltonian path in G, then we denote by $O_1(S(G), P)$ the orientation of $S(G)$ obtained as follows. Say that an edge vertex of X is *touched* if the Hamiltonian P uses the corresponding edge of G and *untouched* otherwise. Partition X as $X = T \cup U$ by distinguishing touched edge vertices from untouched ones. Let $U = \{u_1, \ldots, u_k\}$ and let $\prec$ be the total order on the vertices of $S(G)$ defined by

$$v_1 \prec x^{v_1 v_2} \prec v_2 \prec \cdots \prec v_{n-1} \prec x^{v_{n-1} v_n} \prec v_n \prec u_1 \prec \cdots \prec u_k.$$

The orientation $O_1(S(G), P)$ of $S(G)$ is such that every edge $\{u, v\} \in E(S(G))$ is oriented in D as $\langle u, v \rangle$ if and only if $v \prec u$ (see Example 4.3).

Clearly, this is an acyclic orientation. Orientation $O_2(S(G), P)$ is the one defined analogously, corresponding to the total order

$$v_n \prec x^{v_n v_{n-1}} \prec v_{n-1} \prec \cdots \prec v_2 \prec x^{v_2 v_1} \prec v_1 \prec u_1 \prec \cdots \prec u_k.$$

Example 4.3 Consider the following graph G with vertex set $\{v_1, v_2, \ldots, v_7\}$ having leaves v_1 and v_7. We highlight a Hamiltonian path P in G.

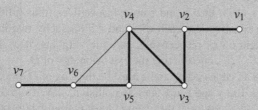

The EAO $O_1(S(G), P)$ is shown below (the edge vertices of $S(G)$ are drawn as diamonds):

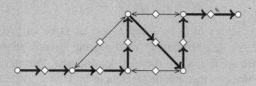

Lemma 4.3 *If G is a graph with exactly two leaves s and t that has a Hamiltonian path P, then both orientations $O_1(S(G), P)$ and $O_2(S(G), P)$ are EAOs of $S(G)$.*

Proof Let $P = (v_1, v_2, \ldots, v_n)$ be a Hamiltonian path in G. Then clearly $v_1 = s$ and $v_n = t$, or vice versa. It suffices to argue that the acyclic orientation $D :=$ $O_1(S(G), P)$ is an EAO of $S(G)$ (the case of $O_2(S(G), P)$ is symmetric). Clearly, D is acyclic. To see that D is extensional, observe that:

- vertex $s = v_1$ is the only vertex with $N^+(s) = \emptyset$;
- every untouched vertex in U, say $x^{uv} \in U$, is the only vertex having $N^+(x^{uv}) = \{u, v\}$;
- every touched vertex in T, say $x^{v_i v_{i+1}} \in T$, is the only vertex having $N^+(x^{v_i v_{i+1}}) = \{v_i\}$;
- every vertex in $V \setminus \{s\}$, say $v_i \in V$ (with $2 \leq i \leq n$), is the only vertex with $N^+(v_i) = \{x^{v_{i-1} v_i}\}$.

⊣

Lemma 4.4 *Let G be a graph. If $S(G)$ is a set graph and D is an EAO of $S(G)$, then D contains a directed path P passing through all vertices of G, and thus G has a Hamiltonian path.*

Proof Let the sink of D be v and let P be a longest directed path in D ending at v. Let $u \in V(G)$ be the endpoint of P other than v. If all vertices of G are on P, we are done. If not, let u' be a vertex of G not on P. Let Q be a longest directed path from u' to v, and let x be the first vertex on Q that belongs to P. (Path Q exists because D is acyclic and has v as unique sink.) Let y and z ($y \neq z$) be the predecessors of x on P and on Q, respectively.

If x is a vertex of G, then y and z are edge vertices (thus different from u and u'). Note that by construction, each of y and z has exactly two incident arcs, one incoming, on P and on Q, respectively, and one outgoing, to x. This implies that $N^+(y) = N^+(z) = \{x\}$, contradicting the extensionality of D.

Otherwise, x is an edge vertex, and x must be the sink of D, since its two incident arcs are incoming. But y and z are again in collision, since from the maximality of the paths and the acyclicity of D they cannot have out-neighbors other than x.   ⊣

We need to refine Lemmas 4.3 and 4.4 as follows.

Lemma 4.5 *If G is a graph with two leaves s and t, then any EAO D of $S(G)$ is isomorphic to either $O_1(S(G), P)$ or $O_2(S(G), P)$, for some Hamiltonian path P of G.*

Proof Let the two leaves of G be s and t, and let D be an EAO of $S(G)$. From Lemma 4.4, there is a directed path P' in D passing through all vertices of G. The endpoints of P' are s and t, since they are leaves in G. We may assume that s is its last vertex, so that s is the sink of D. Denote by $(t = v_n, v_{n-1}, \ldots, v_1 = s)$ the order in which the vertices of G appear on P'. By the construction of $S(G)$ and the fact that P is a directed path, $N^+(x^{v_{i+1}v_i}) = \{v_i\}$, for every $1 \leq i \leq n-1$. Every vertex of D untouched by P is an edge vertex $x^{v_i v_j}$, with $i, j \in \{2, \ldots, n-1\}$. If $x^{v_i v_j}$ has only one out-neighbor in D, say v_i, it is in collision with $x^{v_{i+1}v_i}$. Therefore, $N^+(x^{v_i v_j}) = \{v_i, v_j\}$. Therefore, it also holds that $N^+(v_i) = \{x^{v_i v_{i-1}}\}$, for every $2 \leq i \leq n$.

We have thus obtained that D is isomorphic to either $O_1(S(G), P)$ or $O_2(S(G), P)$, where P is the undirected path underlying P'.   ⊣

Theorem 4.6 *Let G be a graph with exactly two leaves. Then, the number of Hamiltonian paths of G equals twice the number of EAOs of the graph $S(G)$ obtained by applying a single subdivision to each edge of G.*

Proof If G is a graph with two leaves s and t, then every Hamiltonian path in G between s and t induces exactly two EAOs (having either s or t as sink) for $S(G)$, by Lemmas 4.3 and 4.5. Moreover, different Hamiltonian paths of G induce different pairs of such EAOs for $S(G)$. Conversely, by Lemma 4.4, every EAO of $S(G)$ arises this way.   ⊣

A corollary of the above correspondence is our desired NP-completeness and #P-completeness results.

Corollary 4.6 *The set graph recognition problem (Problem 4.1) is NP-complete, and its counting version is #P-complete, even when the input is restricted to bipartite graphs with exactly two leaves.*

Proof Problem 4.1 belongs to NP, since acyclicity and extensionality can be checked in polynomial time; actually, the extensionality of an acyclic digraph can be verified in linear time [30]. The NP-hardness and #P-completeness follow thus from Theorem 4.6. ⊣

Recall that every connected claw-free graph admits an EAO (Corollary 4.2), which implies that Problem 4.1 is trivial for claw-free graphs. Exercise 4.26 asks the reader to show that Problem 4.1 remains NP-complete for the class of graphs in which no two induced claws have an edge in common.

4.5.2 Hyperextensional Orientations

In this section we prove that the following problem is NP-complete:

> **Problem 4.3 (Existence of a hyperextensional orientation)** Given a graph G, decide whether G admits a hyperextensional orientation (an *HEO*, for short).

Our strategy is the same as in the previous section. As done therein, it can also be shown that the counting version of Problem 4.3 is #P-complete, and this is left as an exercise for the reader (Exercise 4.28).

We show that the Hamiltonian paths of a graph G with exactly two leaves are in bijection with pairs of HEOs of the bipartite graph obtained from $S(G)$ by appending a gadget, called G_8 below, to one of its leaves. The purpose of this gadget is to enforce the sink of the HEO to be only one of two vertices.

Given digraphs D_1 and D_2 with disjoint vertex sets, and given vertices $v_i \in V(D_i)$, $i = 1, 2$, we denote by $U(D_1, v_1, v_2, D_2)$ the digraph obtained by taking a copy of D_1 and a copy of D_2 and adding the arc $\langle v_1, v_2 \rangle$. Formally, $U(D_1, v_1, v_2, D_2)$ has

- $V(D_1) \cup V(D_2)$ as vertex set,
- $E(D_1) \cup E(D_2) \cup \{\langle v_1, v_2 \rangle\}$ as the arc relation.

We define this operation analogously for graphs.

Our reduction will encode any graph G having two leaves s and t by the graph $U(S(G), s, a_8, G_8)$, where G_8 is the graph underlying the digraph D_8 shown in Fig. 4.7. Graph G_8 has been chosen to admit exactly two HEOs (see Exercise 4.27), a fact which turns out to be useful in proving #P-completeness in Exercise 4.28.

We start with a preliminary lemma about the hyperextensionality of $U(D_1, v_1, v_2, D_2)$. Recall that, according to Lemma 3.4, any simple hyperextensional digraph has a sink.

Lemma 4.6 *Let D_1 and D_2 be two simple hyperextensional digraphs. If the sink of D_1 is s and D_2 has a source t, then the digraph $U(D_1, s, t, D_2)$ is hyperextensional.*

Proof Let $D = U(D_1, s, t, D_2)$ and let B be the bisimilarity relation over D (thus B is symmetric). We need to prove that $x = y$ whenever xBy (or, equivalently,

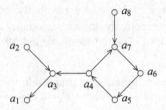

Fig. 4.7 Digraph D_8 is a gadget to force a sink when orientations can have cycles; one of a_1 or a_2 must be a sink in any *extensional* (not necessarily acyclic) orientation of its underlying undirected graph, which we call G_8

yBx) holds. To begin with, Lemma 3.4 implies that D_1 and D_2 are also extensional, and hence, by construction, so is D. We argue by contradiction and consider three cases.

First, assume that xBy holds for distinct $x, y \in V(D_2)$. However, B restricted to $V(D_2)$, that is, the relation $B_2 = \{\langle x, y \rangle : x, y \in V(D_2) \mid xBy\}$, is a bisimulation over D_2. This contradicts the fact that D_2 is hyperextensional.

Second, suppose that $x_0 B y_0$ holds for distinct $x_0 \in V(D_1)$ and $y_0 \in V(D_2)$. By Lemma 3.4 applied to D_1, there is a directed path P from x_0 to s. Take $x_1 \in N^+(x_0)$ so that x_1 is the second vertex on the directed path P (or set $x_1 = t$, if $x_0 = s$). Since $x_0 B y_0$ holds, there exists $y_1 \in N^+(y_0)$, thus $y_1 \neq x_1$, such that $x_1 B y_1$. By repeating the above procedure sufficiently many times while following P, we reach a pair $\langle x_i, y_i \rangle$ (where $i \geqslant 0$) such that $x_i = s$, $y_i \in V(D_2)$ and sBy_i. Since $N_D^+(s) = \{t\}$, there exists a $y_{i+1} \in N^+(y_i)$ so that tBy_{i+1}. Recall that t is a source of D_2, therefore $t \neq y_{i+1}$. This contradicts the previous case.

Finally, suppose that xBy holds for distinct $x, y \in V(D_1)$. We will show that the restriction of B to $V(D_1)$, that is, the relation $B_1 = \{\langle x, y \rangle : x, y \in V(D_1) \mid xBy\}$, is a bisimulation over D_1. The fact that also xB_1y holds for $x \neq y$ will then contradict the hyperextensionality of D_1.

Let xB_1y be an arbitrary pair in B_1, where $x, y \in V(D_1)$ are distinct and $s \notin \{x, y\}$. By the definition of B_1 and by the fact that xBy, then both conditions in the definition of bisimulation hold. In fact, xB_1s or sB_1x cannot hold for $x \in V(D_1) \setminus \{s\}$ because $N^+(s) = \{t\}$, and, by the previous case, there can be no $x' \in N_{D_1}^+(x)$ such that $x'Bt$ or tBx'. Therefore, B_1 is a bisimulation over D_1.   ⊣

Lemma 4.7 *If G is a graph with two leaves s and t, and G has a Hamiltonian path, then the graph $U(S(G), s, a_8, G_8)$ admits an HEO.*

Proof First, let $P = (v_1, \ldots, v_n)$ be a Hamiltonian s-t path in G, and let $D := O_1(S(G), P)$ be the EAO derived from $S(G)$ and P (as defined before Lemma 4.3). Observe that D is a simple digraph. In particular, s is the sink of D. By Lemma 3.5, D is also hyperextensional. As shown in Example 4.4, the orientation D_8 of G_8 is hyperextensional, but then, since a_8 is a source of D_8, by Lemma 4.6 the digraph $U(D, s, a_8, D_8)$ is hyperextensional as well, which proves the claim.   ⊣

Example 4.4 To see that the digraph D_8:

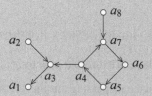

is hyperextensional, we can apply the partition refinement algorithm of [82]. This algorithm maintains, at each step i, a partition π_i of $V(D_8)$ which is coarser than the one, $\pi_\star$, associated with the bisimilarity relation over D_8; that is, every bisimilarity class is contained in some block of π_i. This partition gets iteratively refined, until it becomes the sought coarsest stable partition $\pi_\star$.

In its most basic form, at each iteration the algorithm finds a "splitter" block of π_i (cf. Fig. 3.6), namely, an $S_i \in \pi_i$, such that there exist $P_i \in \pi_i$ and $x, y \in P_i$ such that x has a vertex of S_i as out-neighbor, while y has no out-neighbors in S_i . Then it gets π_{i+1} from π_i by replacing block P_i by the two blocks:

- $\{x \in P_i \mid N^+(x) \cap S_i \neq \emptyset\}$,
- $\{x \in P_i \mid N^+(x) \cap S_i = \emptyset\}$.

Here is an example of execution of this algorithm:

i	π_i	S_i	P_i
1	$[a_1, a_2, a_3, a_4, a_5, a_6, a_7, a_8]$	$[a_1, a_2, a_3, a_4, a_5, a_6, a_7, a_8]$	$[a_1, a_2, a_3, a_4, a_5, a_6, a_7, a_8]$
2	$[a_2, a_3, a_4, a_5, a_6, a_7, a_8][a_1]$	$[a_1]$	$[a_2, a_3, a_4, a_5, a_6, a_7, a_8]$
3	$[a_3][a_2, a_4, a_5, a_6, a_7, a_8][a_1]$	$[a_2, a_4, a_5, a_6, a_7, a_8]$	$[a_2, a_4, a_5, a_6, a_7, a_8]$
4	$[a_3][a_4, a_5, a_6, a_7, a_8][a_2][a_1]$	$[a_3]$	$[a_4, a_5, a_6, a_7, a_8]$
5	$[a_3][a_4][a_5, a_6, a_7, a_8][a_2][a_1]$	$[a_4]$	$[a_5, a_6, a_7, a_8]$
6	$[a_3][a_4][a_5][a_6, a_7, a_8][a_2][a_1]$	$[a_5]$	$[a_6, a_7, a_8]$
7	$[a_3][a_4][a_5][a_6][a_7, a_8][a_2][a_1]$	$[a_6]$	$[a_7, a_8]$
8	$[a_3][a_4][a_5][a_6][a_7][a_8][a_2][a_1]$	–	–

The hyperextensionality of D_8 amounts to the fact that the resulting $\pi_\star$ consists of singletons, which indicates that bisimilarity is the identity relation.

Lemma 4.8 *Let G be a graph. If $U(S(G), s, a_8, G_8)$ admits an HEO, then G has a Hamiltonian path.*

Proof We reason as in the proof of Lemma 4.4. Let D be an HEO of $U(S(G), s, a_8, G_8)$. By construction, D is a simple digraph. Therefore, by Lemma 3.4, D

- is extensional,
- has a (unique) sink $v \in \{a_1, a_2\}$ because it has two leaves, a_1 and a_2, with $N(a_1) = N(a_2)$, and
- from every vertex of D there is a directed path to v.

We claim that D has a directed path passing through all the vertices of G, which thus constitutes a Hamiltonian path for G.

Indeed, let P be a longest directed path in D starting in a vertex of G and ending at v, the sink of D. Let $u \in V(G)$ be the endpoint of P other than v. If all vertices of G are on P, we are done. If not, let u' be a vertex of G not on P. Let Q be a longest directed path from u' to v, and let x be the first vertex on Q that belongs to P. We have that x is a vertex of $S(G)$ since by construction P passes through the edge $\{s, a_8\}$ before reaching $v \in \{a_1, a_2\}$ (recall also that $\{s, a_8\}$ is the only connection between $S(G)$ and G_8). Let y and z ($y \neq z$) be the predecessors of x on P and on Q, respectively.

If x is a vertex of G, then y and z are edge vertices (thus different from u and u'). Note that by construction, each of y and z have exactly two incident arcs, one incoming, on P or on Q, and one outgoing to x. This implies that $N^+(y) = N^+(z) = \{x\}$, contradicting the extensionality of D.

Otherwise, x is an edge vertex, and x must be the sink of D, since its two incident arcs are incoming. This contradicts the fact that the sink of D is a vertex of G_8. $\dashv$

Corollary 4.7 *The problem of deciding whether a graph admits an HEO (Problem 4.3) is NP-complete, even when the input is restricted to bipartite graphs with exactly three leaves.*

Proof Problem 4.3 belongs to NP, since hyperextensionality can by checked in polynomial time, for example, by the algorithm of [82]. The hardness follows from Lemmas 4.7 and 4.8. $\dashv$

Exercises

4.1 Are the two graphs below set graphs?

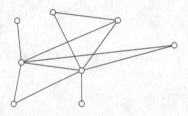 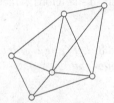

4.2 Prove that the first set in the Ackermann order whose (directed) pointed membership graph does not admit a (directed) Hamiltonian path has Ackermann number 12.

4.3 (*) Determine the first set in the Ackermann order such that the undirected graph underlying its membership graph does not admit a Hamiltonian path.

4.4 Given the complete graph on n vertices K_n, we define the *corona of* K_n, denoted $K_n \odot K_1$, as the graph on vertex set $\{\alpha_1, \ldots, \alpha_n\} \cup \{\beta_1, \ldots, \beta_n\}$, where $\{\alpha_1, \ldots, \alpha_n\}$ is a clique, and each β_i is adjacent only to α_i, $1 \leq i \leq n$. Show that for every $n \geq 3$, $K_n \odot K_1$ is a set graph which does not have a Hamiltonian path. Prove also that there is a unique EAO of $K_n \odot K_1$, up to isomorphism.

4.5 Show that the graph depicted below is not a set graph.

4.6 Let G be a set graph and let $X \subseteq V(G)$. Denote by $Y_1, \ldots, Y_t$ the vertex sets of the connected components of the graph $G - X$. Prove the following inequality:

$$t \leq \max \left| \bigcup_{i=1}^{t} \mathcal{P}\big(N(v_i) \cap X\big) \right|,$$

where the maximum is taken over all t-tuples $\langle v_1, \ldots, v_t \rangle \in Y_1 \times \cdots \times Y_t$. Argue that the graph from Exercise 4.5 does not satisfy this inequality.

4.7 Show that a complete bipartite graph $K_{m,n}$ is a set graph if and only if $|m - n| \leq 1$.

4.8 Give an example of a family of bipartite set graphs with partite sets A and B such that $|A|$ is exponential in $|B|$.

4.9 Let G be a set graph and let $v \in V(G)$ be a sink in an EAO of G. Show that v is not a cut vertex of G.

4.10 Let $G = (V, E)$ be a graph, and let $L(G)$ be its line graph, namely, the graph (V', E') with $V' = E$ and $E' = \{\{x, y\} : x, y \in E \mid x \cap y \neq \emptyset\}$. Show that $L(G)$ is claw-free.

4.11 Adapt the proof of Theorem 4.2 and use Exercise 4.9 above for proving the following stronger result about claw-free graphs. If G is a connected claw-free graph and $v \in V(G)$, then G admits an extensional acyclic orientation with sink v if and only if v is not a cut vertex of G.

4.12 Give a linear-time algorithm for checking whether a graph is a jellyfish graph (recall Definition 4.3).

4.13 Give a linear time algorithm for constructing an EAO orientation of a jellyfish graph (recall Definition 4.3 and Theorem 4.3).

4.14 (*) Give a structural characterization (analogous to what Theorem 4.3 says of unicyclic graphs) of connected set graphs which are *bicyclic*, that is, of the connected graphs G in which the number of edges equals one plus the number of vertices.

4.15 Show that a connected graph G is complete multipartite if and only if G is (P_4, paw)-free.

4.16 Show that a graph is a block graph if and only if it is diamond-free and it does not have any cycle of length at least four as induced subgraph.

4.17 Use the characterization given in Proposition 4.4 for writing an efficient algorithm that decides whether a given block graph is a set graph. What is the complexity of your algorithm?

4.18 Let G be a claw-free graph. Use Corollary 4.2 to argue that the following conditions are equivalent:

(1) G is a set graph.
(2) G is connected and satisfies the cut vertex condition.
(3) G is connected and claw-free.

4.19 (*) A vertex v in a graph G is *simplicial* if $N(v)$ forms a clique. Suppose that $G_1, \ldots, G_k$ are disjoint graphs with simplicial vertices $v_1, \ldots, v_k$, respectively. Let G be the graph obtained by identifying all $v_1, \ldots, v_k$. We say that G is obtained by an *s-gluing* from $G_1, \ldots, G_k$ (the letter "s" stands for "simplicial").

Prove the following structural result describing a way of building all (apple, $K_{2,3}$, dart, $K_{1,1,3}$)-free graphs from claw-free graphs. Let X denote the class of (apple, $K_{2,3}$, dart, $K_{1,1,3}$)-free graphs. Show that a connected graph G belongs to X if and only if one of the following conditions holds:

(i) G is claw-free.
(ii) There exist $k \geqslant 3$ graphs $G_1, \ldots, G_k \in X$ such that G can be obtained by an s-gluing of $G_1, \ldots, G_k$.

4.20 Show that the decomposition of a given connected (apple, $K_{2,3}$, dart, $K_{1,1,3}$)-free graph into claw-free graphs described in Exercise 4.19 can be implemented in polynomial time.

4.21 Show that every set graph G endowed with at least three vertices has a vertex v such that:

1. $G - v$ is a set graph;
2. if $N(v) = \{w\}$, then also $G - \{v, w\}$ is a set graph.

4.22 Prove Corollary 4.5.

4.23 Prove that the Hamiltonian path problem on graphs with exactly two leaves (Problem 4.2 on page 118) is NP-complete, and that its counting version is #P-complete. *Hint.* Given a graph G, construct G^+ having $V(G) \cup \{s_1, s_2, t_1, t_2\}$ as vertex set, and $E(G) \cup \{\{s_1, s_2\}, \{t_1, t_2\}\} \cup \{\{s_2, v\}, \{t_2, v\} : v \in V(G)\}$ as edge set. Use the fact that the counting version of the Hamiltonian path problem is #P-complete.

4.24 Say that an extensional acyclic digraph D is *slim* is the digraph obtained by removing any arc from D is no longer extensional. Show that the problem of deciding whether a graph admits a slim EAO is NP-complete. *Hint.* Show that reduction used in the NP-completeness proof for the set graph recognition problem (Problem 4.1) works also for this problem.

4.25 Show that the problem of deciding whether a given graph admits an EAO in which *reversing* any arc produces either a cycle or a collision is NP-complete. *Hint.* Show that reduction used in the NP-completeness proof for the set graph recognition problem (Problem 4.1) works also for this problem.

4.26 Show that the set graph recognition problem (Problem 4.1) is NP-complete for the class of graphs in which no two induced claws have an edge in common. *Hint.* Observe that the following simplified version of Lemma 4.3 holds: "If G is a graph with a Hamiltonian path, then $S(G)$ admits an HEO", and use the fact that the Hamiltonian path problem remains NP-complete for cubic graphs [42].

4.27 Show that the graph G_8, namely, the undirected graph underlying the digraph shown in Fig. 4.7, admits exactly two HEOs.

4.28 Using the fact from Exercise 4.27 above, show that the counting version of Problem 4.3, namely, finding the number of HEOs of a given graph, is #P-complete.

Chapter 5
Graphs as Transitive Sets

Set Theory is meant to be a *lingua franca* for the various disciplines of mathematics, whose scholars can, in set-theoretic terms, reason uniformly about graphs, integer arithmetic, complex analysis, and whatnot. Undeniably, any skillful mathematician benefits from specific disciplinary knowledge when reasoning inside a realm of mathematics; on the other hand, amalgamation of diverse disciplines into a single "big" theory discloses prospects of high versatility in proof development.

Modeling a mathematical notion in convenient set-theoretic terms is seldom a trivial task, though, and formalization inside Set Theory calls for design choices. This chapter will illustrate this, referring to the class of claw-free graphs thoroughly discussed in the previous chapter.

The "standard" way of representing a graph is the one used up to now: a set V whatsoever paired with a set E of doubletons. When the graph is undirected, each doubleton is a subset $\{x, y\}$ of V; it takes, instead, the form $\{\{x\}, \{x, y\}\}$—under Kuratowski's representation of ordered pairs—when the graph is directed. This modeling is not particularly revealing when it comes to reasoning about some special class of graphs.

Alternative representations of the same mathematical entity often provide different insights on its nature. For example, as we have seen for natural numbers, von Neumann's representation of $\mathbb{N}$ in terms of the successor function $n \mapsto n \cup \{n\}$ outperforms Zermelo's simple-minded representation of $\mathbb{N}$ as the smallest superset of $\{\emptyset\}$ which is closed under the function $n \mapsto \{n\}$: the former representation, in fact, easily generalizes to ordinal numbers and mimics number comparison $<$ directly through $\in$.

Likewise, we will manage to represent connected claw-free graphs as specially constrained *transitive* sets: each element x' of the set T that represents one such

© Springer International Publishing AG 2017
E.G. Omodeo et al., *On Sets and Graphs*, DOI 10.1007/978-3-319-54981-1_5

graph will act as a corresponding vertex x and the edge relationship will be mimicked by membership over T. That is, either $x' \in y'$ or $y' \in x'$ will hold when $\{x', y'\} \subseteq T$ is the set corresponding to an edge $\{x, y\}$ of the original graph. Even though our representation will superimpose an apparently artificial orientation to an undirected graph, we will take advantage precisely of this redundancy to prove with relative ease that every connected claw-free graph has a near-perfect matching[1] and a Hamiltonian cycle in its square.

Some evidence of the usefulness of the approach just outlined is that by relying on the representation of claw-free graphs as transitive sets, we can verify the correctness of the proofs supplied in this chapter with the aid of a programmed proof assistant. With a more conventional representation of graphs, the size of the formalized proofs would have been discouragingly big for such an experiment; on the opposite, thanks to the representation adopted, the formalized proof turns out to be an affordable "*proof pearl*" [18, 80, 81]. The strength reduction in our experiment owes to the fact—shown in this chapter—that studying connected claw-free graphs amounts to reasoning about special sets. Apt to the purpose, the proof checker which supported our experiment is based on Set Theory.

Proof assistants which are endowed with a formal theory of sets are not rare, and they are enjoying a season of increasing popularity and success. Those systems differ in the assumptions made about sets, in whether such assumptions are stated as explicit axioms or built into the computerized proof-manipulation machinery and in the importance attached to a type model. Thus, in order to give a full account of our proof verification experiment, we must also describe our working environment. Part of this chapter is in fact devoted to our proof checker, Ref, *per se*.

In this chapter we must enforce a clear-cut distinction between undirected graphs (to which the properties of connectedness and claw-freeness refer) and the directed ones (to which the properties of acyclicity, extensionality, and weak extensionality refer). Graphs of the two kinds will be treated together, for directed graphs will result from undirected graphs via the orientation techniques under discussion. We will, hence, carefully use the terms *graph*, *digraph*, and *arc* to mean: undirected graph, directed graph, and oriented edge, respectively.

---

[1]In its customary wording, this theorem claims that every claw-free connected graph *with an even number* of vertices has a *perfect* matching; according to our touched-up rewording, one vertex can be left unmatched—needless to say, this will happen when the number of vertices is odd.

5.1 Proofs + Sets = Checkable Proofs

Here we begin to dissect logically the proofs of two claims concerning any connected graph $G = (V, E)$ endowed with at least three vertices:

(A) if G is claw-free, it has a *near-perfect matching* [108, 111], namely, a set $P \subseteq E$ such that

- no two edges in P have a vertex in common;
- at most one vertex is left untouched by P: stated in symbols, $|V \setminus \cup P| \leqslant 1$;

(B) if G is claw-free, then *the square of G has a Hamiltonian cycle C* [60], namely, there exists a set of doubletons

$$C = \{\{x_0, x_1\}, \ldots, \{x_n, x_{n+1}\}\} \cup \{\{x_{n+1}, x_0\}\}$$

$$\subseteq E \cup \{\{u, v\} : \{u, w\} \in E, \{w, v\} \in E\}$$

such that every $x \in V$ occurs in the sequence $x_0, x_1, \ldots, x_n, x_{n+1}$ exactly once.

We will be more formal and meticulous than usual in providing set-theoretic definitions of all notions involved and in developing the said two proofs; as announced above, we can afford doing so because we can rely on specifications written for a proof checker which does, in fact, verify the correctness of our arguments.

The proofs which we will analyze are recent [64]. They are simpler than the ones discovered first, because we can benefit from what we know from Chap. 4—as recalled right below—about convenient orientations of claw-free graphs. The advantage resulting from this view has emerged, even more strikingly, from [110]: a proof of the somewhat deeper theorem [47] that all connected claw-free graphs have a *vertex-pancyclic* square was also attained cheaply through the same orientation of claw-free graphs. But, as yet, this proof has not been formalized in terms suitable for a proof checker.

At a high level of reasoning, the fact that an extensional acyclic orientation can be imposed on the edges of any connected claw-free graph G enables us to supersede G by a *transitive set*; this is why, in preparation for (A) and (B), we will develop formal proofs of two additional facts—already proved in Chaps. 4 and 3, respectively, in more general but in less detailed terms than in the ongoing:

(1) every connected claw-free graph is a set graph;

(2) the vertices of any (weakly) extensional and acyclic digraph can be injectively decorated with sets so that its edges $\{x, y\}$ correspond to the pairs s_x, s_y of decorating sets such that $s_x \in s_y$.

At a low level of reasoning, we will need one more preliminary fact, whose proof is relegated to Appendix A; namely, that in any connected graph G:

(0) there is a vertex $z \in V$ whose removal (with all incident edges) from G does not disrupt the connectedness of G.

Claim (5.1), by decreasing the cardinality of V, plays a technical role in many inductive proofs about connected graphs; in particular, we will exploit it to prove (5.1).

Our aim, in what follows, is to reason about graphs and digraphs *inside* a set theory. This standpoint differs from the approach taken in Sect. 2.3, where we were rooting graph theories directly on first-order predicate calculus: here we will represent graphs as peculiar sets. While postponing any consideration about *di*graphs to Sect. 5.2, let us begin with graphs, formally defined as follows[2]:

$$\mathsf{uGraph}(V, E) \;\leftrightarrow_{\mathrm{Def}}\; \mathsf{Finite}\,(V) \;\&\; E \subseteq \{\{x, y\} : x \in V, \, y \in V \setminus \{x\}\}.$$

Since Set Theory enforces an obvious distinction between doubletons and ordered pairs, we see no advantage in modeling graphs as symmetric digraphs and just will represent graphs and digraphs in different ways. In the present context, in fact, relative advantages and drawbacks of representation choices must be assessed according to criteria related to proof-checking: transparency of argumentation will be valuable anyhow, but we will also strive to avoid layering notions too deeply, in order to avoid intertwining proofs on different topics beyond necessity (albeit this may, at times, be an aid rather than a hindrance for a trained mathematical mind). This will be illustrated in this section through a characterization of connectedness which does not refer to paths (and hence, more or less explicitly, to natural numbers).

Connectedness is a key notion involved in our formalization experiment. A customary way of stating that a graph $G = (V, E)$ is connected is by saying that

---

[2]This definition covers, along with the notion of graph proper, also the case $V = E = \emptyset$.

between any two vertices there is a path (see Sect. 2.3.2). Driven by Panel 5.1 (see also Exercise 5.1), we choose this equivalent way of formalizing the notion:

$$\mathsf{Conn}(V, E) \;\leftrightarrow_{\mathrm{Def}}\; V = \cup E \;\&\; \{p \subseteq E \mid \cup p \cap \cup(E \setminus p) = \emptyset\} \subseteq \{\emptyset, E\};$$

Panel 5.1 Partitions, reachability, and connectedness

We define the *refinement* relation between sets P, Q as follows:

$$P \sqsubseteq Q \;\leftrightarrow_{\mathrm{Def}}\; \cup P = \cup Q \;\&\; \forall p \in P \, \exists! q \in Q \, p \cap q \neq \emptyset\,;$$

in words, P is *finer* than Q (and Q is *coarser* than P) if the members of sets in P are the same as the members of sets in Q and every set belonging to P intersects one and only one set belonging to Q. We also make the definition

$$\mathsf{PartitionOf}(P, X) \;\leftrightarrow_{\mathrm{Def}}\; P \sqsubseteq P \;\&\; \cup P = X\,;$$

plainly, this condition is met iff P is a set of pairwise disjoint nonnull sets—to be called the *blocks* of P—whose union yields X. We then call *disconnected* any set D such that $\{\cup d : d \in D\}$ is a partition:

$$\mathsf{Disconnected}(D) \;\leftrightarrow_{\mathrm{Def}}\; \mathsf{PartitionOf}\left(\left\{\cup d : \; d \in D\right\}, \cup \cup D\right).$$

After defining

$$\sqcap \mathcal{Q} \;=_{\mathrm{Def}}\; \left\{\bigcap_{x \in b \in Q \in \mathcal{Q}} b : x \in \cup \cup \mathcal{Q}\right\},$$

one gets the finest disconnected partition of any set E such that $\emptyset \notin E$ as $\sqcap \mathcal{Q}$, where $\mathcal{Q}$ is the set—to which $\{E\} \setminus \{\emptyset\}$ belongs—of all disconnected partitions of E.

Definition: The *reachability partition* of a set $E \not\ni \emptyset$ is that disconnected partition of E such that no finer disconnected partition of E exists. We say that E is *connected* when this partition has fewer than two blocks.

---

that is, no vertex is isolated and no set p of edges other than $\emptyset$ and E is vertex disjoint from its complement $E \setminus p$ (see Example 5.1). Here we deliberately leave out of consideration the graph consisting of a single vertex (on the other hand, a graph with two vertices and one edge, or with no vertices at all, falls under our definition). To separate concerns, this definition does not insist that V and E form a graph: e.g., any set a satisfies $\mathsf{Conn}(a, \{a\})$, even when a is not a doubleton and hence it does not qualify as an edge.

Example 5.1 A partition of the set of edges of a connected graph is shown below; the blocks of this partition, p and q = E \ p, are not vertex disjoint. No white-vertex removal disrupts the connectedness of G.

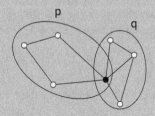

Non-cut vertices are now easy to characterize:

$$\mathsf{NonCut}(z, V, E) \leftrightarrow_{\mathrm{Def}} z \in V \ \& \ \mathsf{Conn}(V \setminus \{z\}, \{a \in E \mid z \notin a\});$$

that is, z is a non-cut vertex if removal of z along with all edges incident to z leads to a connected graph.

We postpone to Appendix A.1 the formal proof of this restatement of claim (5.1):

Theorem 5.1 *Every connected graph with at least two edges has a non-cut vertex:*

$$\Big(\mathsf{uGraph}(V, E) \ \& \ \mathsf{Conn}(V, E) \ \& \ E \nsubseteq \{\mathbf{arb}(E)\}\Big) \to \exists z \, \mathsf{NonCut}(z, V, E).$$

(Here $\mathbf{arb}(E)$ denotes an edge e_0 arbitrarily drawn from E—if $E = \emptyset$, then $e_0 = \emptyset$. The quantifier-free condition $E \nsubseteq \{\mathbf{arb}(E)\}$ thus rules out the possibilities that E be empty or consist of e_0 alone.)

A characterization of connectedness, less slick than the one adopted above but in whose terms the existence of non-cut vertices would be easier to prove, is presented in Exercises 5.12 and 5.13. (Our rationale for not adopting it has been that it is overly knowledgeable.)

5.2   Formalizing Our Mathematical Roadmap

This section continues to pave the way to proofs of the classical results on claw-free graphs referred to as (5.1) and (5.1) at the beginning of Sect. 5.1. Those proofs, before we can subject them to automated verification, must be recast more formally than we intend to do right away; only then they will constitute an implementation ready for automated verification. Here we outline the main phases of our enterprise in terms more convenient for the reader than for our automated assistant.

As a general rule, it is a virtuous proof-development practice to prepare an accurate draft of a challenging proof before undertaking its implementation with a proof checker. Indeed, by sketching the argumentation line at the level of ordinary mathematical discourse, one can better elicit the underlying insight and identify useful preliminary lemmas and reusable proof components. Some aspects—hopefully marginal—of the proofs may later be worked out differently; but this way of proceeding is aimed at ensuring that once fully formalized the proof will still retain transparency, to the benefit of the mathematically inclined reader. By reflecting this recommended practice into the organization of this chapter, we wish to convey to our readers a taste of our work and of our proof-development method.

Our next target will be: Show how to impose a very special orientation on the edges of a connected graph $G = (V, E)$. The resulting structure will be a *digraph* $D = (V, A)$.

Throughout this chapter, we will designate the oriented pair $\langle x, y \rangle$ as $[x, y]$ and its components as $p^{[1]} = x$ and $p^{[2]} = y$, respectively; see Fig. 5.1.[3] As usual, $\mathbf{dom}\,(F)$ will indicate the set $\{\, p^{[1]} : p \in F \,\}$ of first components of the ordered pairs belonging to a set F.

Figure 5.2 shows the definitions of graph-theoretic notions relevant to the proof-checking experiment on which we report and introduces the notions of mapping ('Svm'), transitive set, and finitude and the *recursive* property of hereditary finitude. It brings to light, once more, the salient role of abstraction terms of the kind here called *setformers* in set-based specifications: e.g., the setformer

$$\{\{p^{[1]}, p^{[2]}\} : p \in A \mid p = [p^{[1]}, p^{[2]}]\}$$

collects all doubletons (or singletons) which result from the ordered pairs in A when the positions of their components are purposely forgotten.

DEF pair$_0$: [Kuratowski's pair] $[\mathsf{X},\mathsf{Y}]$ $=_{\mathrm{Def}}$ $\{\{\mathsf{X}\},\{\mathsf{X},\mathsf{Y}\}\}$

DEF pair$_1$: [1st pair-component] $\mathsf{P}^{[1]}$ $=_{\mathrm{Def}}$ $\mathbf{arb}(\{\mathsf{x} : \mathsf{s} \in \mathsf{P}, \mathsf{x} \in \mathsf{s} \mid \mathsf{s} = \{\mathsf{x}\}\})$

DEF pair$_2$: [2nd pair-component] $\mathsf{P}^{[2]}$ $=_{\mathrm{Def}}$ $\mathbf{arb}(\{\mathsf{y} : \mathsf{d} \in \mathsf{P}, \mathsf{y} \in \mathsf{d} \mid \mathsf{P} = \{\{\mathsf{y}\}\} \vee \mathsf{d} \backslash \{\mathsf{y}\} \in \mathsf{P}\})$

DEF prod: [Cartesian product] $\mathsf{S} \times \mathsf{T}$ $=_{\mathrm{Def}}$ $\{[\mathsf{x},\mathsf{y}] : \mathsf{x} \in \mathsf{S}, \mathsf{y} \in \mathsf{T}\}$

Fig. 5.1 Definitions of the ordered pair, of its projections, and of Cartesian product

---

[3]By slightly deviating from the notation employed so far, in this chapter we are conforming to the manner formulae are pretty-printed by our proof checker Ref. For example, we will often write $\langle \exists x \in t \mid \varphi \rangle$ instead of $\exists s \in t\ \varphi$ and will do similarly with universal restricted quantifiers.

DEF acyclic: [Acyclicity]   Acyclic$(V,A) \leftrightarrow_{\text{Def}}$
$$\left\langle \forall w \subseteq V \,|\, w \neq \emptyset \to \left\langle \exists t \in w \,|\, \emptyset = \{ y \in w \,|\, [t,y] \in A \} \right\rangle \right\rangle$$

DEF xtens$_0$: [Extensionality]   Extensional$(V,A) \leftrightarrow_{\text{Def}}$
$$\left\langle \forall x \in V, y \in V, \exists z \,|\, ([x,z] \in A \leftrightarrow [y,z] \in A) \to x = y \right\rangle$$

DEF xtens$_1$: [Weak extensionality]   WExtensional$(V,A) \leftrightarrow_{\text{Def}}$
$$\text{Extensional}(V \cap \mathbf{dom}(A \cap (V \times V)), A \cap (V \times V))$$

DEF orien : [Orientation of a graph]   Orientates$(A,V,E) \leftrightarrow_{\text{Def}}$
$$E \cap \{\{x,y\} : x \in V, y \in V \setminus \{x\}\} = \left\{ \left\{ p^{[1]}, p^{[2]} \right\} : p \in A \,|\, p = \left[p^{[1]}, p^{[2]} \right] \right\}$$

DEF maps$_1$: [Map domain, i.e. first components of pairs in map] $\mathbf{dom}(F) =_{\text{Def}}$
$$\left\{ p^{[1]} : p \in F \right\}$$

DEF maps$_2$: [Map restriction] $F_{|S} =_{\text{Def}}$
$$\left\{ p \in F \,|\, p^{[1]} \in S \right\}$$

DEF maps$_3$: [Value of single-valued function] $F|X =_{\text{Def}}$
$$\mathbf{arb}\left(F_{|\{X\}}\right)^{[2]}$$

DEF maps$_4$: [Multi-image, i.e. second components of pairs in map] $\mathbf{img}(F) =_{\text{Def}}$
$$\left\{ p^{[2]} : p \in F \right\}$$

DEF maps$_5$: [Map predicate]   Is_map$(F) \leftrightarrow_{\text{Def}}$
$$\left\langle \forall p \in F \,|\, p = \left[p^{[1]}, p^{[2]} \right] \right\rangle$$

DEF maps$_6$: [Single-valued map]   Svm$(F) \leftrightarrow_{\text{Def}}$
$$\text{Is\_map}(F) \;\&\; \left\langle \forall p \in F, q \in F \,|\, p^{[1]} = q^{[1]} \to p = q \right\rangle$$

DEF unionset: [Family of all members of members of a set] $\bigcup(S) =_{\text{Def}}$
$$\{u : v \in S, u \in v\}$$

DEF pow: [Family of all subsets of a given set] $\mathcal{P}(S) =_{\text{Def}}$
$$\{x : x \subseteq S\}$$

DEF transitivity: [Transitive set]   Trans$(T) \leftrightarrow_{\text{Def}}$
$$\{y \in T \,|\, y \not\subseteq T\} = \emptyset$$

DEF Finite : [Finitude]   Finite$(S) \leftrightarrow_{\text{Def}}$
$$\left\langle \forall g \in \mathcal{P}(\mathcal{P}(S)) \setminus \{\emptyset\}, \exists m \,|\, g \cap \mathcal{P}(m) = \{m\} \right\rangle$$

DEF HerFin: [Hereditary finitude]   HerFin$(S) \leftrightarrow_{\text{Def}}$
$$\text{Finite}(S) \;\&\; \left\langle \forall x \in S \,|\, \text{HerFin}(x) \right\rangle$$

Fig. 5.2 Four properties refer to digraphs, all others to generic sets

The first definition in Fig. 5.2 specifies the property of a digraph (V, A) in which every nonnull set w of vertices has a *local sink*, namely, a $t \in w$ devoid of outgoing arcs $[t, y]$ with $y \in w$. This forbids walks of the form $[x_0, x_1], [x_1, x_2], \ldots$ *(ad inf.)* consisting of arcs; see Exercise 2.13. In the most common situation, in which A is a finite set, this simply amounts to forbidding cycles $[x_0, x_1], [x_1, x_2], \ldots, [x_k, x_0]$ of arcs. From now on the meaning of the **arb** operator, which—as mentioned above— picks from each set $w \neq \emptyset$ an $\mathbf{arb}(w) = t \in w$, will be constrained similarly: by requiring that $\emptyset = \{y \in w \,|\, y \in t\}$, we will use **arb** to witness that $\in$ is a well-founded relationship, thus excluding cycles $x_0 \in x_1 \in \cdots \in x_k \in x_0$. (We also put $\mathbf{arb}(\emptyset) = \emptyset$.)

The second definition in Fig. 5.2 introduces extensionality: a digraph enjoys this property if no two vertices in it have the same out-neighborhood. The next definition introduces *weak* extensionality in rather cautious terms, namely, without presupposing that A consists of pairs $[x, y]$ with $x, y \in V$: this is why this definition treats only the elements of $A \cap (V \times V)$ as arcs; moreover, the requirement that two vertices cannot have the same out-neighborhood here regards only nonsink vertices. The fourth definition states that $\mathsf{Orientates}(A, V, E)$ holds when the edges in E are the doubletons resulting from the pairs in A; to keep the statement simple, here we do not insist that V and E form a graph; hence we cannot take the inclusion $\{\{x, y\} : x \in V, y \in V \setminus \{x\}\} \subseteq E$ for granted.

Six subsequent definitions in Fig. 5.2 regard maps and map-related operations such as the value $F \upharpoonright x$ resulting from application of F to an operand x. This notation is a bit unusual and deserves some justification: to enforce a useful distinction, we denote by $\gamma(x)$—e.g., $\mathcal{P}(x)$—the application of a *global* function γ to an argument x ("global" meaning that the domain of γ consists of all sets), while denoting by $F \upharpoonright x$—or, often, simply by $F x$—the application to x of a *map* F (typically single valued), viewed as a set of pairs. In the light of these definitions and under the assumption $A \subseteq V \times V$, sources and sinks of the digraph $D = (V, A)$, and the out-neighbors of a vertex x, can be characterized very straightforwardly:

s is a source:	$s \in V \setminus \mathrm{img}(A)$,
s is a sink:	$s \in V \setminus \mathrm{dom}(A)$,
s is an out-neighbor of x :	$s \in \{\mathsf{p}^{[2]} : \mathsf{p} \in A_{\restriction \{x\}}\} \left(= \mathrm{img}\left(A_{\restriction \{x\}}\right)\right)$.

The penultimate definition in Fig. 5.2 states that a set S is finite if every nonnull set g consisting of subsets of S has an $\subseteq$-minimal element m.

5.2.1 Acyclic Orientations and Extensionalization

We will explain here how to convert an arbitrary graph into a weakly extensional acyclic digraph. Specifically, the claim which will be proved is

Theorem 5.2 *Let s be a vertex of the graph $G = (V, E)$; then an orientation A can be assigned to the edges E so that the resulting digraph $D = (V, A)$ is weakly extensional and acyclic, and s is a source (namely, a vertex with no incoming arcs) in D. This claim can be formalized in slightly more generic terms:*

$$(\mathsf{Finite}(V) \ \& \ s \in V) \rightarrow \exists A(\mathsf{Orientates}(A, V, E) \ \& \ \mathsf{Acyclic}(V, A)$$
$$\& \ \mathsf{WExtensional}(V, A) \ \& \ s \notin \mathrm{img}(A)).$$

Proof Arguing by contradiction, suppose that there is a counterexample; then, by exploiting the finiteness hypothesis, take a *minimal* counterexample V_1, s_1, E_0.

Fig. 5.3 Extending a weakly extensional acyclic orientation of $(V_0, E_0 \cap \{\{x, y\} : x \in V_0, y \in V_0 \setminus \{x\}\})$ in which a neighbor t_1 of s_1 acts as a source, by making s_1 a source

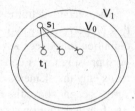

THM acyclicity$_0$: [Adjunction of an outer vertex to a digraph cannot disrupt acyclicity]
   $V \times V \supseteq A \,\&\, X \notin V \,\&\, V \supseteq S \,\&\, \text{Acyclic}(V, A) \rightarrow \text{Acyclic}(V \cup \{X\}, A \cup (\{X\} \times S))$

THM acyclicity$_1$: [Reduction of the set of arcs of a digraph preserves its acyclicity]
   $\text{Acyclic}(V, A) \,\&\, V' \subseteq V \,\&\, A' \subseteq A \rightarrow \text{Acyclic}(V', A')$

THM acyclicity$_2$: [Acyclic digraphs are devoid of self-loops and of symmetrical arcs]
   $\text{Acyclic}(V, A) \,\&\, \{Y, X\} \subseteq V \,\&\, [X, Y] \in A \rightarrow [Y, X] \notin A \,\&\, X \neq Y$

THM acyclicity$_4$: [Every acyclic graph has sinks and sources]
   $\text{Acyclic}(V, A) \,\&\, \text{Finite}(V) \,\&\, V \neq \emptyset \rightarrow$
   $\quad \langle \exists s \in V, t \in V \mid \emptyset = \{y \in V \mid [s, y] \in A \vee [y, t] \in A\} \rangle$

THM acyclicity$_5$: [No triangle inside an acyclic digraph]
   $\text{Acyclic}(V, A) \,\&\, \{X, Y, Z\} \subseteq V \,\&\, \{[X, Y], [Y, Z]\} \subseteq A \rightarrow [Z, X] \notin A$

THM acyclicity$_6$: [Adjunction of an inner vertex to a digraph cannot disrupt acyclicity]
   $V \times V \supseteq A \,\&\, X \notin V \,\&\, V \supseteq S \,\&\, \text{Acyclic}(V, A) \rightarrow \text{Acyclic}(V \cup \{X\}, A \cup (S \times \{X\}))$

Fig. 5.4 Properties of acyclicity proved by means of Ref

We are supposing that there is no acyclic, weakly extensional orientation of the graph $(V_1, E_0 \cap \{\{x, y\} : x \in V_1, y \in V_1 \setminus \{x\}\})$ having s_1 as a source, whereas, for every $V_0 \subsetneq V_1$, one can orient $(V_0, E_0 \cap \{\{x, y\} : x \in V_0, y \in V_0 \setminus \{x\}\})$ by an acyclic and weakly extensional $A_0 \subseteq V_0 \times V_0$, for any vertex $t \in V_0$, so that t plays the role of a source. Let, in particular, $V_0 = V_1 \setminus \{s_1\}$. Unless s_1 is an isolated vertex, an acyclic and weakly extensional orientation of V_0 exists that has as a source a chosen neighbor t_1 of s_1 (see Fig. 5.3). However, that orientation could trivially be extended into a weakly extensional acyclic orientation of the graph with vertices V_1 so that s_1 becomes a source; this contradiction shows that s_1 cannot have neighbors in V_1, which is also untenable: any orientation for V_0, in fact, works also as an orientation for V_1 and, as such, has each isolated vertex of V_1—in particular s_1—as a source. ⊣

Figure 5.4 lists various properties enjoyed by acyclicity whose proofs are left to the reader as Exercise 5.7. Of the items in that list, THMs acyclicity$_1$, acyclicity$_2$, and acyclicity$_4$ will enter in the proofs related to the decoration of $D = (V, A)$, which will be discussed next (see Sect. 5.2.2); THMs acyclicity$_0$, acyclicity$_1$, acyclicity$_5$, and acyclicity$_6$ will enter more or less directly in the proof that every connected claw-free graph is a set graph (i.e., it admits an EAO).

5.2.2 Injective Digraph Decorations

Alongside with the issue of getting a weakly extensional acyclic orientation of any given graph $G = (V, E)$, an important technical issue is how to decorate the resulting digraph $D = (V, A)$ by sets, so that its arcs mirror membership: assuming that $V \neq \emptyset$, we will now see how to define a function mski on the set V of vertices so as to satisfy the following conditions relative to D:

- for each vertex w which is not a sink, mski sends w to the multi-image, under mski, of the out-neighborhood of w:

$$\left(\forall w \in \mathbf{dom}\,(A)\right)\left(\mathsf{mski}\!\restriction\! w = \mathbf{img}\left(\mathsf{mski}_{|\mathbf{img}(A_{|\{w\}})}\right)\right);$$

- mski sends different vertices to different values;
- mski sends every vertex to a finite set and sends one vertex to $\emptyset$.

Remark 5.1 The identifier "mski" wants to recall Andrzej Mostowski, the logician who introduced the collapsing function, cf. Lemma 3.1. Injective decorations *à la* Mostowski of acyclic graphs have already surfaced a few times in this book; see, e.g., Examples 1.8, 3.10, and 3.23. Much as we will do now, we have repeatedly adapted the decoration technique to digraphs which we did not require to be extensional, unlike the "collapsing lemma" just cited did.   ⊣

The construction of mski can be carried out in many ways, and we opt for the rather simple technique (see Example 5.2) of using the global function

$$\mathsf{tag}(w) = \begin{cases} \text{if} & w \in \mathbf{dom}\,(A) \,\cup\, \{\mathbf{arb}\,(V \backslash \mathbf{dom}\,(A))\} \\ \text{then} & \emptyset \\ \text{else} & \{\{V\} \cup (V \backslash \{w\})\} \quad \text{fi} \end{cases}$$

to define mski recursively over V, by taking advantage of the finiteness of V and of the acyclicity of D:

$$\mathsf{mski}\!\restriction\! x = \{\mathsf{mski}\!\restriction\! p^{[2]} : p \in A_{|\{x\}} \mid p^{[2]} \in V\} \cup \mathsf{tag}(x).$$

This will automatically enforce the first condition we want to impose on mski, for all $x \in V$; moreover, mski readily turns out to be a function sending each vertex to a finite set and sending the sink $\mathbf{arb}\,(V \backslash \mathbf{dom}\,(A))$ to $\emptyset$.

Example 5.2 Consider the weakly extensional acyclic digraph with vertex set $V = \{a, b, c, d, e\}$ and set of arcs A, drawn here below.

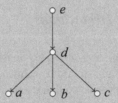

By definition, every nonsink vertex receives the tag $\emptyset$ because it belongs to $\mathbf{dom}\,(A)$. We assume that a is the vertex $a = \{\mathbf{arb}\,(V \backslash \mathbf{dom}\,(A))\}$, thus $\mathsf{tag}(a) = \emptyset$. Both sinks b and c belong to $V \backslash \mathbf{dom}\,(A)$, and hence

- $\mathsf{tag}(b) = \{\{a, b, c, d, e\}, a, c, d, e\}$
- $\mathsf{tag}(c) = \{\{a, b, c, d, e\}, a, b, d, e\}$

Below, next to each vertex $x \in V$, we show the set $\mathsf{mski} \upharpoonright x$ decorating x.

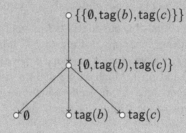

The injectivity of mski is obvious over the set $V \setminus \mathbf{dom}\,(A)$ of all sinks (because, there, tag is injective and mski and tag take the same values); from the sinks it easily extends to all other vertices: cardinality considerations (see Exercise 5.9) show in fact that a value collision between a sink and an internal vertex is impossible, and the weak extensionality assumption prevents collisions to occur between internal vertices. It follows from the injectivity of mski that $u \in \mathsf{mski} \upharpoonright x$ is satisfied if and only if either $u = \mathsf{mski} \upharpoonright y$ and $[x, y] \in A$ or u is the set—if any—satisfying $\{u\} = \mathsf{tag}(x)$. Note that the second case of this alternative vanishes when the digraph has just one sink, i.e., it is extensional.

We leave the verification that mski has all desired properties to the reader, as Exercise 5.9. *Weak* extensionality has been presupposed till now; additional properties of mski can be proved if we call for slightly more, namely, for extensionality, see Exercise 5.10.

5.2.3 Claw-Free Graphs Mirrored into Transitive Finite Sets

Our next, richer construction will associate with each connected claw-free graph $G = (V, E)$ an injection f from V onto a transitive, hereditarily finite set v_G so that $\{x, y\} \in E$ if and only if either $f x \in f y$ or $f y \in f x$.

The main new notion entering into play is rendered formally as follows:

$$\mathsf{ClawFreeG}(V, E) \;\leftrightarrow_{\mathrm{Def}}$$
$$\forall w, x, y, z \, (\{w, x, y, z\} \subseteq V \;\&\; \{\{w, y\}, \{y, x\}, \{y, z\}\} \subseteq E \rightarrow$$
$$(x = z \vee w \in \{z, x\} \vee \{x, z\} \in E \vee \{z, w\} \in E \vee \{w, x\} \in E)).$$

Here, the *definiens* requires that no subgraph of G induced by four vertices has the shape of a "Y" (see the image of $K_{1,3}$ recalled from p. 103 at the top of Fig. 5.6).

We will exploit Theorem 5.1 on the existence of non-cut vertices in order to get the following downgraded version of Theorem 4.2. Our forthcoming proof will retrace the one of Theorem 4.2, while now providing more details and also making explicit use of those formal predicates $\mathsf{uGraph}(\cdot, \cdot), \cdots, \mathsf{Extensional}(\cdot, \cdot)$ that we have introduced short ago (see Sect. 5.1 and Fig. 5.2).

Theorem 5.3 *Suppose that $G = (V, E)$ is a connected and claw-free graph; then an orientation A can be assigned to the edges E so that the resulting digraph $D = (V, A)$ is extensional and acyclic:*

$$(\mathsf{uGraph}(V, E) \;\&\; \mathsf{Conn}(V, E) \;\&\; \mathsf{ClawFreeG}(V, E)) \rightarrow \exists A \, (A \subseteq V \times V$$
$$\&\; \mathsf{Orientates}(A, V, E) \;\&\; \mathsf{Acyclic}(V, A) \;\&\; \mathsf{Extensional}(V, A)).$$

Proof Arguing by contradiction, assume that there is a counterexample $G_2 = (V_2, E_2)$ to the claim. Then, thanks to the finiteness assumption built in our definition of uGraph, we can take a minimal counterexample $G_1 = (V_1, E_1)$ with $V_1 \subseteq V_2$

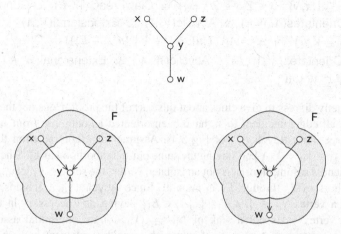

Fig. 5.5 The forbidden orientations of a *claw* (as usually intended—see graph *on the top*) inside a *claw-free set* (as formally defined in Fig. 5.5 for our proof-checking experiment)

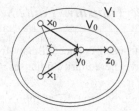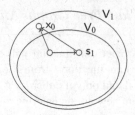

Fig. 5.6 The main cases in the proof of Theorem 5.3. On *the left*, the acyclic digraph $D_0 = (V_1 \setminus \{x_0\}, A_0)$ has no sink adjacent to x_0 through E_1; on *the right*, the contrary case

and $E_1 = E_2 \cap \{\{x, y\} : x \in V_1, y \in V_1\}$. Note that V_1 cannot be a singleton, else a contradiction would arise: the null set of arcs would in fact be an extensional acyclic orientation of G_1; the possibility that E_1 be a singleton is discarded with equal ease.

By Theorem 5.1, we can hence consider a non-cut vertex x_0 of G_1. Now consider the graph $G_0 = (V_0, E_0)$ induced by G_1 on the strict subset $V_0 = V_1 \setminus \{x_0\}$ of the set of vertices. This graph inherits the claw-freeness property; it is in fact plain that

$$\big(\mathsf{ClawFreeG}(V, E) \ \& \ W \subseteq V \ \to \ \mathsf{ClawFreeG}(W, \{e \in E \mid e \subseteq W\})\big)$$

holds in general; therefore, the minimality of V_1 ensures us that we can obtain an extensional acyclic orientation A_0 of this induced graph.

We first deal with the case when the sink of the acyclic digraph $D_0 = (V_0, A_0)$ is not adjacent to x_0 through E_1 (see Fig. 5.6, left). In this case we orient the edges incident to x_0 as outgoing from x_0, to get an extensional acyclic orientation A_1 for G_1. Note that the neighbors of x_0 through E_1 are $\{t \in V_1 \mid \{x_0, t\} \in E_1\}$, and hence we are putting $A_1 = A_0 \cup (\{x_0\} \times \{t \in V_1 \mid \{x_0, t\} \in E_1\})$.

Here we resort to the following auxiliary lemma which holds for W, V, u, A, E, Z and A' whatsoever (for the purposes of the case at hand, these must be instantiated as $W = V_1, V = V_0 \setminus \{x_0\}, u = x_0, A = A_0, E = E_1, Z = V_2$, and $A' = A_1$):

$$\big(W = V \cup \{u\} \ \& \ u \notin V \ \& \ \{s \in V \mid A_{|\{s\}} = \emptyset \ \& \ \{s, u\} \in E\} = \emptyset$$
$$\& \ E \subseteq \{\{x, y\} : x \in Z, y \in Z \setminus \{x\}\} \ \& \ \mathsf{ClawFreeG}(W, E) \ \& \ \mathsf{Conn}(W, E)$$
$$\& \ \mathsf{Orientates}(A, V, E) \ \& \ \mathsf{Acyclic}(V, A) \ \& \ \mathsf{Extensional}(V, A)$$
$$\& \ A \subseteq V \times V \ \& \ A' = A \cup (\{u\} \times \{t \in V \mid \{u, t\} \in E\})\big)$$
$$\to \ \big(\mathsf{Orientates}(A', W, E) \ \& \ \mathsf{Acyclic}(W, A') \ \& \ \mathsf{Extensional}(W, A')$$
$$\& \ A' \cdot \subseteq W \times W\big).$$

Let us briefly digress to give clues about this part of the proof. Consider the digraph in which all edges incident to x_0 have been oriented as outgoing from x_0; this is acyclic (see THM $\mathsf{acyclicity}_0$ of Fig. 5.4). Assume for a contradiction that there exists an $x_1 \in V_0 = V_1 \setminus \{x_0\}$ having the same out-neighborhood as x_0. Since, by the connectedness assumption, x_0 is not an isolated vertex, the set $\{t \in V_1 \mid \{x_0, t\} \in E_1\}$ of neighbors of x_1 through E_1 is nonnull. Since $\mathsf{Acyclic}(V_0, A_0)$ holds, we can consider a vertex $y_0 \in \{t \in V_1 \mid \{x_0, t\} \in E_1\}$ having no successors in common with x_0. Vertex y_0 is not a sink of $D_0 = (V_0, A_0)$ by our initial assumption; thus there exists a successor z_0 of y_0 which is neither adjacent to x_0 nor to x_1, in consequence of the choice of y_0, of the fact that x_0 and x_1 have the same out-neighbors, and of THM acyclicity$_5$ of Fig. 5.4. Since also x_0 and x_1 are not adjacent, by THM acyclicity$_2$ of Fig. 5.4, it follows that the set $\{x_0, x_1, y_0, z_0\}$ is a claw of the graph (V_1, E_2), a contradiction.

Next we deal with the case when the sink s_1 of D_0 is adjacent to x_0 through E_1 (see Fig. 5.6, right). Here we resort to the following auxiliary lemma, which holds for W, V, u, s, A, E and Z whatsoever (for our purposes, these must be instantiated as $W = V_1$, $V = V_0$, $u = x_0$, $s = s_1$, $A = A_0$, $E = E_2$ and $Z = V_2$):

$$\begin{aligned}
\big(W = V \cup \{u\} \ \& \ u \notin V \ \& \ s \in V \ \& \ \{y \in V \mid [s, y] \in A\} = \emptyset \\
\& \ s \in \{t \in W \mid \{u, t\} \in E\} \ \& \ E \subseteq \{\{x, y\} : x \in Z, y \in Z \setminus \{x\}\} \\
\& \ \mathsf{Orientates}(A, V, E) \ \& \ \mathsf{Acyclic}(V, A) \ \& \ \mathsf{Extensional}(V, A) \\
\& \ A \subseteq V \times V \big) \rightarrow \exists A' \big(A' \subseteq W \times W \ \& \ \mathsf{Orientates}(A', W, E) \\
\& \ \mathsf{Acyclic}(W, A') \ \& \ \mathsf{Extensional}(W, A') \big).
\end{aligned}$$

Concretely, this amounts to constructing A_1 from A_0 by orienting all edges incident to x_0 as incoming to x_0: thus an acyclic digraph $D_1 = (V_1, A_1)$ results (see THM acyclicity$_6$ of Fig. 5.4); moreover, D_1 is extensional because s_1 has x_0 as its sole out-neighbor, whereas every other vertex in V_0 has at least one vertex in V_0 as out-neighbor.

In both cases we have been able to extend A_0 into an acyclic extensional orientation of G_1 and hence have found a contradiction, as we aimed at.   ⊣

Let us exploit the orientation, just found, of a connected claw-free graph G in order to decorate its vertices injectively as explained in Sect. 5.2.1. The multi-image v_G of the decoration $f = \mathsf{mski}$ is a transitive finite set (see Exercise 5.10), over which membership mimics E as recollected below:

Theorem 5.4 (Reflection theorem [64]) *If $G = (V, E)$ is a connected claw-free graph, then a finite transitive set v_G and a bijection $f : V \longrightarrow v_G$ can be found so that the biimplication*

$$\{x, y\} \in E \ \leftrightarrow \ \big(f\, x \in f\, y \vee f\, y \in f\, x\big)$$

holds for x, y in V. (An example of this is shown in Example 5.3.)

Example 5.3 The image below shows a connected claw-free graph, on the left. On the right, it shows an extensional acyclic orientation of it and, near each vertex of the oriented version of the graph, the set associated to it by the Mostowski decoration.

(continued)

Example 5.3 (continued)
   Notice that this claw-free graph is the same seen in Example 1.11, but drawn "upside-down." Its extensional acyclic orientation shown above is different from the one in Example 1.11. (Recall from Corollary 4.6 that, given an arbitrary graph, finding the number of all its extensional acyclic orientations is computationally difficult, namely, #P-complete.)

It makes sense, in the light of this mirroring, to study the properties of connected claw-free graphs by reasoning about "membership digraphs" whose vertices are hereditarily finite sets and whose arcs reflect the membership relation between them. In sight of this, we supply in Fig. 5.7 definitions of $\in$-claws and claw-free *sets*. In the second of these, the assumption that S is transitive is omitted and left pending to be introduced explicitly in the pertaining theorems.

An $\in$-*claw* is thereby defined to be a pair y, F of sets such that:

1. F has at least three elements,
2. no element of F belongs to any other element of F,
3. either y belongs to all elements of F or there is a $w \in y$ such that y belongs to all elements of $F \setminus \{w\}$.

Accordingly, a *claw-free set* will be one which does not include an $\in$-claw. For that, it suffices that it does not contain an $\in$-claw y, F, with $|F| = 3$, like the ones shown in Fig. 5.5.

On the basis of these definitions, one easily proves (see Exercise 5.14 below) the monotonicity of claw-freeness, along with one slightly less obvious property:

THM clawFreeness$_a$ *Subsets of claw-free sets are claw-free:*

$$\text{ClawFree}(S) \ \& \ T \subseteq S \ \rightarrow \ \text{ClawFree}(T) \,;$$

DEF claw: [Pair characterizing an $\in$-claw, possibly endowed with more than 3 elements]
   MembClaw$(Y,F) \ \leftrightarrow_{\text{Def}} \ F \cap \bigcup F = \emptyset \ \&$
      $\langle \exists x,z,w \, | \, F \supseteq \{x,z,w\} \ \& \ x \neq z \ \& \ w \notin \{x,z\} \ \& \ \{w\} \cap Y \supseteq \{v \in F \, | \, Y \notin v\} \rangle$
DEF clawFreeness: [Claw-freeness, for a membership digraph]
   ClawFree$(S) \ \leftrightarrow_{\text{Def}} \ \langle \forall y \in S, e \subseteq S \, | \, \neg\text{MembClaw}(y,e) \rangle$

Fig. 5.7 Adaptation of graph-theoretic notions to membership digraphs

THM clawFreeness$_b$ *In a claw-free set, any potential* $\in$*-claw must have a bypass:*

$$(\, \mathsf{ClawFree}(S) \; \& \; S \supseteq \{y, x, z, w\} \; \& \; y \in x \cap z$$
$$\& \; w \in y \; \& \; x \notin z \cup \{z\} \; \& \; z \notin x \,) \; \rightarrow \; w \in x \cup z.$$

Preparatory to Sect. 5.2.4, let us hint at our view of a set S as being a digraph $D(S)$ with sources $S \setminus \cup S$. Relevant for what is to follow, we will focus on the *pivots* of $D(S)$, which we define to be the elements of $(\cup S) \setminus \cup(S \cap \cup S)$—simplifiable into $(\cup S) \setminus \cup \cup S$ when S is transitive; in graph-theoretic terms, $y \in S$ is a pivot of $D(S)$ if y is an out-neighbor of a source of $D(S)$, but is not at the end of any directed path included in S whose length exceeds 1. The salient property of a pivot y is that if x, z are in-neighbors of y, then neither $x \in z$ nor $z \in x$ holds, thanks to the claim stated in Exercise 5.3 below, whose contrapositive ensures, when $y \notin \cup(S \cap \cup S)$, the incomparability of x and z.

It hence turns out that in a claw-free set the in-neighbors of a pivot $y \in x \in S$ form a set $\{x, z\}$, possibly singleton. This is straightforward: should y have three in-neighbors x, z, and w, the pair $y, \{x, z, w\}$ would then be an $\in$-claw.

The reader is invited, at this point, to do Exercises 5.14 and 5.3.

5.2.4 Perfect Matchings, Hamiltonian Cycles, and Claw-Freeness

This section revisits two classical results on connected claw-free graphs and provides new proofs exploiting the reflection result Theorem 5.4, by which such graphs have been assimilated to special transitive sets and their edges have become membership arcs. How can an orientation artificially imposed on the edges be of any help in proving those results? To see an analogous situation, recall the classical "mutilated chessboard" problem:

> A chess board can be tiled by 32 dominoes, each covering two squares. If two diagonally opposite squares are removed, can the remaining 62 squares be tiled by dominoes? [91]

Abstractly speaking, the colors of the squares on the chessboard should have no bearing whatsoever on the solution of the problem; but, in practice, they offer a crucially useful hook to the human mind[4]:

> No. Each domino covers a white square and a black square, so a tiled area must have equal numbers of both colours. The mutilated board cannot be tiled because the two removed squares have the same colour. [91]

---

[4]The "negative solution" to the tiling-problem instance recalled here has its positive counterpart in Gomory's theorem, telling us that regardless of where one white and one black square are deleted from an ordinary chessboard, the remaining board can always be tiled by 31 dominoes.

Similarly, thanks to Theorem 5.4, our proofs on claw-free graphs will be shorter than the classical ones and relatively easy to grasp.

Let us start by recalling, from Sect. 2.3.2, two graph-theoretic definitions which we will specify later on in the language of our proof checker. These are presupposed for results which we will revisit in purely set-theoretic terms. A *perfect matching* in a graph G is a subset M of its edges such that no two edges in M have a common endpoint and every vertex of G is endpoint of some edge in M. A *Hamiltonian cycle* in a graph G is a subset C of its edges which forms a cycle such that every vertex of G is endpoint of an edge in C.

In addition, we need the notion of *square* of a graph G, denoted G^2, and defined as the graph with the same vertex set as G in which two vertices are adjacent if either they are adjacent in G, or they have a common neighbor in G. See Example 5.4.

Example 5.4 Let G be the following connected claw-free graph:

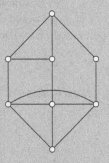

Below on the left, we draw with bold edges a Hamiltonian cycle in G^2; wiggly edges are edges of G^2 not present in G. On the right, we draw with bold edges a perfect matching in G.

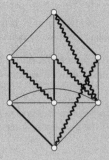 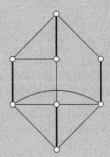

We are now ready for outlining, in terms convenient for the reader, the two proofs concerning connected claw-free graphs whose checkable implementation is under the spotlight of this chapter. The following result first appeared in [108, 111]. We report here below its alternative proof, from [64]. Recall also that we already discussed this proof while illustrating it on a concrete graph in Example 1.12.

Theorem 5.5 *If G is a connected claw-free graph with $2n$ vertices, then G has a perfect matching.*

(We leave it to the reader as Exercise 5.11 a comparison between the claim of Theorem 5.5 and its reformulation shown at the beginning of Sect. 5.1.)

Proof We reason by induction on n. If $n > 1$, let D be an acyclic orientation of G having a unique sink. Let $x_0, x_1, \ldots, x_r$ be a directed path of maximum length in D, and let $x = x_0$ and $y = x_1$. Observe that no two vertices of $N^-(y)$ are adjacent, due to the "*pivotal*" choice of y. Since G is claw-free, it follows that $|N^-(y)| \leqslant 2$, for otherwise $\{y\} \cup N^-(y)$ would contain an induced claw in G.

If $N^-(y) = \{x\}$, then $G - \{x, y\}$ is connected, else D would have two sinks. From the inductive hypothesis, $G - \{x, y\}$ has a perfect matching, which, with the adjunction of the edge $\{x, y\}$, constitutes a perfect matching for G.

If $N^-(y) = \{x, z\}$, then, similarly, $G - \{x, z\}$ is connected. Let $\{y, w\}$ be an edge of the perfect matching of $G - \{x, z\}$, obtained from the inductive assumption. Since G is claw-free, assume without loss of generality that $\{x, w\} \in E(G)$. Obtain a perfect matching for G from the one for $G - \{x, z\}$ by replacing the edge $\{y, w\}$ by the edges $\{x, w\}$ and $\{z, y\}$.   ⊣

In [60] it was proved that a connected claw-free graph with at least three vertices has a Hamiltonian square. We state below the slightly stronger result of [64] together with its proof given therein.

Theorem 5.6 *If G is a connected claw-free graph with at least three vertices, and $S \subseteq V(G)$ is the set of sources of an acyclic orientation of G endowed with exactly one sink, then G^2 has a Hamiltonian cycle C such that for every $t \in S$, at least one edge of C incident to t belongs to $E(G)$.*

Proof Arguing as in the preceding proof, unless G has three or four vertices, we select a "pivotal" pair x, y, so that $|N^-(y)| \leqslant 2$; moreover, $G - N^-(y)$ has at least three vertices and is connected, since otherwise D would have two sinks. In applying the inductive hypothesis to $H = G - N^-(y)$, take the orientation induced by D, so that y is a source, and consider a Hamiltonian cycle C of H^2 containing an edge $yw \in E(G)$.

If $N^-(y) = \{x\}$, notice that $\{x, w\} \in E(G^2)$. Obtain a Hamiltonian cycle in G^2 by replacing the edge $\{y, w\}$ in C by the path (y, x, w) (so that $\{x, y\} \in E(G)$). If $N^-(y) = \{x, z\}$, due to the claw-freeness of G, at least one of the edges $\{x, w\}$ or $\{z, w\}$, say $\{x, w\}$, belongs to G. Moreover, $\{z, x\} \in E(G^2)$. Obtain a Hamiltonian cycle in G^2 by replacing the edge $\{y, w\}$ in C by the path (y, z, x, w) (so that $\{y, z\}, \{x, w\} \in E(G)$).   ⊣

Our formal specification of the above-stated theorems will refer to claw-free and transitive sets, instead of to claw-free graphs. This change of viewpoint is legitimatized by the reflection result Theorem 5.4. Specifically, our proofs will refer to special acyclic digraphs $D(S)$, each supported by a set S. The vertices of $D(S)$ are the sets belonging to S, and for u, w in S, there is an arc from u to w if and only if $w \in u$. Consistently with Sect. 3.4, we will call $D(S)$ a *membership digraph* when the set S is transitive: this requirement entails the extensionality of $D(S)$, namely, that distinct vertices have different out-neighborhoods. $D(S)$ has the virtue that the in-/out-neighborhood of any w in S coincides, respectively, with $\{u \in S \mid w \in u\}$ and with $w \cap S$; here $w \cap S$ simplifies into w when S is transitive. As said earlier, the sources of $D(S)$ form the set difference $S \setminus \bigcup S$. In our arguments about $D(S)$, the orientation of edges can be left as implicit; moreover, the unique-sink assumption readily follows from two facts: $\emptyset$ is the sole set devoid of elements, by the extensionality axiom; moreover, $\emptyset$ belongs to every transitive nonnull set, by the foundation axiom (see Exercise 2.4).

5.3   Set-Based Proof Verification

A convenient computerized system for reasoning about the entities of our discourse—graphs of many kinds—is the proof-checker **Referee** [72, 105], which we here call **Ref** for brevity and which we will describe in Sect. 5.3.1.[5] Upon receiving script files named *scenarios*, which contain fully formalized specifications of definitions, theorems, and proofs, **Ref** examines them and establishes whether or not they are logically correct. This system, consistently with its foundation which is the Zermelo-Fraenkel theory enriched with a universal choice axiom, ultimately represents every entity in the user's universes of discourse as a set; the framework it provides offers infinite sets also, but these are not relevant for our present purposes. **Ref** served as our assistant in the preparation of the subject matter of this chapter in the way described in the ongoing. We will resume the description of our experiment in Appendix A, where we will keep a closer eye on technical details.

We will study graphs as peculiar sets, but will sometimes extend to generic sets—even infinite sets—the import of properties, e.g., connectedness (see Panel 5.1), which are normally referred to graphs. The insight which, by proceeding this way, we will get into the discrete nature of many arguments, will make them shareable with our proof assistant.

---

[5]The proof-checker **Referee** is also known by its fancier name **ÆtnaNova**.

5.3.1   The *Referee* System

5.3.1.1   Ref's Basic Definition-Handling and Proof-Checking Abilities

Our proof-checker Ref (which stands for "Referee") processes script files, named *(proof) scenarios* to establish whether or not they are formally correct. A scenario, typically written by a working mathematician or computer scientist, consists of definitions, theorem statements, detailed proofs of the theorems, and "theories" (see below); one can also intermix comments with these syntactical entities. Figure 5.8 shows a tiny scenario in pretty-printed format, Fig. 5.9 shows another one in the keyboard-oriented format in which it actually gets submitted to Ref. After checking

---

DEF unionset: [Family of all members of members of a set] $\bigcup X =_{\text{Def}} \{u : v \in X, u \in v\}$

THM un_6: [Unionset of adjunction] $\bigcup(X \cup \{Y\}) = Y \cup \bigcup X$. PROOF:
Suppose_not(x_0, y_0) $\Rightarrow$ *Stat0* : $\bigcup(x_0 \cup \{y_0\}) \neq y_0 \cup \bigcup x_0$

$\langle$a$\rangle \hookrightarrow Stat0 \Rightarrow$ a $\in \bigcup(x_0 \cup \{y_0\}) \not\hookrightarrow$a $\in y_0 \cup \bigcup x_0$
 | Arguing by contradiction, let x_0, y_0 be a counterexample, so that in either one of
 | $\bigcup(x_0 \cup \{y_0\})$ and $y_0 \cup \bigcup x_0$ there is an a not belonging to the other set. Taking the def-
 | inition of $\bigcup$ into account, by monotonicity we must exclude the possibility that a $\in$
 | $\bigcup x_0 \setminus \bigcup(x_0 \cup \{y_0\})$; through variable-substitution, we must also discard the possibility that
 | a $\in \bigcup(x_0 \cup \{y_0\}) \setminus \bigcup x_0 \setminus y_0$.
Set_monot $\Rightarrow$ $\{u : v \in x_0, u \in v\} \subseteq \{u : v \in x_0 \cup \{y_0\}, u \in v\}$
Suppose $\Rightarrow$ *Stat1* : a $\in \{u : v \in x_0 \cup \{y_0\}, u \in v\}$ & a $\notin \{u : v \in x_0, u \in v\}$ & a $\notin y_0$
$\langle v_0, u_0, v_0, u_0 \rangle \hookrightarrow Stat1 \Rightarrow$ false;   Discharge $\Rightarrow$ AUTO
Use_def($\bigcup$) $\Rightarrow$ *Stat2* : a $\notin \{u : v \in x_0 \cup \{y_0\}, u \in v\}$ & a $\in y_0$
 | The only possibility left, namely that a $\in y_0 \setminus \bigcup(x_0 \cup \{y_0\})$, is also manifestly absurd. This
 | contradiction leads us to the desired conclusion.
$\langle y_0, a \rangle \hookrightarrow Stat2 \Rightarrow$ false;   Discharge $\Rightarrow$ QED

THM un_{13}: [Less-one lemma for unionset]
      $\bigcup M = T \setminus \{C\}$ & $S = T \cup X \cup \{V\}$ & $Y \in S \cap \{C, V\} \rightarrow$
                       $\langle \exists d \,|\, \bigcup(M \cup \{X \cup \{Y\}\}) = S \setminus \{d\} \rangle$. PROOF:

Suppose_not(m, t, c, s, x, v, y) $\Rightarrow$ *Stat0* : $\neg \langle \exists d \,|\, \bigcup(m \cup \{x \cup \{y\}\}) = s \setminus \{d\} \rangle$
                       & $\bigcup m = t \setminus \{c\}$ & $s = t \cup x \cup \{v\}$ & ($y = v \vee (c = y$ & $y \in s$))
 | For, supposing the contrary, $\bigcup(m \cup \{x \cup \{y\}\})$ would differ from each of $s \setminus \{s\}$,
 | $s \setminus \{c\}$, and $s \setminus \{v\}$, the first of which equals s. Thanks to THM un_6, we can rewrite
 | $\bigcup(m \cup \{x \cup \{y\}\})$ as $x \cup \{y\} \cup \bigcup m$; but then the decision algorithm for a fragment of
 | set theory known as 'multi-level syllogistic with singleton' yields an immediate contradic-
 | tion.
$\langle s \rangle \hookrightarrow Stat0 \Rightarrow$ $\bigcup(m \cup \{x \cup \{y\}\}) \neq s$
$\langle c \rangle \hookrightarrow Stat0 \Rightarrow$ $\bigcup(m \cup \{x \cup \{y\}\}) \neq s \setminus \{c\}$
$\langle v \rangle \hookrightarrow Stat0 \Rightarrow$ $\bigcup(m \cup \{x \cup \{y\}\}) \neq s \setminus \{v\}$
$\langle m, x \cup \{y\} \rangle \hookrightarrow Tun_6 \Rightarrow$ AUTO
EQUAL $\Rightarrow$ *Stat1* : $x \cup \{y\} \cup \bigcup m \neq s \setminus \{c\}$ & $x \cup \{y\} \cup \bigcup m \neq s \setminus \{v\}$ & $x \cup \{y\} \cup \bigcup m \neq s$
(*Stat0, Stat1*)Discharge $\Rightarrow$ QED

Fig. 5.8 Tiny scenario for Ref, regarding the unionset operation

```
Def unionset: [Family of all members of members of a set]
    Un(X) := {u: v in X, u in v}
--
THEORY imageOfDoubleton(f(X),x0,x1)
END imageOfDoubleton
--
ENTER_THEORY imageOfDoubleton
--
Theorem imageOfDoubleton:
    [The image of a doubleton is either a doubleton or a singleton]
    ({f(v): v in 0 } = 0) & ({f(v): v in {x0} } = {f(x0)}) &
    ({f(v): v in {x0,x1} } = {f(x0),f(x1)}). Proof:
Suppose_not() ==> AUTO
--
-- Ref has the built-in ability to reduce ${f(v): v in 0}$ to $0$ and
-- ${f(v): v in {x0}}$ to ${f(x0)}$; hence we are left with only the
-- doubleton to consider. Let $c$ belong to one of ${f(v): v in {x0,x1}}$
-- and  ${f(x0),f(x1)}$ but not the other. After excluding, through
-- variable-substitution, the case $c notin {f(v): v in {x0,x1}}$,
-- we easily discard both possibilities $c=f(x0)$ and $c=f(x1)$, through
-- variable-substitution and equality propagation.
--
    SIMPLF ==> Stat1: {f(v): v in {x0,x1} } /= {f(x0),f(x1)}
    c-->Stat1 ==> (c in {f(v): v in {x0,x1}}) neq (c in {f(x0),f(x1)})
    Suppose ==> Stat2: c notin {f(v): v in {x0,x1}}
        x0-->Stat2 ==> AUTO
        x1-->Stat2 ==> AUTO
    Discharge ==> Stat3: (c in {f(v): v in {x0,x1}}) &
                          (c notin {f(x0),f(x1)})
    xp-->Stat3 ==> (xp in {x0,x1}) & (f(xp) /= f(x0)) & (f(xp) /= f(x1))
    Suppose ==> xp = x0
        EQUAL ==> false
    Discharge ==> xp = x1
    EQUAL ==> false
    Discharge ==> QED
--
ENTER_THEORY Set_theory
```

Fig. 5.9 Another tiny scenario for Ref, in keyboard-oriented *input format*

a scenario for syntactic validity, Ref verifies that the proofs are compliant with the version of set theory built into it.

The deductive system underlying Ref—mainly first-order, but with an important second-order feature to be discussed in Sect. 5.3.1.2—is a variant of the Zermelo-Fraenkel set theory with axioms of foundation and universal choice (see Remark 2.2). This is apparent from the very syntax of the language in which scenarios are written, which extends the usual language of first-order predicate logic with constructs reflecting the underlying theory of sets: we can, e.g., as shown by most of the abbreviating definitions in Fig. 5.2, exploit the very flexible set abstraction construct

$$\{ term : iterators \mid condition \}$$

presented in Sect. 2.2 to specify many familiar operations and relations over sets. For example, the setformer $\{u : v \in S, u \in v\}$ is used to define the unionset global operation $\cup S$. Other built-in set constructs are the Boolean dyadic operators $\cap, \setminus, \cup$, the "elementary set" constructor $\{S_1, \ldots, S_n\}$ (by means of which one can build, e.g., $\emptyset, \{x\}, \{u, \{\emptyset, u\}\}, \{v, w, \{y\}, z\}$, etc.), the pair-forming construct $[\ell, r]$, and the

THM fin$_7$: [Finite, nonnull sets have sources] Finite(F) & F$\neq\emptyset\to$F\$\bigcup$F$\neq\emptyset$. PROOF:
Suppose_not(f$_1$) $\Rightarrow$ AUTO
|| Arguing by contradiction, suppose that there are counterexamples to the claim. Then, by
|| exploiting finite induction, we can pick a minimal counterexample, f$_0$.
APPLY $\langle$fin$_\Theta$: f$_0\rangle$ finiteInduction$(s_0\mapsto f_1, P(S)\mapsto (S\neq\emptyset$ & S\$\bigcup$S$=\emptyset))\Rightarrow$
     Stat0 : $\langle\forall s|s\subseteq f_0\to$Finite(s) & (s$\neq\emptyset$ & s\$\bigcup$s$=\emptyset\leftrightarrow$s$=f_0$)$\rangle$
Loc_def $\Rightarrow$ a$=$**arb**(f$_0$)
$\langle$f$_0\rangle\hookrightarrow$Stat0 $\Rightarrow$ Stat1 : Finite(f$_0$) & a $\in$ f$_0$ & f$_0$\$\bigcup$f$_0=\emptyset$
|| Momentarily supposing that f$_0=\{$a$\}$, one gets $\bigcup$f$_0\not\subseteq$a, because $\bigcup$f$_0\subseteq$a would imply
|| f$_0$\$\bigcup$f$_0\supseteq\{a\}$\a and hence would imply the emptiness of $\{$a$\}$\a, whence the manifest ab-
|| surdity a $\in$ a follows. But, on the other hand, $\bigcup\{$a$\}\subseteq$a trivially holds; therefore we must
|| exclude that f$_0$ is a singleton $\{$a$\}$.
Suppose $\Rightarrow$ f$_0=\{$a$\}$ & $\bigcup$f$_0\not\subseteq$a
     EQUAL $\Rightarrow$ $\bigcup\{$a$\}\not\subseteq$a
     Use_def($\bigcup$) $\Rightarrow$ $\{$u : v $\in\{$a$\}$,u $\in$ v$\}\not\subseteq$a
SIMPLF $\Rightarrow$ false;   Discharge $\Rightarrow$ AUTO
|| Due to our minimality assumption, the strict nonnull subset f$_0$\$\{$**arb**(f$_0$)$\}$ of f$_0$ cannot be
|| a counterexample to the claim; therefore it has sources and hence f$_0$\$\bigcup$(f$_0$\$\{$**arb**(f$_0$)$\})\neq\emptyset$.
$\langle$f$_0$\$\{a\}$,a$\rangle\hookrightarrow$Tun$_6(\star)\Rightarrow$ $\bigcup$(f$_0$\$\{a\}\cup\{a\})=\bigcup$(f$_0$\$\{a\})\cup$a & f$_0$\$\{a\}\cup\{a\}=f_0$
$\langle$f$_0$\$\{a\}\rangle\hookrightarrow$Stat0($\star$) $\Rightarrow$ f$_0$\$\bigcup$(f$_0$\$\{a\})\neq\emptyset$
|| Since **arb**(f$_0$) does not intersect f$_0$, the inequality just found conflicts with the equal-
|| ity f$_0$\$(\bigcup$(f$_0$\$\{$**arb**(f$_0$)$\})\cup$**arb**(f$_0$)$)=\emptyset$ which one gets from THM un$_6$ through equality
|| propagation.
EQUAL $\Rightarrow$ f$_0$\$(\bigcup$(f$_0$\$\{a\})\cup$a$)=\emptyset$
Discharge $\Rightarrow$ QED

Fig. 5.10 An application of finite induction in Ref

monadic operator **arb** embodying the axiom of universal choice and ensuring the well-foundedness of $\in$. (The operator **arb** has been mentioned in connection with Theorem 5.1 and Lemma A.1 and briefly explained in Sect. 5.2.)

   Set theory also reflects into the semantics of many of the 15 or so mechanisms constituting the inferential armory of Ref: for example, the inclusion

$$\{u : v \in x_0, u \in v\} \subseteq \{u : v \in x_0 \cup \{y_0\}, u \in v\}$$

can be proved in a single step (cf. Fig. 5.8), as an application of the inference mechanism named SET_MONOT; likewise (cf. Fig. 5.10), one can infer

$$\{u : v \in \{a\}, u \in v\} = a$$

in a single step, through the mechanism named SIMPLF, which unravels the setformer $\{u : v \in \{a\}, u \in v\}$ first into $\{u : u \in a\}$ and then into a. Collectively, the inference mechanisms embody almost every feature of the axioms of ZFC: the only axiom of set theory which Ref must maintain as an explicit assumption is, in fact, the one stating that there exist infinite sets.

   Definitions often introduce abbreviating notation such as the unionset operation in the example made short ago; sometimes they bring into play recursive notions such as the one of hereditary finitude seen in Fig. 5.2. In fact, as a benefit originating from the assumption that set membership is a well-founded relation, Ref offers a built-in form of definition recurring over the members of one operand of the predicate or global function being defined. To see which recursive schemes are permitted, cf. [105, pp. 215–217]; clues on how to overcome the limitations of the built-in form of recursion are given in [79].

Proofs are formed by two-portion lines: the second portion of each line, separated by the sign $\Rightarrow$ from the first and at times carrying an identifying label of the form *Statlabel*, is the *assertion* being derived; the first portion is the *hint*, referencing the basic inference mechanism which enables that derivation in Ref. For example, the hint Use_def($\cup$) suggests that one is expanding previous occurrences of the symbol $\cup$ inside the proof by the appropriate definition; EQUAL triggers a simple equality-propagation activity; etc. Besides referencing an inference mechanism, each hint may supply it with auxiliary parameters (e.g., SUPPOSE_NOT(x_0, y_0)), including the context of preceding statements in which it should operate (e.g., (*Stat0, Stat1*)DISCHARGE draws a contradiction from the conjunction of the assertions labeled *Stat0* and *Stat1*).

In the style of natural deduction, the scaffold of a proof consists of SUPPOSE/ DISCHARGE matching pairs. Each SUPPOSE hint introduces a statement S to be exploited provisionally until a manifest contradiction will arise, leading—through the matching DISCHARGE—to the conclusion $\neg S$. SUPPOSE_NOT is a variant of SUPPOSE that occurs, exclusively and always, as the first inference step in Ref proofs; its purpose is to open an argument by contradiction, and its parameters are distinct constants local to a proof, which correspond in number and in positions to the distinct unquantified variables appearing in the claim T of the theorem being proved. Such theorem variables (whose first letters are always capitalized for emphasis) are in fact understood to be universally bound; and the corresponding constants, replacing them throughout the argument, are meant to constitute a counterexample to T. Needless to say, within a proof DISCHARGE hints must match SUPPOSE and SUPPOSE_NOT hints in the same balanced way in which closed parentheses match open parentheses within an arithmetic expression. A statement of the form DISCHARGE$\Rightarrow$ QED, invariably appearing at the end of the proof, hence counterbalances the initial SUPPOSE_NOT: it indicates that a contradiction was derived from the assumed existence of a counterexample, thus leading to the desired conclusion.

Two variable-instantiating mechanisms are available in Ref (both uniform, in the sense that all occurrences of the same variable are replaced by the same term). One of these replaces existentially bound occurrences of variables by previously unused constants: among others, it acts as a sub-mechanism of the SUPPOSE_NOT— which, by implicitly negating the claim, converts its universally bound variables into existentially bound variables. The other mechanism replaces universally bound occurrences of variables by arbitrary terms: among others, it acts as a sub-mechanism in *theorem citations*, which have the form

$$\langle t_1, \ldots, t_m \rangle \hookrightarrow T \text{theoremName} \Rightarrow \quad \cdots$$

as illustrated by the citation of THM un_6 within the proof of THM un_{13}. In *statement citations*, which have the form

$$\langle t_1, \ldots, t_m \rangle \hookrightarrow Stat\text{label} \Rightarrow \quad \cdots :,$$

instantiation can affect universally—as well as existentially—bound variables, and the two kinds of instantiation can be intermixed in a single step. Many trivial logical equivalences, such as

$$s \not\subseteq t \;\leftrightarrow\; (\exists x \in s \,|\, x \notin t)$$
$$y \in \{t : x \in s \,|\, P(x)\} \;\leftrightarrow\; (\exists x \in s \,|\, P(x) \ \& \ y = t)$$

and the like, enable instantiation of variables even in statements where quantifiers are not explicitly visible. For example, x occurs as a universally bound variable in $\{x \in s \,|\, P(x)\} = \emptyset$, and, accordingly, its instantiation as t yields $\neg(t \in s \ \& \ P(t))$. Similarly, introducing a brand new constant e in the citation of an inequality $s_1 \neq s_2$ makes sense: it yields $e \in s_1 \leftrightarrow e \notin s_2$.

As an example, consider the hint $\langle v_0, u_0, v_0, u_0 \rangle \hookrightarrow Stat1$ in Fig. 5.8. This calls for instantiation of the four bound variables appearing in the setformers of *Stat1*: the constants v_0 and u_0, created in order to instantiate the existentially bound variables of the first setformer, must then be exploited to instantiate the universally bound variables of the other setformer; the overall effect will be the (tacit) derivation, from *Stat1*, of the formula

$$v_0 \in x_0 \cup \{y_0\} \ \& \ u_0 \in v_0 \ \& \ a = u_0 \ \& \ (a \neq u_0 \vee v_0 \notin x_0 \vee u_0 \notin v_0) \ \& \ a \notin y_0,$$

whose unsatisfiability ELEM, the mechanism to be discussed next, can detect.

ELEM is the most central of all Ref's inference mechanisms. Its use is, often, tacitly combined with other forms of inference; for example, although it is not called explicitly into play by any of the hints in Figs. 5.8 and 5.10, ELEM plays a role there in yielding the contradictions needed for the final Discharges of the proofs of THM un_{13} and THM fin_7. ELEM implements an enhanced variant of *multi-level syllogistic* [37, 38], a decision algorithm which determines whether a given unquantified set-theoretic formula involving only individual variables (which designate sets) and a restricted collection of set operators is satisfiable.[6] By means of that algorithm, the Ref verifier can identify many cases in which a conjunction constructed by negating a statement S of a proof and conjoining a selection of earlier statements is unsatisfiable, so that S follows from the preceding context. Should some of the constructs appearing in this context be unamenable to Ref's built-in syllogistic, a preprocessing phase replaces all parts of the context whose lead operators cannot be treated by the decision algorithm by new variables designating

---

[6]In essence, the decision algorithm mentioned here is a solver for Problem 3.4, discussed earlier.

either sets (when they occur as terms) or propositions (when they occur as sub-
formulae), replacing syntactically identical (or recognizably equal) parts by the
same variable.

Occasionally an assertion is represented laconically by the keyword AUTO, when
no ambiguity or obscurity can ensue from this. Thus, in Fig. 5.8:

- When AUTO occurs in the proof of THM un_6, it stands for the negation of the
  overall assertion labeled *Stat1*. In fact, the Discharge hint introducing this AUTO
  matches the Suppose hint introducing that assertion—an assertion which has led
  to a contradiction, as is the task of Discharge to ascertain.
- When AUTO occurs in the proof of THM un_{13}, it stands for the formula

$$\cup(m \cup \{x \cup \{y\}\}) = x \cup \{y\} \cup \cup m$$

resulting from the replacement of X by m and of Y by $x \cup \{y\}$ within the claim
of THM un_6. This instantiated citation is what the hint $\langle m, x \cup \{y\}\rangle \hookrightarrow Tun_6$
calls for, since, as explained earlier, X and Y behave in THM un_6 as universally
quantified variables.

The case when the assertion of a proof line is sharply determined by the lines
and the hint that precede it in the proof is rather rare, though. For example, it is
entirely a matter of taste—in view of the trivial fact $s \notin s$—whether to derive
$\cup(m \cup \{x \cup \{y\}\}) \neq s$ or $\cup(m \cup \{x \cup \{y\}\}) \neq s\backslash\{s\}$ as the second step in the
second proof of that in Fig. 5.8; but plainly, it is the latter that would more directly
result from the substitution $\langle s \rangle \hookrightarrow Stat0$, requested there, of s for the universally
bound variable d of the assertion *Stat0*.

Such indeterminacy in the outcome of an inference step stems, principally, from
the pervasiveness of ELEM as a "behind-the-scenes" inference mechanism. A hint
may, in fact, sometimes—as in the example just made, and in all those cases
when one can resort to the keyword AUTO—designate a specific assertion which,
however, is shrouded in a "penumbra" of easy consequences of its conjunction
with the context preceding it in the proof; then, from among those consequences,
derivable via ELEM, the Ref's user is left free to pick one or another. Additional
fuzziness often comes from the hint itself: while, on the one hand, the hint
Use_def$(\cup(x_0 \cup \{y_0\}))$ readily prompts the assertion $\cup(x_0 \cup \{y_0\}) = \{u : v \in
x_0 \cup \{y_0\}, u \in v\}$, on the other hand, the hint Use_def$(\cup)$ might refer to one or
another occurrence of the symbol $\cup$ inside the proof, or to multiple occurrences of
it at once; hence it is appropriate that the Ref's user reduce the degree of ambiguity
by explicitly providing an assertion instead of resorting to an uninformative AUTO.

The main reference about Ref is the monograph [105]; an online manual
is available at URL http://aetnanova.units.it/Ref_user_manual.html; other relevant
references are [20, 39, 72, 79].

Fig. 5.11 Interface of a
THEORY used in connection
with **img** (·) and **dom** (·)

$$\begin{array}{|l|}\hline \text{THEORY pointwise}\big(g(Y), f(X)\big) \\ \quad \big\langle \forall x \mid f(x) = \{g(y) : y \in x\} \big\rangle \\ \Rightarrow \\ \quad \big\langle \forall x \mid x = \emptyset \leftrightarrow f(x) = \emptyset \big\rangle \\ \quad \big\langle \forall s,t \mid s \subseteq t \rightarrow f(s) \subseteq f(t) \big\rangle \\ \quad \big\langle \forall x,y,z \mid x = y \cup z \rightarrow f(x) = f(y) \cup f(z) \big\rangle \\ \text{END pointwise} \\ \hline \end{array}$$

5.3.1.2 Proof Encapsulation in Ref

> *Beyond this, definitions serve to "instantiate," that is, to introduce the objects whose special properties are crucial to an intended argument. Like the selection of crucial lines, points, and circles from the infinity of geometric elements that might be considered in a Euclidean argument, definitions of this kind often carry a proof's most vital ideas.*

> (J. T. Schwartz, [105, p. 9])

The proof-checker Ref has a second-order construct named THEORY (cf. [79] and [105, pp. 19–25]), aimed at proof reuse, akin to a mechanism for parameterized specifications of the Clear specification language [17]. Besides providing theorems of which it holds the proofs, a THEORY has the ability to instantiate "objects whose special properties are crucial to an intended argument." Like procedures of a programming language, Ref's THEORYs have input formal parameters, in exchange for whose actualization they supply useful information. Actual input parameters must satisfy a conjunction of statements, called the *assumptions* of the THEORY. (In the example shown in Fig. 5.9, THEORY imageOfDoubleton has three parameters—a global function $f(\cdot)$ and two sets x_0 and x_1—but no assumptions concerning them.)

As an introductory example, consider the THEORY pointwise displayed in Fig. 5.11, which expects as parameters two functions $f(\cdot)$ and $g(\cdot)$ between which the assumed relationship

$$f(X) = \{ g(y) : y \in X \}$$

must hold for every set X. We can, in view of the definition maps$_1$ of Fig. 5.2, pass **dom** (X) and $Y^{[1]}$ as actual parameters to $f(X)$ and to $g(Y)$, respectively; **img** (X) and $Y^{[2]}$ qualify as actual parameters too, thanks to another definition of Fig. 5.2. Both of the following applications hence are viable:

$$\begin{array}{|l|}\hline \text{APPLY } \langle\,\rangle \text{ pointwise}\big(g(Y) \mapsto Y^{[2]}, f(X) \mapsto \textbf{img}\,(X)\big) \Rightarrow \quad \cdots \cdots \\ \text{APPLY } \langle\,\rangle \text{ pointwise}\big(g(Y) \mapsto Y^{[1]}, f(X) \mapsto \textbf{dom}\,(X)\big) \Rightarrow \quad \cdots \cdots \\ \hline \end{array}$$

THEORY finiteInduction$(s_0, P(S))$
    Finite(s_0) & $P(s_0)$
$\Rightarrow (\mathrm{fin}_\Theta)$
    ‖ In exchange for the given s_0, which has been assumed to be finite and to satisfy
    ‖ property P, this theory returns an $\subseteq$-minimal finite set fin_Θ which also satisfies P.
    $\langle \forall S \mid S \subseteq \mathrm{fin}_\Theta \rightarrow \mathrm{Finite}(S)\ \&\ (P(S) \leftrightarrow S = \mathrm{fin}_\Theta) \rangle$
END finiteInduction

Fig. 5.12 A finite induction mechanism

Before being in the position of invoking our THEORY in this manner, we will have
derived inside it the claims which are displayed in Fig. 5.11 below the '$\Rightarrow$' sign. As
a result of the applications indicated above, Ref will accept as proved claims the
monotonicity and additivity properties of **dom** and **img**, respectively:

$$\langle \forall x \mid x = \emptyset \leftrightarrow \mathbf{dom}(x) = \emptyset \rangle\ \&\ \langle \forall s, t \mid s \subseteq t \rightarrow \mathbf{dom}(s) \subseteq \mathbf{dom}(t) \rangle$$
$$\&\ \langle \forall x, y, z \mid x = y \cup z \rightarrow \mathbf{dom}(x) = \mathbf{dom}(y)\ \cup \mathbf{dom}(z) \rangle$$

and

$$\langle \forall x \mid x = \emptyset \leftrightarrow \mathbf{img}(x) = \emptyset \rangle\ \&\ \langle \forall s, t \mid s \subseteq t \rightarrow \mathbf{img}(s) \subseteq \mathbf{img}(t) \rangle$$
$$\&\ \langle \forall x, y, z \mid x = y \cup z \rightarrow \mathbf{img}(x) = \mathbf{img}(y)\ \cup \mathbf{img}(z) \rangle.$$

A THEORY usually encapsulates the definitions of entities related to the input
parameters, and it supplies, along with some consequences of the assumptions,
theorems talking about these internally defined entities, which the THEORY returns
as output parameters.[7] After having been derived by the user once and for all inside
the THEORY, the consequences of the assumptions, as well as the claims involving
the output parameters, are available to be exploited repeatedly.

Another simple, yet very important example is the THEORY finiteInduction
displayed in Fig. 5.12, which receives a finite set s_0 along with a property P such
that $P(s_0)$ holds and returns a "minimal witness" of P, i.e., a finite set fin_Θ satisfying
$P(\mathit{fin}_\Theta)$ none of whose strict subsets t satisfies $P(t)$.

To see this device at work, consider the proof, displayed in Fig. 5.10, of the claim
that $F \setminus \cup F \neq \emptyset$ holds for every finite nonnull set F. To carry through an argument
by contradiction, in this case, we must resort to a *minimal* counterexample to the
claim: at the outset—through the by now familiar Suppose_not mechanism—we
simply assume that

$$\mathsf{Finite}(f_1)\ \&\ f_1 \neq \emptyset\ \&\ f_1 \setminus \cup f_1 = \emptyset,$$

---

[7]As a visible countersign, the formal output parameters of a THEORY must carry the Greek letter
Θ as a subscript.

but then we want to add the minimality requirement. All that we must do, to shift from f_1 to a minimal f_0, is to actualize the triplet s_0, $P(S)$, fin_Θ of parameters of the THEORY finiteInduction as f_1, $S \neq \emptyset$ & $S \setminus \cup S = \emptyset$, and f_0, in the way shown by the APPLY line of the proof.

We take the opportunity, since the construct **arb** has now shown up again (see Sect. 5.2), to recapitulate the properties of this important set operation which reflects the well-foundedness of membership and whose meaning is built into the ELEM inference mechanism: $X \mapsto \mathbf{arb}(X)$ is a function such that, for every set X,

$$\left(\mathbf{arb}(X) \in X \ \& \ X \cap \mathbf{arb}(X) = \emptyset\right) \vee \left(X = \emptyset \ \& \ \mathbf{arb}(X) = \emptyset\right).$$

Awareness of the semantics of **arb** sometimes slows down the proof-checking process; one can disable it locally by the flag "$(\star)$," as shown twice in Fig. 5.10, to avoid this drawback.

5.3.1.3   An Indication on the Effectiveness of Ref

To convey a rough idea of how effective Ref's inferential armory is, we report that fewer than 400 Ref's inference lines are needed to prove all ancillary laws shown in Figs. 5.13 and 5.14, which regard Cartesian product and other map-related constructs introduced above (see Fig. 5.2), along with the internals of the THEORYs pointwise and finiteInduction (see Figs. 5.11 and 5.12) and with half-a-dozen minor lemmas that all of these presuppose.

5.3.2   Proof-Checking on Graphs as Transitive Sets

5.3.2.1   Reflecting Connected Claw-Free Graphs into Transitive Sets

As explained in Sects. 5.2.1 and 5.2.2, our formalization experiment includes a basic edge-to-membership translation which we can effect by:

 (i) converting an arbitrary graph into a weakly extensional acyclic digraph,
(ii) decorating the resulting digraph by sets, so that its arcs mirror membership.

This overall task, and its subtask (5.3.2.1), culminates in the two THEORYs shown in Fig. 5.15. In particular, the THEORY finMostowskiDecoration implements (5.3.2.1), while Theorem 5.2, statable in Ref as

THM xtensionalization$_0$. Finite(V) & $S \in V \rightarrow \langle \exists d \mid$ Orientates(d, V, E) & Acyclic(V, d)
        & WExtensional(V, d) & $S \notin$ img(d) $\rangle$,

is the key result corresponding to (5.3.2.1), which makes the THEORY finGraphRepr easily obtainable from the other THEORY.

THM range_1: [Additivity of range] $(X = Y \cup Z \rightarrow \text{img}(X) = \text{img}(Y) \cup \text{img}(Z))$ &
$$(\text{img}(X) = \emptyset \leftrightarrow X = \emptyset)$$

THM domain_1: [Additivity of domain] $(X = Y \cup Z \rightarrow \text{dom}(X) = \text{dom}(Y) \cup \text{dom}(Z))$ &
$$(\text{dom}(X) = \emptyset \leftrightarrow X = \emptyset)$$

THM domain_2: [Elements of domain, of range] $P \in S \rightarrow P^{[1]} \in \text{dom}(S)$ & $P^{[2]} \in \text{img}(S)$

THM domain_3: [Typical el'ts of domain, of range] $[X, Y] \in S \rightarrow X \in \text{dom}(S)$ & $Y \in \text{img}(S)$

THM restr_0: [The restriction of any (set or) map is included in it] $F_{|A} \subseteq F$

THM restr_1: [The domain of a restriction is included in the restraining set] $P \in F_{|A} \rightarrow P^{[1]} \in A$

THM restr_2: [Each pair in a map belongs to the shoot of an element of its domain]
$$P = [X, Y] \rightarrow (P \in F_{|\{Z\}} \leftrightarrow Z = X \ \& \ [Z, Y] \in F)$$

THM restr_3: [The elements of a map shoot are pairs of fixed first component]
$$\text{Is\_map}(F) \ \& \ P \in F_{|\{Z\}} \rightarrow P = \left[Z, P^{[2]}\right] \ \& \ P \in F$$

THM image_0: [The image, under $\emptyset$, of any entity is $\emptyset$] $Z = \emptyset \rightarrow \text{Svm}(Z) \ \& \ Z \restriction X = \emptyset$

THM image_1: [In a map, each element brings in an operand-image pair]
$$X \in \text{dom}(F) \ \& \ \text{Is\_map}(F) \rightarrow [X, F \restriction X] \in F_{|\{X\}} \cap F$$

THM image_2: [Concerning set-union applied to an operand] $X \notin \text{dom}(F) \rightarrow (F \cup G) \restriction X = G \restriction X$

THM image_3: [Application to an entity outside map's domain] $X \notin \text{dom}(F) \rightarrow F \restriction X = \emptyset$

THM image_4: [Applying a single-valued map] $\text{Svm}(F) \ \& \ P \in F \rightarrow P = \left[P^{[1]}, F \restriction P^{[1]}\right]$

THM image_5: [Form of a single-valued map] $\text{Svm}(F) \leftrightarrow F = \{[x, F \restriction x] : x \in \text{dom}(F)\}$

THM svm_0: [Every subset of a single valued maps is single valued] $\text{Svm}(F) \ \& \ G \subseteq F \rightarrow \text{Svm}(G)$

THM svm_1: [Union of domain-disjoint single-valued maps]
$$\text{Svm}(F) \ \& \ \text{Svm}(G) \ \& \ \text{dom}(F) \cap \text{dom}(G) = \emptyset \rightarrow \text{Svm}(F \cup G)$$

THM singletonMap_0: [Domain and range of a singleton]
$$\text{dom}(\{P\}) = \left\{P^{[1]}\right\} \ \& \ \text{img}(\{P\}) = \left\{P^{[2]}\right\}$$

THM singletonMap_1: [Singleton maps are single valued]
$$(F \subseteq \{[X, Y]\} \rightarrow \text{Svm}(F)) \ \& \ \text{dom}(\{[X, Y]\}) = \{X\} \ \& \ \text{img}(\{[X, Y]\}) = \{Y\}$$

THM singletonMap_2: [Singleton map application] $F = \{[X, Y]\} \rightarrow F \restriction X = Y$

THM singletonMap_3: [Transplant of singleton sub-map]
$$\text{Svm}(F) \ \& \ [X, Y] \in F \ \& \ Z \notin \text{dom}(F) \ \& \ G = F \setminus \{[X, Y]\} \cup \{[Z, Y]\} \rightarrow$$
$$\text{Svm}(G) \ \& \ \text{dom}(G) = \text{dom}(F) \setminus \{X\} \cup \{Z\} \ \& \ \text{img}(G) = \text{img}(F)$$

THM part_whole_1: [The part is smaller than the whole]
$$\text{Svm}(H) \ \& \ \text{Finite}(H) \ \& \ \text{img}(H) \supseteq \text{dom}(H) \rightarrow \text{img}(H) = \text{dom}(H)$$

Fig. 5.13 Laws concerning map-related constructs

The THEORY finGraphRepr does not, by itself, play a role in our study on claw-free graphs, but so does its specialization to the case of connected claw-free graphs, namely, the THEORY herfinCCFGraphRepr shown in Fig. 5.16, which implements Theorem 5.4 by relying on a Ref-verified proof of Theorem 5.3.

$$\text{THM cartesian}_0.\ X \in S \times T \leftrightarrow X = \left[X^{[1]}, X^{[2]}\right]\ \&\ X^{[1]} \in S\ \&\ X^{[2]} \in T$$

$$\text{THM cartesian}_1.\ S \neq \emptyset \rightarrow \mathbf{img}\,(S \times T) = T$$

$$\text{THM cartesian}_2.\ T \neq \emptyset \rightarrow \mathbf{dom}\,(S \times T) = S$$

$$\text{THM cartesian}_3:\ [\text{Annichilator of Cartesian product}]\ S \times \emptyset = \emptyset\ \&\ \emptyset \times S = \emptyset$$

$$\text{THM cartesian}_6:\ [\text{All subsets of a Cartesian product are maps}]\ F \subseteq S \times T \rightarrow \mathsf{Is\_map}(F)$$

Fig. 5.14 Laws concerning the Cartesian product

$$\text{THEORY finMostowskiDecoration}(v_0, d_0)$$
$$v_0 \times v_0 \supseteq d_0\ \&\ \mathsf{Finite}(v_0)\ \&\ \mathsf{Acyclic}(v_0, d_0)\ \&\ \mathsf{WExtensional}(v_0, d_0)$$
$$\Rightarrow (\mathsf{mski}_\Theta)$$
$$\mathsf{Svm}(\mathsf{mski}_\Theta)\ \&\ \mathbf{dom}\,(\mathsf{mski}_\Theta) = v_0$$
$$\langle \forall w \mid w \in \mathbf{dom}\,(d_0)\ \rightarrow\ \mathsf{mski}_\Theta \restriction w = \left\{\mathsf{mski}_\Theta \restriction p^{[2]} : p \in d_{0 \restriction \{w\}}\right\}\ \&\ \mathsf{mski}_\Theta \restriction w \neq \emptyset \rangle$$
$$(v_0 \neq \emptyset \rightarrow \emptyset \in \mathbf{img}\,(\mathsf{mski}_\Theta))\ \&\ \langle \forall y \mid y \in \mathbf{img}\,(\mathsf{mski}_\Theta) \rightarrow \mathsf{Finite}(y) \rangle$$
$$\langle \forall x, y \mid \{x, y\} \subseteq v_0\ \&\ \mathsf{mski}_\Theta \restriction x = \mathsf{mski}_\Theta \restriction y \rightarrow x = y \rangle$$
$$\langle \forall y, x \mid y \in v_0\ \rightarrow\ (\mathsf{mski}_\Theta \restriction y \in \mathsf{mski}_\Theta \restriction x \leftrightarrow [x, y] \in d_0) \rangle$$
$$\text{END finMostowskiDecoration}$$

$$\text{THEORY finGraphRepr}(v_0, e_0)$$
$$\mathsf{uGraph}(v_0, e_0)$$
$$\Rightarrow (\mathsf{wski}_\Theta)$$
$$\mathsf{Svm}(\mathsf{wski}_\Theta)\ \&\ \mathbf{dom}\,(\mathsf{wski}_\Theta) = v_0$$
$$(v_0 \neq \emptyset \rightarrow \emptyset \in \mathbf{img}\,(\mathsf{mski}_\Theta))\ \&\ \langle \forall y \mid y \in \mathbf{img}\,(\mathsf{wski}_\Theta) \rightarrow \mathsf{Finite}(y) \rangle$$
$$\langle \forall x, y \mid \{x, y\} \subseteq v_0\ \&\ \mathsf{wski}_\Theta \restriction x = \mathsf{wski}_\Theta \restriction y \rightarrow x = y \rangle$$
$$\langle \forall x, y \mid \{x, y\} \subseteq v_0 \rightarrow ((\mathsf{wski}_\Theta \restriction y \in \mathsf{wski}_\Theta \restriction x \lor \mathsf{wski}_\Theta \restriction x \in \mathsf{wski}_\Theta \restriction y) \leftrightarrow \{x, y\} \in e_0) \rangle$$
$$\langle \forall x \mid \mathsf{wski}_\Theta \restriction x \cap \mathbf{img}\,(\mathsf{wski}_\Theta) \neq \emptyset \rightarrow \mathsf{wski}_\Theta \restriction x \subseteq \mathbf{img}\,(\mathsf{wski}_\Theta) \rangle$$
$$\text{END finGraphRepr}$$

Fig. 5.15 THEORYs on Mostowski's decoration and on graph representation

$$\text{THEORY herfinCCFGraphRepr}(v_0, e_0)$$
$$\mathsf{uGraph}(v_0, e_0)\ \&\ \mathsf{ClawFreeG}(v_0, e_0)\ \&\ \big(\mathsf{Conn}(v_0, e_0) \lor v_0 = \{\mathbf{arb}(v_0)\}\big)$$
$$\Rightarrow (\mathsf{trans}_\Theta)$$
$$\mathsf{Svm}(\mathsf{trans}_\Theta)\ \&\ \mathbf{dom}\,(\mathsf{trans}_\Theta) = v_0$$
$$\langle \forall x, y \mid \{x, y\} \subseteq v_0\ \&\ \mathsf{trans}_\Theta \restriction x = \mathsf{trans}_\Theta \restriction y \rightarrow x = y \rangle$$
$$\langle \forall x, y \mid \{x, y\} \subseteq v_0 \rightarrow ((\mathsf{trans}_\Theta \restriction y \in \mathsf{trans}_\Theta \restriction x \lor \mathsf{trans}_\Theta \restriction x \in \mathsf{trans}_\Theta \restriction y) \leftrightarrow \{x, y\} \in e_0) \rangle$$
$$v_0 \neq \emptyset \rightarrow \emptyset \in \mathbf{img}\,(\mathsf{mski}_\Theta)$$
$$\{y \in \mathbf{img}\,(\mathsf{trans}_\Theta) \mid y \not\subseteq \mathbf{img}\,(\mathsf{trans}_\Theta)\} = \emptyset$$
$$\mathsf{HerFin}(\mathbf{img}\,(\mathsf{trans}_\Theta))$$
$$\text{END herfinCCFGraphRepr}$$

Fig. 5.16 THEORY representing a connected claw-free graph as a transitive, hereditarily finite set

$$
\boxed{
\begin{array}{l}
\text{THEORY finAcycLabeling}(v_0, d_0, h(s,x)) \\
\quad \text{Finite}(v_0) \;\&\; \text{Acyclic}(v_0, d_0) \\
\Rightarrow (\text{lab}_\Theta) \\
\quad \text{Svm}(\text{lab}_\Theta) \;\&\; \mathbf{dom}\,(\text{lab}_\Theta) = v_0 \\
\quad \Big\langle \forall x \in v_0 \mid \text{lab}_\Theta \restriction x = h\Big(\big\{ \text{lab}_\Theta \restriction p^{[2]} : p \in d_{0\restriction\{x\}} \mid p^{[2]} \in v_0 \big\}, x\Big)\Big\rangle \\
\text{END finAcycLabeling}
\end{array}}
$$

Fig. 5.17 A THEORY recursively assigning a tag to each vertex of an acyclic graph

We do not provide here a detailed account on how to carry out task (5.3.2.1), namely, the Ref-based implementation of THEORY finMostowskiDecoration; nevertheless, we show in Fig. 5.17 the interface of a THEORY whose development should precede the said implementation. Exercise 5.8 asks the reader to explain why the latter is a useful preparatory step for the former.

5.3.2.2 Proving Properties of Connected Claw-Free Graphs

Down-to-Earth Notions for Our Experiment and a Crucial Auxiliary Theory

Along with many ancillary laws (such as the ones listed in Fig. 5.18 and the ones cited in Exercise 5.14), two instantiating mechanisms play a key role in our proof-pearl scenario. One relates to the finiteness of the graphs under study here: this assumption conveniently reflects into the induction principle discussed in Sect. 5.3.1.2 (see Fig. 5.12).

The other THEORY (Fig. 5.19) more specifically reflects our claw-freeness and transitivity assumptions; it factors out a mathematical insight which is common to the two main proofs on which we are reporting. Essentially—after a formal statement of the final claim of Sect. 5.2.3—it says that in a transitive claw-free set $s_0 \nsubseteq \{\emptyset\}$, we can always select a pivot y_Θ and its in-neighborhood $\{x_\Theta, z_\Theta\}$. Along with $y_\Theta, x_\Theta, z_\Theta$, this THEORY returns the set $t_\Theta = s_0 \setminus \{x_\Theta, z_\Theta\} = \{v \in s_0 \mid y_\Theta \notin v\}$, strictly included in s_0; in its turn, t_Θ is proved to be claw-free and transitive (Fig. 5.20).

For an intuition of how the quadruple $x_\Theta, y_\Theta, z_\Theta, t_\Theta$ can be obtained, referring to the classical notion of *rank* recursively definable (cf. Fig. 3.1) as

$$
\text{rank}(s) =_{\text{Def}} \begin{cases} 0 & \text{if } s = \emptyset \\ \max\{\, \text{rank}(t) + 1 : t \in s \,\} & \text{otherwise,} \end{cases}
$$

observe that a transitive set s_0 not included in $\{\emptyset\}$ must have rank $r \geqslant 2$ and hence must have elements x_Θ, y_Θ such that $y_\Theta \in x_\Theta$ and $\text{rank}(y_\Theta) = r - 2$.

Although a recursive definition such as the one of rank just seen is supported by Ref, we preferred to avoid it, judging that a digression on numbers (definable *à la* von Neumann in Ref) would have been out of focus relative to our goal.

THM un_0: [Unionset operation yielding null result] $\bigcup X = \emptyset \leftrightarrow X \subseteq \{\emptyset\}$
THM un_2: [Unionset of set-adjunction] $X \cup \{Y\} = Z \rightarrow \bigcup Z = \bigcup X \cup Y$
THM un_4: [Unionset of single-/double-ton] $\{X,Y\} = Z \rightarrow \bigcup Z = X \cup Y$
THM un_8: [Additivity of monadic union] $X \cup Y = Z \rightarrow \bigcup X \cup \bigcup Y = \bigcup Z$
THM un_9: [Unionset as an upper bound] $(X \in S \rightarrow X \subseteq \bigcup S)$ & $(\langle \forall y \in S \,|\, y \subseteq X \rangle \rightarrow \bigcup S \subseteq X)$
THM un_{10}: [Monotonicity of unionset, special cases] $\bigcup(X \setminus \{Y\}) \supseteq \bigcup X \setminus Y$ & $\bigcup X \supseteq \bigcup(X \setminus \{Y\})$
THM un_{11}: [Unionset of a set obtained through removal followed by adjunction]
$\qquad \bigcup M \supseteq P$ & $Q \cup R = P \cup S \rightarrow \bigcup(M \setminus \{P\} \cup \{Q,R\}) = \bigcup M \cup S$
THM un_{12}: [Incomparability of pre-pivotal elements]
$\qquad Y \in X$ & $X \in Z$ & $X,Z \in S \rightarrow Y \in \bigcup(S \cap \bigcup S)$
THM un_{13}: [Less-one lemma for unionset]
$\qquad \bigcup M = T \setminus \{C\}$ & $S = T \cup X \cup \{V\}$ & $\big(Y = V \vee (C = Y$ & $Y \in S)\big) \rightarrow$
$\qquad\qquad\qquad \langle \exists d \,|\, \bigcup(M \cup \{X \cup \{Y\}\}) = S \setminus \{d\} \rangle$
THM tr_1: [The unionset of a transitive set is included in it] $\mathrm{Trans}(T) \leftrightarrow T \supseteq \bigcup T$
THM tr_2: [For a transitive set, elements are also subsets] $\mathrm{Trans}(T)$ & $X \in T \rightarrow X \subseteq T$
THM tr_3: [Trapping phenomenon for trivial sets] $\mathrm{Trans}(S)$ & $X,Z \in S$ & $X \notin Z$ & $Z \notin X$ &
$\qquad S \setminus \{X,Z\} \subseteq \{\emptyset,\{\emptyset\}\} \rightarrow S \subseteq \{\emptyset,\{\emptyset\},\{\{\emptyset\}\},\{\emptyset,\{\emptyset\}\}\}$
THM tr_4: [Peddicord's lemma] $\mathrm{Trans}(T)$ & $S \subsetneq T$ & $A = \mathbf{arb}(T \setminus S) \rightarrow A \subseteq S$ & $A \in T \setminus S$
THM tr_5: [$\emptyset$ belongs to any nonnull transitive set t, $\{\emptyset\}$ also does if $t \not\subseteq \{\emptyset\}$, and so on]
$\qquad \mathrm{Trans}(T)$ & $N \in \{\emptyset,\{\emptyset\},\{\emptyset,\{\emptyset\}\}\}$ & $T \not\subseteq N \rightarrow$
$\qquad\qquad\qquad N \subseteq T$ & $(N \in T \vee (N = \{\emptyset,\{\emptyset\}\}$ & $\{\{\emptyset\}\} \in T))$
THM tr_6: [Source removal from a transitive set does not disrupt transitivity]
$\qquad \mathrm{Trans}(S)$ & $S \supseteq T$ & $(S \setminus T) \cap \bigcup S = \emptyset \rightarrow \mathrm{Trans}(T)$
THM fin_0: [Monotonicity of finiteness] $Y \supseteq X$ & $\mathrm{Finite}(Y) \rightarrow \mathrm{Finite}(X)$
THM fin_7: [Finite, nonnull sets, have sources] $\mathrm{Finite}(F)$ & $F \neq \emptyset \rightarrow F \setminus \bigcup F \neq \emptyset$

Fig. 5.18 Various laws on the basic set-theoretic constructs $\bigcup$, Trans, and Finite

Fig. 5.19 A key quadruple associated with a claw-free set

THEORY pivotsForClawFreeness(s_0)
$\quad$ClawFree(s_0) & Trans(s_0) & Finite(s_0)
$\quad s_0 \not\subseteq \{\emptyset\}$
$\Rightarrow (x_\Theta, y_\Theta, z_\Theta, t_\Theta)$
$\quad \langle \forall x \in s_0, y \in x \setminus \bigcup\bigcup s_0 \,|\, \langle \exists z \,|\, \{v \in s_0 \,|\, y \in v\} = \{x,z\} \rangle \rangle$
$\quad \{x_\Theta, y_\Theta, z_\Theta\} \subseteq s_0$ & $\{x_\Theta, z_\Theta\} = \{v \in s_0 \,|\, y_\Theta \in v\}$
$\quad x_\Theta \notin z_\Theta$ & $z_\Theta \notin x_\Theta$ & $y_\Theta \in x_\Theta \cap z_\Theta \setminus \bigcup\bigcup s_0$
$\quad y_\Theta \in t_\Theta \setminus \bigcup t_\Theta$ & $t_\Theta = s_0 \setminus \{x_\Theta, z_\Theta\}$ & $t_\Theta = \{v \in s_0 \,|\, y_\Theta \notin v\}$
$\quad$ClawFree(t_Θ) & Trans(t_Θ)
END pivotsForClawFreeness

As a surrogate for the rank notion, we conceal inside this THEORY the definition of the *frontier* of a set s: this consists of those elements s to which a pivot of s belongs:

DEF frontier: [Frontier of a set] $\mathrm{front}(S) =_{\mathrm{Def}} \{x \in S \,|\, x \cap S \setminus \bigcup(S \cap \bigcup S) \neq \emptyset\}$.

Fig. 5.20 A key quadruple
associated with a transitive
claw-free set s_0

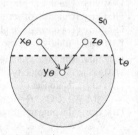

Admittedly, there is no deep rationale behind our decision to encapsulate this definition in the THEORY pivotsForClawFreeness along with the two theorems about the frontier to be seen in the coming paragraphs: these statements do not really depend upon the theory which we are discussing and which makes use of them, but they are rather technical and, for the time being, we do not know whether they will turn out to be useful elsewhere.

Aided by this notion, we get x and y by drawing arbitrarily the former from $\mathsf{front}(s_0)$, the latter from $x_\Theta \setminus \cup \cup s_0$. This presupposes, of course, a proof that $\mathsf{front}(s_0) \neq \emptyset$, a fact simply following from the more general proposition

$$\text{THM front}ier_1.\ \mathsf{Finite}(S \cap \cup S)\ \&\ S \cap \cup S \neq \emptyset \rightarrow \mathsf{front}(S) \neq \emptyset,$$

applicable to s_0 thanks to the assumption $s_0 \not\subseteq \{\emptyset\}$ of the THEORY at hand.

It is worth noting that the proof of THM $\mathsf{frontier}_1$ depends, albeit indirectly with our treatment, on the mechanism of finite induction. As a matter of fact, it relies on THM fin_7 whose proof, as we have seen, exploits the THEORY finiteInduction. Phrased in words, the argument proceeds as follows. By THM fin_7, when $s \cap \cup s$ is finite and nonnull, we can pick a $y \in s \cap \cup s \setminus \cup(s \cap \cup s)$, so that $y \in s$, $y \notin \cup(s \cap \cup s)$, but $y \in x$ holds for some $x \in s$. Supposing that $\mathsf{front}(s)$ is empty and x is as said, $x \notin \{x \in s \mid x \cap s \setminus \cup(s \cap \cup s) \neq \emptyset\}$ must hold, and hence no y' belonging to $x \cap s$ can lie outside $\cup(s \cap \cup s)$, which contradicts what we know about y. This contradiction leads us to the desired conclusion.

To conclude the development of the THEORY pivotsForClawFreeness, one must show that $t_\Theta = \{v \in s_0 \mid y_\Theta \notin v\}$ is transitive, as follows from

$$\text{THM front}ier_2.\ \mathsf{Trans}(S)\ \&\ X \in \mathsf{front}(S)\ \&\ Y \in X \backslash \cup \cup S\ \&\ T = \{z \in S \mid Y \notin z\}$$
$$\rightarrow \mathsf{Trans}(T)\ \&\ T \subseteq S\ \&\ X \notin T\ \&\ Y \in T \backslash \cup T,$$

in view of THM tr_3 from Fig. 5.18.

Preparatory Lemmas

Since an edge of a graph is represented as membership between two sets, we define a *matching* to be a set of disjoint doubletons $\{x, y\}$, one of whose elements belongs to the other.

DEF matching: [set of disjoint membership pairs]

$$\text{Matching(M)} \leftrightarrow_{\text{Def}} \langle \forall p \in M, \exists x \in p, y \in x, \forall q \in M \mid x \in q \vee y \in q \rightarrow \{x, y\} = q \rangle.$$

The following theorems about matchings admit straightforward proofs.

THM matching$_0$: [The null set is a matching] Matching($/$)0
THM matching$_2$: [All subsets of a matching are matchings]
$\quad$ Matching(M) & M $\supseteq$ N $\rightarrow$ Matching(N)
THM matching$_3$: [Bottom-up assembly of a finite matching]
$\quad$ Matching(M) & X $\notin \bigcup$M & Y $\notin \bigcup$M & Y $\in$ X $\rightarrow$
$\qquad\qquad$ Matching(M $\cup$ {{X,Y}})
THM matching$_4$: [Deviated matching] Matching(M) &
$\{$Y,W$\} \in$ M & X $\notin \bigcup$M & Z $\notin \bigcup$M & Y $\in$ Z & Y$\neq$X & X$\neq$Z & W $\in$ X $\rightarrow$
$\qquad\qquad$ Matching(M\{{Y,W}} $\cup$ {{Y,Z},{X,W}})

The last two of these reflect our proof strategy: THM matching$_3$ claims that we can extend a matching by insertion of a doubleton of new sets, while THM matching$_4$ states conditions under which we can break a pair $\{y, w\}$ of a matching M into two doubletons $\{y, z\}$ and $\{x, w\}$ (see Fig. 5.21).

Next come our definitions pertaining to Hamiltonian cycles. These notions must refer to the edges in the square of a claw-free set, which will be formalized as doubletons. In order to define a Hamiltonian cycle, we can avoid speaking of sequences of vertices of a graph and refer only to subsets of edges forming a cycle. This is done in two steps: we define Hank(H) to hold, for an H such that $\emptyset \notin H$, if every element $x \in e \in H$ is a member of another element $q \neq e$ of H. Roughly speaking, this says that every endpoint of an edge of H has degree at least 2 in H, but notice that for the time being we are not insisting that H is formed by doubletons. Next, we define Cycle(C) to hold if Hank(C) holds and C is inclusion-minimal with this property (see Example 5.5).

Fig. 5.21 Two strategies for extending a matching

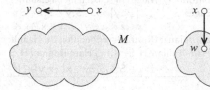

Example 5.5 A hank H and a cycle C:

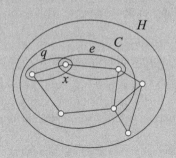

DEF $cycle_0$: [Collection of edges whose endpoints have degree greater than 1]

$\quad$ $\mathsf{Hank}(H)$ $\leftrightarrow_{\text{Def}}$ $\emptyset \notin H$ & $\langle \forall e \in H \mid e \subseteq \bigcup(H\setminus\{e\}) \rangle$

DEF $cycle_1$: [Cycle (unless null)]

$\quad$ $\mathsf{Cycle}(C)$ $\leftrightarrow_{\text{Def}}$ $\mathsf{Hank}(C)$ & $\langle \forall d \subseteq C \mid \mathsf{Hank}(d)\ \&\ d \neq \emptyset \rightarrow d = C \rangle$

Let us briefly digress to show that whenever C is a nonnull subset of edges of a graph G and $\mathsf{Cycle}(C)$ holds, the subgraph $G[C]$ of G induced by the edges of C is a cycle in the customary sense.

We argue first that $G[C]$ contains a cycle, i.e., that there exist $\{x,y\} \in C$ and $x_1,\ldots,x_k \in \bigcup C$, $k \geq 1$, such that $\{x,x_1\}, \{x_1,x_2\}, \ldots, \{x_{k-1},x_k\}, \{x_k,y\} \in C$. To find such elements of $\bigcup C$, let $P = (x,x_1,\ldots,x_t)$, $t \geq 2$, be a longest path in $G[C]$. Since $\mathsf{Hank}(C)$ holds, there exists $\{x,y\} \in C \setminus \{\{x,x_1\}\}$. Our first claim immediately follows since the maximality of P implies that $y \in P$ and thus that y is equal to some x_{k+1}, with $1 \leq k \leq t-1$.

Let, now, D denote the set $\{\{x,y\},\{x,x_1\},\{x_1,x_2\},\ldots,\{x_k,y\}\}$. From the choice of $x_1,\ldots,x_k$, $\mathsf{Hank}(D)$ holds. The initial assumption $\mathsf{Cycle}(C)$ and the fact that $\emptyset \neq D \subseteq C$ imply $C = D$, showing thus that C is a cycle in the customary sense.

DEF $hamiltonian_1$: [Hamiltonian cycle, in graph without isolated vertices]

$\quad$ $\mathsf{Hamiltonian}(H,S,E)$ $\leftrightarrow_{\text{Def}}$ $\mathsf{Cycle}(H)$ & $\bigcup H = S$ & $H \subseteq E$

DEF $hamiltonian_2$: [Edges in squared membership]

$\quad$ $\mathsf{sqEdges}(S)$ $=_{\text{Def}}$ $\{\{x,y\} : x \in S, y \in S, z \in S \mid$
$\qquad x \in y \vee (x \in z\ \&\ z \in y) \vee (z \in x \cap y\ \&\ x \neq y)\}$

DEF $hamiltonian_3$: [Restraining condition for Hamiltonian cycles]

$\quad$ $\mathsf{SqHamiltonian}(H,S)$ $\leftrightarrow_{\text{Def}}$ $\mathsf{Hamiltonian}(H,S,\mathsf{sqEdges}(S))$ &
$\qquad \langle \forall x \in S\setminus\bigcup S, \exists y \in x \mid \{x,y\} \in H \rangle$

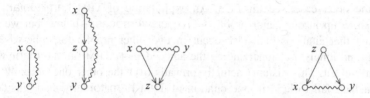

Fig. 5.22 Four orientations of a path of length 1 or 2 between two vertices x and y; the last of these is not taken into account by our definition sqEdges

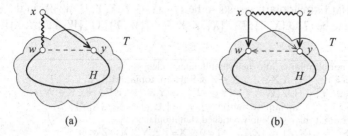

Fig. 5.23 Two enrichment strategies for Hamiltonian cycles in squares of claw-free graphs

Given a graph (S, E), we say that $H \subseteq E$ is a *Hamiltonian cycle* of it if $\mathsf{Cycle}(H)$ holds, and each vertex v of S is *covered* by an edge a of H, in the sense that $v \in a$. Given a set S, we characterize the set of *square edges* of S by allowing only three of the four possible membership alignments of two sets x, y whose distance in the graph $G(S)$ underlying S is 1 or 2 (see Fig. 5.22).

These three configurations suffice in a proof of the announced theorem. To complete our setup, we need the notion of SqHamiltonian, which describes a Hamiltonian cycle H of a set S reflecting the claim of Theorem 5.6: in the first place, we require H to be Hamiltonian in the square of the underlying graph $G(S)$; secondly, H must cover each source of S by an edge of $G(S)$.

The following theorems about Hamiltonian cycles are easily proved.

The last two of these will serve as base case for the proof we are after, namely, the case when a transitive set s has three or four elements. In particular, if $s = \{x, y, z\}$ is a transitive tripleton, then its elements are $x = \emptyset, y = \{\emptyset\}, z = \{\{\emptyset\}\} \vee z = \{\emptyset, \{\emptyset\}\}$, and x, z form a square edge; therefore, $\{x, y\}, \{y, z\}, \{z, x\}$ form a hank and then clearly a cycle, because hanks of cardinality 1 or 2 do not exist. When s has 4 elements and is transitive, it must contradict $s \subseteq \{\emptyset, \{\emptyset\}\}$ as well as the consequent of THM hamiltonian$_4$, whose contrapositive hence tells us again that the square of s has a Hamiltonian cycle. When s has five elements or more, then, mimicking the proof of Theorem 5.6 seen above, we will proceed differently, depending on whether the selected pivot of s belongs to a single element of s or to two: THM hamiltonian$_2$ will serve us when s has two such predecessors, and THM hamiltonian$_1$ will settle the other case; see Fig. 5.23. An exploitation of THM hamiltonian$_1$ will hence show up in our major proof of THM clawFreeness$_1$ below, but we must point out that citations of it also occur in two other proofs; for, otherwise, the second (and longish) disjunct ending the antecedent of THM hamiltonian$_1$ would seem irrelevant. This disjunct actually prevails over the other one once: $W \in Y$ may, in fact, fail to hold in one citation of THM hamiltonian$_1$ made inside the proof of THM hamiltonian$_2$. This citation is possible thanks to the fact that the two enrichment strategies for Hamiltonian cycles are conceptually the same. Actually, the proof of THM hamiltonian$_2$ basically consists of just two citations of THM hamiltonian$_1$: in the first step, we take the tuple $\langle S, T, X, Y, H, W, K \rangle$ about which THM hamiltonian$_1$ speaks to be $\langle T \cup \{X\}, T, X, Y, H, W, \emptyset \rangle$; in the second, we take it to be $\langle T \cup \{X, Z\}, T \cup \{X\}, Z, Y, H \setminus \{\{W, Y\}\} \cup \{\{W, X\}, \{X, Y\}\}, X, W \rangle$.

THM hamiltonian$_1$: [Enriched Hamiltonian cycles]

$S = T \cup \{X\}$ & $X \notin T$ & $Y \in X$ & SqHamiltonian(H, T) &
$\{W, Y\} \in H$ & $(W \in Y \lor (Y \in W$ & $K \neq Y$ & $\{W, K\} \in H$ & $K \in W)) \to$
  SqHamiltonian$(H \setminus \{\{W, Y\}\} \cup \{\{W, X\}, \{X, Y\}\}, S)$

THM hamiltonian$_2$: [Doubly enriched Hamiltonian cycles]

$S = T \cup \{X, Z\}$ & $\{X, Z\} \cap T = \emptyset$ & $X \neq Z$ & $Y \in X \cap Z$ &
SqHamiltonian(H, T) & $\{W, Y\} \in H$ & $W \in Y \cap X \to$
  SqHamiltonian$(H \setminus \{\{W, Y\}\} \cup \{\{W, X\}, \{X, Z\}, \{Z, Y\}\}, S)$

THM hamiltonian$_3$: [Trivial Hamiltonian cycles]

$S = \{X, Y, Z\}$ & $X \in Y$ & $Y \in Z \to$ SqHamiltonian$(\{\{X, Y\}, \{Y, Z\}, \{Z, X\}\}, S)$

THM hamiltonian$_4$. [Any nontrivial transitive set whose square is devoid of
    Hamiltonian cycles must strictly comprise certain sets]

Trans(S) & $S \not\subseteq \{\emptyset, \{\emptyset\}\}$ & $\neg \langle \exists h \mid$ SqHamiltonian$(h, S) \rangle \to$
  $S \neq \{\emptyset, \{\emptyset\}, \{\{\emptyset\}\}\}$ & $S \neq \{\emptyset, \{\emptyset\}, \{\emptyset, \{\emptyset\}\}\}$ &
  $S \neq \{\emptyset, \{\emptyset\}, \{\{\emptyset\}\}, \{\emptyset, \{\emptyset\}\}\}$ & $S \supseteq \{\emptyset, \{\emptyset\}\}$ &
  $(\{\{\emptyset\}\} \in S \lor \{\emptyset, \{\emptyset\}\} \in S)$

Hamiltonicity and Perfect Matching

We will now examine in detail our formal reconstruction of Theorem 5.6, as readjusted for membership graphs and certified correct with Ref, as THM clawFreeness$_1$.

Assuming the contrary, let s_1 be a finite transitive claw-free set with at least three elements, i.e., $s_1 \not\subseteq \{\emptyset, \{\emptyset\}\}$, which has no Hamiltonian cycles in its square (step 1). Through the finiteInduction THEORY, pick an inclusion-minimal finite transitive nontrivial claw-free set s_0 likewise lacking such a cycle (steps 2, 3).

Nontrivial claw-free transitive sets have Hamiltonian squares

THM clawFreeness$_1$.
Finite(S) & Trans(S) & ClawFree(S) & S $\not\subseteq$ {$\emptyset$, {$\emptyset$}} $\to$ $\langle\exists h \mid$ SqHamiltonian(h, S)$\rangle$. PROOF:
1 Suppose_not(s$_1$) $\Rightarrow$ AUTO

2 APPLY $\langle$fin$_\Theta$: s$_0\rangle$ finiteInduction$\big($s$_0 \mapsto$ s$_1$,

   P(S) $\mapsto$ (Trans(S) & ClawFree(S) & S $\not\subseteq$ {$\emptyset$,{$\emptyset$}} & $\neg\langle\exists h \mid$ SqHamiltonian(h,S)$\rangle$)$\big)$ $\Rightarrow$
    Stat1 : $\langle\forall$s $\mid$ s $\subseteq$ s$_0 \to$ Finite(s) & (Trans(s) & ClawFree(s) & s $\not\subseteq$ {$\emptyset$, {$\emptyset$}} &
            $\neg\langle\exists h \mid$ SqHamiltonian(h, s)$\rangle \leftrightarrow$ s = s$_0$)$\rangle$

3 $\langle$s$_0\rangle\hookrightarrow$Stat1 $\Rightarrow$ Stat2 : $\neg\langle\exists h \mid$ SqHamiltonian(h, s$_0$)$\rangle$ &
                  Finite(s$_0$) & Trans(s$_0$) & ClawFree(s$_0$) & s$_0 \not\subseteq$ {$\emptyset$, {$\emptyset$}}

4 APPLY (x$_\Theta$: x, y$_\Theta$: y, z$_\Theta$: z, t$_\Theta$: t) pivotsForClawFreeness(s$_0 \mapsto$ s$_0$) $\Rightarrow$

  {v $\in$ s$_0 \mid$ y $\in$ v} = {x, z} & x, y, z $\in$ s$_0$ & y $\in$ x $\cap$ z $\setminus \cup\cup$s$_0$ &

  y $\in$ t$\setminus\cup$t & t = s$_0\setminus$\{x, z\} & t = {u $\in$ s$_0 \mid$ y $\notin$ u} &

  Trans(t) & ClawFree(t) & x $\notin$ t & x $\notin$ z & z $\notin$ x
5 Suppose $\Rightarrow$ t $\subseteq$ {$\emptyset$, {$\emptyset$}}
6   $\langle$s$_0$, x, z$\rangle\hookrightarrow$Ttr$_3$ $\Rightarrow$ s$_0 \subseteq$ {$\emptyset$, {$\emptyset$}, {{$\emptyset$}}, {$\emptyset$, {$\emptyset$}}}
7 $\langle$s$_0\rangle\hookrightarrow$Thamiltonian$_4$ $\Rightarrow$ false;   Discharge $\Rightarrow$ AUTO
8 $\langle$t$\rangle\hookrightarrow$Stat1 $\Rightarrow$ Stat9 : $\langle\exists h \mid$ SqHamiltonian(h, t)$\rangle$
9 $\langle$h$_0\rangle\hookrightarrow$Stat9 $\Rightarrow$ SqHamiltonian(h$_0$, t)
10 Use_def(Hamiltonian(h$_0$, t, sqEdges(t))) $\Rightarrow$ AUTO
11 Use_def(SqHamiltonian) $\Rightarrow$ Stat11 : ($\forall$x $\in$ t$\setminus\cup$t, $\exists$y $\in$ x $\mid$ {x, y} $\in$ h$_0$) &
                Cycle(h$_0$) & $\cup$h$_0$ = t & h$_0 \subseteq$ sqEdges(t)
12 $\langle$y, w$\rangle\hookrightarrow$Stat11 $\Rightarrow$ w $\in$ y & {w, y} $\in$ h$_0$
13 Suppose $\Rightarrow$ ⟩ x = z
14 $\langle$s$_0$, t, x, y, h$_0$, w, $\emptyset\rangle\hookrightarrow$Thamiltonian$_1$ $\Rightarrow$
           SqHamiltonian(h$_0\setminus$ {{w, y}} $\cup$ {{w, x}, {x, y}}, s$_0$)
15 $\langle$h$_0\setminus$ {{w, y}} $\cup$ {{w, x}, {x, y}}$\rangle\hookrightarrow$Stat2 $\Rightarrow$ false;   Discharge $\Rightarrow$ x $\neq$ z
16 $\langle$s$_0$, y$\rangle\hookrightarrow$Ttr$_2$ $\Rightarrow$ w $\in$ s$_0$
17 $\langle$s$_0$, y, x, z, w$\rangle\hookrightarrow$TclawFreeness$_b$ $\Rightarrow$ w $\in$ x $\cup$ z
18 Suppose $\Rightarrow$ w $\in$ x
19 $\langle$s$_0$, t, x, z, y, h$_0$, w$\rangle\hookrightarrow$Thamiltonian$_2$ $\Rightarrow$
           SqHamiltonian(h$_0\setminus$ {{w, y}} $\cup$ {{w, x}, {x, z}, {z, y}}, s$_0$)
20 $\langle$h$_0\setminus$ {{w, y}} $\cup$ {{w, x}, {x, z}, {z, y}}$\rangle\hookrightarrow$Stat2 $\Rightarrow$ false
21 Discharge $\Rightarrow$ w $\in$ z
22 $\langle$s$_0$, t, z, x, y, h$_0$, w$\rangle\hookrightarrow$Thamiltonian$_2$ $\Rightarrow$
           SqHamiltonian(h$_0\setminus$ {{w, y}} $\cup$ {{w, z}, {z, x}, {x, y}}, s$_0$)
23 $\langle$h$_0\setminus$ {{w, y}} $\cup$ {{w, z}, {z, x}, {x, y}}$\rangle\hookrightarrow$Stat2 $\Rightarrow$ false;   Discharge $\Rightarrow$ QED

The THEORY pivotsForClawFreeness can be applied to s_0 (step 4): we thereby pick an element x from the frontier of s_0 and an element y of x which is pivotal relative to s_0. This y will have at most two in-neighbors (one of the two being x) in s_0. We denote by z an in-neighbor of y in s_0, possibly such that $z \neq x$. Observe, among others, that neither one of x, z can belong to the other.

If the removal of x, z from s_0 leads to a set t included in {$\emptyset$, {$\emptyset$}} (step 5), then by THM tr$_3$ we get $s_0 \subseteq$ {$\emptyset$, {$\emptyset$} , {{$\emptyset$}} , {$\emptyset$, {$\emptyset$}}}. This leads us to a contradiction, in light of THM hamiltonian$_4$ (step 7). Therefore, t is not trivial and the inductive hypothesis applies to it (step 8): thanks to that hypothesis, we can find a Hamiltonian cycle h_0 for t (step 9).

Recalling the definitions of Hamiltonian and sqHamiltonian (steps 10, 11), since y is a source of $t = \cup h_0$, there is an edge $\{y, w\}$ in h_0, with $w \in y$ (step 12).

If $x = z$, the set $h_1 = h_0 \setminus \{\{y, w\}\} \cup \{\{x, y\}, \{x, w\}\}$ is a Hamiltonian cycle for s_0, by THM hamiltonian$_1$ (step 14). This conflicts with the minimality of s_0 (step 15): in fact $\{x, w\}$ is a square edge, since $w \in y$ and $y \in x$ both hold.

On the other hand, if $x \neq z$, claw-freeness implies, via THM clawFreeness$_b$, that either $w \in x$ or $w \in z$ must hold (step 17). Assume the former (step 18), and put $h_2 = h_0 \setminus \{\{y, w\}\} \cup \{\{y, z\}, \{z, x\}, \{x, w\}\}$, where $\{x, z\}$ is a square edge and $\{x, w\}$ and $\{y, z\}$ are genuine edges incident to the sources x, z. By THM hamiltonian$_2$, h_2 is a Hamiltonian cycle for s_0 (step 19), and we are again facing a contradiction (step 20). The case $w \in z$ is entirely symmetric (steps 21, 22, 23), which proves the initial claim.

The result on the existence of a perfect matching is usually restricted to graphs whose set of vertices has an even cardinality, as we have done in our Theorem 5.5; but here, since numbers pop up only in this place, we omit the evenness constraint: transitive, claw-free sets admit a "near-perfect matching," that is to say, a matching which does not cover at most one of its elements. In Ref:

> THM clawFreeness$_2$: [Every claw-free transitive set has a near-perfect matching]
> Finite(S) & Trans(S) & ClawFree(S) $\rightarrow$ ⟨∃m, y | Matching(m) & S \ {y} $= \cup$m⟩

(The trick, since we are not insisting that y must belong to S, is that when S has an even cardinality, we can pick any y lying outside S, e.g., $y = S$.)

Our formal reconstruction of Theorem 5.5 is almost twice as long as the proof about Hamiltonicity just detailed, but closely resembles it. No surprise, a kinship between the proofs of Theorems 5.6 and 5.5 already emerged at the informal level of discussion in Sect. 5.2.4 and actually gave us the rationale for isolating a reusable construction inside the THEORY pivotsForClawFreeness (cf. Fig. 5.19). Thanks to this design choice, the two formal proofs have kept their close analogy, and it seems pointless to supply again here many explanations, which anyway are offered as comments inside the text of the proof available at the URL

   aetnanova.units.it/scenarios/GraphsAsTransitiveSets/.

5.4 Conclusions About Our Proof-Checking Experiment

In the transfers of techniques and results across the realms of graphs and sets, claw-free graphs turned out to occupy a preeminent place. As discussed in this chapter, a set-theoretic interpretation of edges as membership pairs leads to plain proofs of two classical results about connected claw-free graphs. We have taken a natural and concrete step by formalizing these two set-inspired proofs in the proof-checker Ref, which ultimately represents every entity in the user's universe of discourse as a set.

The material treated in what precedes (and also, in part, relegated to Appendix A) has, in fact, revolved around an experiment leading to a computer-verified proof of the Milanič–Tomescu reflection theorem—actually of a special case of it, Theorem 5.4. By proving that theorem, we gained a change of perspective on connected claw-free graphs: we could replace graphs of that kind by (hereditarily finite) transitive claw-free *sets*, thereby producing proofs, far easier than the classical ones, of two propositions regarding them.

To take advantage of the set-theoretic foundation of Ref, we exploited set equivalents of the graph-theoretic notions involved in our experiment: edge, source, square, etc. We have eased some proofs by resorting to weak counterparts of well-established notions such as cycle, claw-freeness, longest directed path, etc. Thanks to these and to plain definitions of genuine graph-theoretic notions such as connectedness, we were in fact able to implement most proofs without big effort.

- In our formal development, we could have defined a transitive set to be *claw-free* if none of the four non-isomorphic membership renderings of a claw are induced by any quadruple of its elements; but actually, it sufficed to forbid two out of these four (see Fig. 5.5) to get the desired proofs. This explains why our results are easier to achieve but in some respects are more general. To see the difference, observe that the graph in Fig. 5.24, once suitably oriented, can be handled by our theorems, whereas its Hamiltonicity and its perfect matchings are not seen by the traditional results [60, 108, 111].
- The graph "squares" about which our Hamiltonicity proof speaks are actually poorer in edges than the standard ones, since we allow only three out of the four membership alignments, cf. Fig. 5.22.
- We have addressed issues regarding graphs, which we see as pre-algorithmic and, as such, application oriented. Nonetheless, our results are so close to the foundations of mathematics that we found no reason to introduce numbers or to speak about paths, and we were able to avoid recursion even in the determination of the pivots, as explained on p. 160 in connection with Fig. 5.19.
- The graphs that can be represented as membership digraphs form a broad class of graphs, which includes, in partial overlap with connected claw-free graphs, all graphs endowed with a Hamiltonian path [64], as seen in Chap. 4.

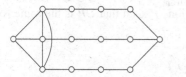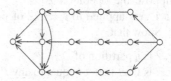

Fig. 5.24 A graph endowed with a claw, but admitting an orientation compatible with our set-theoretic definition of claw-freeness (cf. Fig. 5.5)

As mentioned in Sect. 5.3.2.1 (see, in particular, Fig. 5.15), we have also proved with Ref a representation result referring to a graph whatsoever. This other result lies at a more fundamental level than the representation, through membership digraphs, of graphs belonging to special classes (viz., connected claw-free graphs and graphs endowed with a Hamiltonian path). Its experimental setup and the proof techniques involved are pretty much the same as for the other case study, but the intermediate acyclic digraph now turns out to be *weakly* extensional instead of just extensional; hence it would be modeled more naturally through a set with atoms than through one belonging to von Neumann's renowned *cumulative hierarchy* (see Sect. 3.2). However, cf. [24, p. 54]:

> Even in this case, one might still wish to prevent the existence of unrestricted atoms. In any case, for the "genuine" sets, Extensionality holds and the other sets are merely harmless curiosities.

To get rid of such "harmless curiosities" as atoms, we had to design a technique— the one presented in Sect. 5.2.2—which, in the end, would remain hidden inside our THEORY finMostowskiDecoration. Nevertheless, we wanted our technique to be as light as possible, because sooner or later we will need similar techniques to handle more challenging situations, involving—as we expect—infinite graphs. The reader will judge whether we have achieved our goal parsimoniously enough.

Our representation theorems exploit sets demandingly: not only have we gone beyond the conventional view that the edges of a graph/digraph simply are doubletons/ordered pairs, but also, as just recalled, we have eliminated atoms from our sets. Also, we have required that the set representing a claw-free graph be transitive. Putting heavy restraints in the formulation of representation theorems is essential in order that a verifier well versed only about first principles can indeed serve as a proof assistant in specific domains.

Our results about graphs are constructive and, indeed, pre-algorithmic. Even the two propositions on the orientability of graphs discussed above (namely, Theorems 5.2 and 5.3) are based on two algorithms of which, in a very definite sense, they prove the correctness.

Exercises

5.1 Suppose the operation $\bigsqcap \mathscr{Q} =_{\text{Def}} \{\bigcap_{x \in b \in Q \in \mathscr{Q}} b : x \in \cup \cup \mathscr{Q}\}$ as defined in Panel 5.1 gets applied to a set $\mathscr{Q}$ of partitions P such that $\cup P$ is the same for all $P \in \mathscr{Q}$. Show that:

(1) $\bigsqcap \mathscr{Q}$ is a partition of $\cup \cup \mathscr{Q}$;
(2) $\bigsqcap \mathscr{Q}$ is finer than or equal to any $P \in \mathscr{Q}$;
(3) $\bigsqcap \mathscr{Q}$ is the coarsest of all partitions of $\cup \cup \mathscr{Q}$ which are finer than or equal to each $P \in \mathscr{Q}$. Moreover,
(4) $\bigsqcap \mathscr{Q}$ is a disconnected partition of E when $\mathscr{Q} \neq \emptyset$ consists of disconnected partitions of E.

Conclude that E has a finest disconnected partition whenever $\emptyset \notin E$.

5.2 Recall Kuratowski's definition of the ordered set-pair and the definitions, seen in Fig. 5.1, of global conjugated projections $P \mapsto P^{[1]}$ and $P \mapsto P^{[2]}$. Check the validity of the following two laws:

THM pair$_0$: [Unambiguity of the pairing function] $[X, Y]^{[1]} = X$ & $[X, Y]^{[2]} = Y$
THM pair$_1$: [Both projections extract $\emptyset$ from $\emptyset$] $P = \emptyset \rightarrow P^{[1]} = \emptyset$ & $P^{[2]} = \emptyset$

Also show that there is plenty of pairs P, Q of sets (where either P or Q is not of the form $[X, Y]$) such that $P^{[1]} = Q^{[1]}$, $P^{[2]} = Q^{[2]}$, and $P \neq Q$.

5.3 Prove the following implication for all sets x, y, z and S:

$$(y \in x \ \& \ x \in z \ \& \ x, z \in S) \rightarrow y \in \cup(S \cap \cup S)$$

(cf. p. 145 and THM un$_{12}$ in Fig. 5.18).

5.4 Check manually the validity of the laws in Fig. 5.13, referring to the definitions of the map-related constructs **dom**, **img**, Svm, $\upharpoonright$, etc., as given in Fig. 5.2.

5.5 Check the validity of the laws in Fig. 5.18, referring to the definitions of the constructs $\cup$, Trans, Finite—and hence, indirectly, of $\mathcal{P}$—as given in Fig. 5.2.

5.6 Check manually the validity of the laws in Fig. 5.14, referring to the definition of the Cartesian power $\times$, as given in Fig. 5.1.

5.7 Prove the properties of acyclicity listed in Fig. 5.4.

5.8 Explain how the developer of THEORY finMostowskiDecoration can take advantage of the availability of the THEORY finAcycLabeling as described in Fig. 5.17.

5.9 Prove in detail the following claims, where $V, A,$ mski are as in Sect. 5.2.2:

- $(\text{Svm}(H) \ \& \ \text{Finite}(H) \ \& \ \text{img}(H) \supseteq \text{dom}(H)) \rightarrow \text{img}(H) = \text{dom}(H)$;
- $(\{x, y\} \subseteq V \ \& \ \text{mski}\upharpoonright y \in \text{mski}\upharpoonright x) \rightarrow x \in \text{dom}(A)$;
- $(\{x, y\} \subseteq V \ \& \ \text{mski}\upharpoonright x = \text{mski}\upharpoonright y) \rightarrow x = y$;
- $(\{x, y\} \subseteq V \ \& \ \text{mski}\upharpoonright y \in \text{mski}\upharpoonright x) \rightarrow x \in \text{dom}(A)$;
- $y \in V \rightarrow (\text{mski}\upharpoonright y \in \text{mski}\upharpoonright x \leftrightarrow [x, y] \in A)$.

5.10 Show that **img**(mski) is a transitive, hereditarily finite set when $D = (V, A)$ is an *extensional* acyclic graph.

5.11 Show that the claim of Theorem 5.5 and the claim (5.1) shown at the beginning of Sect. 5.2 are equivalent.

5.12 Prove that

$$(\text{uGraph}(V, E) \ \& \ \text{Conn}(V, E) \ \& \ V \neq \emptyset) \rightarrow \text{HasSpaningTree}(V, E),$$

where the following definitions apply:

$$\text{HankFree}(T) \leftrightarrow_{\text{Def}} \forall E\big((E \subseteq T \ \& \ E \neq \emptyset) \to$$
$$\exists a\big(a \in E \ \& \ a \not\subseteq \cup(E \setminus \{a\})\big)\big),$$
$$\text{Is\_tree}(V,\,T) \leftrightarrow_{\text{Def}} \text{Conn}(V,\,T) \ \& \ \text{HankFree}(T) \ \& \ T \neq \emptyset,$$
$$\text{HasSpanningTree}(V,\,E) \leftrightarrow_{\text{Def}} \exists T\big(\text{Is\_tree}(V,\,T) \ \& \ T \subseteq E\big).$$

5.13 As a continuation of Exercise 5.12, prove the following analogue of Theorem 5.1

$$\Big(\text{HasSpanningTree}(V,\,E) \ \& \ E \subseteq \big\{\, \{x,y\} : x \in V, \ y \in V \setminus \{x\} \,\big\} \ \&$$
$$V \neq \{\text{arb}\,(V)\}\Big) \to \exists z \in V \ \text{HasSpanningTree}(V \setminus \{z\},\, \{a \in E \,|\, z \notin a\}).$$

5.14 Prove the claims THMs clawFreeness$_a$, clawFreeness$_b$ stated on pp. 144–145.

Part III
Sets as Graphs

Chapter 6
Counting and Encoding Sets

In all that precedes, we have established that sets and hypersets correspond naturally to particular classes of graphs. In what is to come, starting in this chapter with counting and encoding problems, we will address natural questions about sets for whose treatment the set-to-graph correspondence can be of use. We first consider a classical counting problem about the number of transitive sets of a given cardinality n. A most elegant way to assess this number is by counting extensional acyclic graphs in various ways, thereby expressing the sought value via a recursive formula, as a function of n and, perhaps, other parameters.

In addition to being straightforward, the graph-theoretic approach to set-counting problems leads to solutions somewhat more effectively than previously proposed techniques. Essentially, it simplifies things because it enables us to analyze the structure of sets by just *visiting* their membership graphs. As is to be expected, the crucial character of this visit is the *order* in which it is performed: always coherent with the membership relation and thus with the Ackermann order.

Given the success of the Ackermann order—and therefore, ultimately, of $\mathbb{N}_A$—in the well-founded case, our next goal will be to extend and adapt it to the non-well-founded case. This task is not trivial since the lack of well-foundedness on ϵ requires to come up with an ordering that reflects the inner sequence of steps performed on a membership graph to establish the equality relation. The graph-theoretic view and the splitting technique for computing the maximum bisimulation are the key: they just need to be applied to the entire universe of finite membership graphs. Hence, we begin by reconsidering the Ackermann order and showing that the splitting technique can be used to compute the Ackermann order and $\mathbb{N}_A$. Then we prove that, ordering both sets *and* proper hypersets, it can be applied also to the universe of hypersets. This application is illustrated by defining an extension of the mapping $\mathbb{N}_A$ to a function $\mathbb{Q}_A$ mapping hypersets to *dyadic* rational numbers, i.e., rational numbers whose binary expansion requires finitely many digits.

The mapping $\mathbb{N}_A$ can be seen as a *characteristic function*, in the sense that, together with the Ackermann order, it provides us with a complete description of the membership relation on a set. For example, the binary expansion of a given

© Springer International Publishing AG 2017
E.G. Omodeo et al., *On Sets and Graphs*, DOI 10.1007/978-3-319-54981-1_6

$\mathbb{N}_A(h)$, say $\mathbb{N}_A(h) = 100101$, tells us that h is the set whose elements are the first, third, and sixth set in the Ackermann order. Our mapping $\mathbb{Q}_A$ will be characterized by a similar trait: $\mathbb{Q}_A(h) = 100101.011$ will be the code of a hyperset h consisting of the first, third, and sixth well-founded set *and* of the second and third proper *hyper*set. Moreover, $\mathbb{Q}_A$ and $\mathbb{N}_A$ will coincide on well-founded sets.

6.1   Counting Sets as Graphs

Most combinatorial counting problems are usually split in two categories. In the first category, the input consists of a single object, and we are interested in counting all of its substructures that enjoy a certain property. For example, in Sect. 4.5.1 the input consisted of a single undirected graph, and we were interested in counting its extensional acyclic orientations. Another classical example is the problem of finding the number of truth assignments that satisfy a given Boolean formula.

In the second category, the input typically consists of a single number n, and we are interested in finding the number of "objects" of "size" n, where "object" and "size" depend on the problem at hand. For example, objects can be trees with vertex set $\{1, \ldots, n\}$ and their size can be their number of vertices, n.

In this section we take well-founded sets as objects. If we take the size of a well-founded set to be its cardinality, then the counting problem is not interesting, since there are an infinite number of well-founded sets with a given positive cardinality. Likewise, if we take the size of a set to be its rank, then Ackermann's bijection between the elements of HF_n and $\{0, \ldots, \beth(n+1) - 1\}$ gives us the answer (see Lemma 3.2 and Exercise 3.8).

We instead adopt, as size of a well-founded set x, the cardinality of its transitive closure $\mathsf{trCl}(x)$. Moreover, in order to simplify some technical details, we will count only *transitive* well-founded sets x (thus $x = \mathsf{trCl}(x)$) of cardinality n. Exercises 6.1 and 6.2 argue that the former problem can in fact be reduced to the latter.

▍ **Problem 6.1 (Counting transitive sets)**   Given an integer n, find the number s_n of transitive well-founded sets of cardinality n.

The results of this section are mainly from [94]. Problem 6.1 was first solved by Peddicord in 1962 [92].

The most basic strategy for solving Problem 6.1 would be to list, or to *enumerate*, all transitive sets of size n. On the one hand, this is not a trivial task without some combinatorial insight. In fact, we give one such insight in Sect. 6.3.1, where we encode transitive sets with n elements by particular tuples of subsets of an $(n-1)$-element set; enumerating these tuples solves this enumeration problem. On the other hand, this is a very impractical process for obtaining the value s_n, since s_n grows exponentially with n [112]; see also Table 6.1.

Table 6.1 The first ten values of s_n, the solution to Problem 6.1

n	s_n
1	1
2	1
3	2
4	9
5	88
6	1802
7	75598
8	6421599
9	1097780312
10	376516036188

We would hence like to derive an algebraic formula for s_n. In the rest of this section, we study three methods for doing this, by exploiting some combinatorial decompositions of well-founded sets.

The first decomposition of a transitive well-founded set x is the one in which we single out the elements of $\text{trCl}(x) = x$ of maximum rank, that is, of rank $\text{rank}(x)-1$. Recall from our discussion after Definition 3.1 that the rank $\text{rank}(x)$ of a hereditarily finite set x can be recursively defined as

$$\text{rank}(x) = \max_{y \in x} (\text{rank}(y) + 1), \qquad \text{where } \text{rank}(\emptyset) = 0. \tag{6.1}$$

After giving this decomposition, we then interpret it from a more intuitive graph-theoretic point of view. We will then continue with this graph-theoretic perspective for the rest of the section.

First, we put in a nutshell the counting problem at hand:

Problem 6.2 (Counting transitive sets by rank)   Given an integer n, find the number s_n^r of transitive well-founded sets x such that $|x| = n$, and x has r elements of maximum rank, $\text{rank}(x) - 1$.

Clearly, having all $s_n^1, \ldots, s_n^n$, we can solve Problem 6.1 by taking $s_n = \sum_{r=1}^{n} s_n^r$. Observe also that $s_n^n = 0$, unless $n = 1$, in which case $s_1^1 = 1$. The following theorem gives the desired recurrence relation for s_n^r.

Theorem 6.1 *The solution to Problem 6.2 satisfies the following recurrence relation, for $n \geqslant 2$ and all $r \in \{1, \ldots, n-1\}$:*

$$s_n^r = \sum_{k=1}^{n-r} s_{n-r}^k \binom{(2^k - 1)2^{n-r-k}}{r}, \qquad \text{where } s_1^1 = 1.$$

Proof A transitive set x with n elements, out of which r have maximum rank $\text{rank}(x) - 1$, can be partitioned as $x = \{y_1, \ldots, y_r\} \cup x'$, where $y_1, \ldots, y_r$ are the

Fig. 6.1 Illustration of the second, graph-theoretic proof of Theorem 6.1

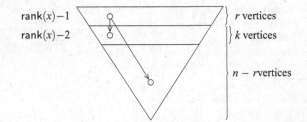

elements of x of rank $\mathsf{rank}(x) - 1$, and $x' = x \setminus \{y_1, \ldots, y_r\}$. Since $x' \subset x$, we have that also x' is transitive. The cardinality of x' is $n - r$ and its number of elements of maximum rank can be any integer $k \in \{1, \ldots, n - r\}$.

Since x is transitive, we have that each y_i is a subset of x'. Each y_i must have at least one element which is of maximum rank in x', while there are no constraints on the other elements of y_i. Thus, each y_i can be one of the $(2^k - 1)2^{n-r-k}$ possible such subsets of x'. Therefore, the number of total sets of the form $\{y_1, \ldots, y_r\}$ is $\binom{(2^k-1)2^{n-r-k}}{r}$, as claimed in the statement of the theorem.   ⊣

Recall from Sect. 3.4 that transitive well-founded sets are in bijection with extensional acyclic graphs, up to isomorphism. Recall also that the rank of a well-founded set x is the length of a longest membership path from its point to $\emptyset$. Therefore, finding an expression for s_n^r is equivalent to counting the non-isomorphic extensional acyclic graphs with n vertices out of which r have maximum rank. We can rewrite the above proof of Theorem 6.1 in these graph-theoretic terms (see also Fig. 6.1).

Proof (of Theorem 6.1; graph-theoretic version) An extensional acyclic graph D on n vertices out of which r have maximum rank can be obtained by adding r new vertices to an extensional acyclic graph D' on $n - r$ vertices out of which k have maximum rank $(1 \leq k \leq n - r)$, such that in D only the new r vertices have maximum rank.

There are $(2^k - 1)2^{n-r-k}$ total candidates for each set of out-neighbors of these r vertices, since at least one out-neighbor must be chosen from the set of k vertices of maximum rank of D', and there is no restriction concerning the remaining $n - r - k$ vertices of D'. Since the sets of out-neighbors of the new r vertices must be pairwise distinct, there are $\binom{(2^k-1)2^{n-r-k}}{r}$ ways of adding these r vertices to D'.   ⊣

The vertices of maximum rank in an extensional acyclic graph are also sources, but the converse does not hold. We now decompose extensional acyclic graphs by sources. Our first solution is based on two ingredients: first, on a count of the extensional acyclic graphs with vertex set $\{1, \ldots, n\}$, which we obtain here by inclusion-exclusion, and, second, on the fact that the membership graph of a transitive set x of cardinality n is isomorphic to exactly $n!$ extensional acyclic graphs with vertex set $\{1, \ldots, n\}$ (recall Lemma 2.1 and Sect. 3.4). Our second solution directly counts non-isomorphic extensional acyclic graphs with a given number of sources.

The problem we are now confronted with is:

Problem 6.3 (Counting extensional acyclic graphs) Given an integer n, find the number e_n of extensional acyclic graphs with vertex set $\{1, \ldots, n\}$.

Theorem 6.2 gives a recurrence relation for e_n using the inclusion-exclusion principle. The same idea works also for counting directed acyclic graphs, as Exercise 6.7 asks the reader to show.

Theorem 6.2 *The solution to Problem* 6.3 *satisfies the following recurrence relation:*

$$e_n = \sum_{k=1}^{n-\lceil \log_2 n \rceil} (-1)^{k+1} \binom{n}{k} (2^{n-k} - n + k)_k e_{n-k}, \quad e_0 = 1.$$

Proof Clearly, $e_1 = 1$ and in any acyclic graph there is at least one vertex v with $N^-(v) = \emptyset$. For all $i \in \{1, \ldots, n\}$, let E_i stand for the set of all extensional acyclic graphs with vertex set $\{1, \ldots, n\}$ and with the property that $N^-(i) = \emptyset$. By the inclusion-exclusion principle, we have

$$e_n = |E_1 \cup E_2 \cup \cdots \cup E_n| = \sum_{k=1}^{n} (-1)^{k+1} \sum_{1 \leq i_1 < \cdots < i_k \leq n} |E_{i_1} \cap \cdots \cap E_{i_k}|.$$

Next, we will see that

$$\sum_{1 \leq i_1 < \cdots < i_k \leq n} |E_{i_1} \cap \cdots \cap E_{i_k}| = \binom{n}{k} (2^{n-k} - n + k)_k e_{n-k},$$

where we have used the falling factorial notation $(x)_k = x(x-1) \cdots (x-k+1)$. Indeed, the number of extensional acyclic graphs with $n - k$ vertices forming the set $\{1, \ldots, n\} \setminus \{i_1, \ldots, i_k\}$ is equal to e_{n-k}. Vertices $i_1, \ldots, i_k$ have the property that $N^-(i_1) = \cdots = N^-(i_k) = \emptyset$; hence they can be joined only by out-going arcs with the remaining $n - k$ vertices. Since the resulting graph must be extensional, no two of the k sources can have the same out-neighbors and, additionally, no source can have the same out-neighbors as a vertex among the remaining $n - k$. Hence, $|E_{i_1} \cap \cdots \cap E_{i_k}|$ is equal to $(2^{n-k} - (n-k)) \cdots (2^{n-k} - (n-k) - (k-1)) e_{n-k}$; as there exist $\binom{n}{k}$ ways to choose $i_1, \ldots, i_k$, we get the above expression.

In conclusion, $e_n = \sum_{k=1}^{n} (-1)^{k+1} \binom{n}{k} (2^{n-k} - n + k)_k e_{n-k}$, and by definition, the non-null terms of this sum are for the argument of the falling factorial greater or equal to k, that is, $2^{n-k} \geq n$ or, equivalently, for $k \leq n - \lceil \log_2 n \rceil$. ⊣

Having obtained s_n we can now use the fact that each transitive set of cardinality n has been counted $n!$ times when obtaining the value s_n because extensional acyclic graphs are rigid (recall Lemmas 2.1 and 3.6), and obtain the following alternative solution to Problem 6.1.

Theorem 6.3 *The solution to Problem* 6.1 *satisfies the following recurrence relation:*

$$s_n = \sum_{k=1}^{n-\lceil \log_2 n \rceil} (-1)^{k+1} \binom{2^{n-k}-n+k}{k} s_{n-k}, \quad e_0 = 1.$$

Proof From Lemmas 2.1 and 3.6, we have $s_n = e_n/n!$. Substituting e_{n-k} by $(n-k)!\, s_{n-k}$ in the recursive expression of e_n from Theorem 6.2, we get

$$s_n = \frac{1}{n!} \sum_{k=1}^{n-\lceil \log_2 n \rceil} (-1)^{k+1} \frac{n!}{k!(n-k)!} (2^{n-k}-n+k)_k (n-k)!\, s_{n-k},$$

which yields the desired result.   ⊣

We now directly count non-isomorphic extensional acyclic graphs, with a given number of sources.

▌ **Problem 6.4 (Counting extensional acyclic graphs by sources)** Given integers n and s, find the number $c_{n,s}$ of non-isomorphic extensional acyclic graphs with n vertices, out of which s are sources.

Cast back in set-theoretic terms, Problem 6.4 asks for the cardinality of the set

$$\Big\{ x \in \mathsf{HF} \mid \mathrm{trCl}\,(x) = x \wedge |x| = n \wedge |\{y \in x \mid \text{ for all } z \in x, y \notin z\}| = s \Big\}.$$

Theorem 6.4 *The solution to Problem* 6.4 *satisfies the following recurrence relation:*

$$c_{n,s} = \frac{1}{s} \Big((2^{n-s}-(n-1))c_{n-1,s-1} + \sum_{k=0}^{n-s-1} \binom{s+k}{k+1} 2^{n-1-(s+k)} c_{n-1,s+k} \Big),$$

where $c_{1,1} = 1$ *and where we define* $c_{n,0}$ *as* 0, *for all* $n \geqslant 1$.

Proof Clearly, $c_{1,1} = 1$, and for all $n \geqslant 1$ and $c_{n,0} = 0$. An extensional acyclic graph on $n \geqslant 2$ vertices and s sources can be obtained from one on $n-1$ vertices by the addition of a source in two ways (see also Fig. 6.2).

First, a source can be added to a graph on $n-1$ vertices and $s-1$ sources. This will be connected to some of the $n-1-(s-1)$ vertices that are not sources, such that in the resulting graph its set of out-neighbors is non-null and is different from the set of out-neighbors of the $n-2$ vertices that are not sinks. There are $(2^{n-s}-(n-1))c_{n-1,s-1}$ ways to add this new source.

Second, a new source can be added to a graph on $n-1$ vertices and $s+k$ sources, for $k = 0,\ldots,n-s-1$, by connecting the new source with exactly $k+1$ already existing sources. This new source can also have arcs toward the remaining $n-1-$

Fig. 6.2 Illustration of the proof of Theorem 6.4. **(a)** First case. **(b)** Second case

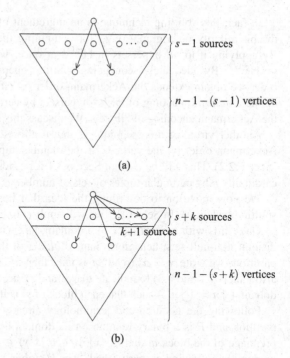

(a)

(b)

$(s + k)$ vertices. In this case, the new source is not a sink since at least one already existing source is among its elements. Moreover, it will certainly have the set of out-neighbors different from any set of out-neighbors of the remaining $n - 1$ vertices. There are $\binom{s+k}{k+1} 2^{n-1-(s+k)} c_{n-1,s+k}$ ways to add this vertex.

In the above process each extensional acyclic graph on n vertices and s sources has been obtained *exactly s* times, by the addition of *each one* of its s sources to exactly *one* extensional acyclic graph on $n - 1$ vertices, hence the factor $1/s$ in the expression of $c_{n,s}$. ⊣

6.2 A New Look at the Ackermann Order

In Sect. 3.3 we introduced the Ackermann encoding of hereditarily finite sets HF by natural numbers. By ordering HF sets according to their encodings, we obtained the Ackermann order on hereditarily finite sets, therein denoted by $\prec$.

The Ackermann encoding, and thus also the Ackermann order, seemed to rely on the fact that we are dealing with well-founded sets. However—following a path first described in [25]—here we revisit the Ackermann order and show that it can be obtained in a rather different manner, where well-foundedness is not necessary. This will be done through a *splitting technique*, first presented in [82] and subsequently refined in many different ways (e.g., [30, 43, 93]).

In fact, this splitting technique is an ingredient of an algorithm for computing the bisimilarity on a graph—e.g., for computing the equality relation on $\mathsf{HF}^{1/2}$. By applying it to *all* hypersets in $\mathsf{HF}^{1/2}$ at once, we manage to obtain an order on $\mathsf{HF}^{1/2}$. By adequately sequencing splitting operations, we also show that the order we obtain extends the Ackermann order on HF. Based on this, we can then define a *natural* encoding of $\mathsf{HF}^{1/2}$ hypersets by certain numbers, in analogy with the Ackermann encoding of HF sets. We discuss this in Sect. 6.3.2.

Another virtue of this technique is that it allows us to efficiently compute the Ackermann order on the vertices of the membership graph of a well-founded set (Sect. 6.2.2). This will be used in Sect. 6.3.1 for quickly encoding transitive sets of cardinality n by particular tuples of sets of numbers.

We now move on to describing the idea that lies behind the abovementioned splitting technique. In Sect. 6.2.1 we will apply it to HF.

One starts with a *partition* π of a domain, i.e., with a set consisting of pairwise disjoint non-null sets, henceforth named *blocks* of the partition. When comparing partitions we write $\pi \sqsubseteq \pi'$ (read "π is *finer* than π'" or "π' is *coarser* than π"; see also Panels 2.1 and 3.3) to indicate that π and π' are partitions—splitting the same domain $\bigcup \pi = \bigcup \pi'$—such that each block of π is included in a block of π'.

Following the notation and terminology already used in Panel 3.3, if π is a partition and R is a binary relation on its domain $\bigcup \pi$, the set $R \mathbin{\raise1pt\hbox{$\triangleleft$}} q$ denotes the preimage of the block q, that is, $\{x \mid \exists y \in q \; \langle x, y \rangle \in R\}$. Central to our subsequent work is the following property (read "π is *R-stable*"):

$$\pi \mathrel{\widehat{=}} R \leftrightarrow_{\mathrm{Def}} \forall p, q \in \pi \left(\emptyset \in \left\{ p \cap R \mathbin{\raise1pt\hbox{$\triangleleft$}} q, p \setminus R \mathbin{\raise1pt\hbox{$\triangleleft$}} q \right\} \right)$$

stating that π cannot be further *refined* using R. Otherwise stated, stability requires that a block p be included in $R \mathbin{\raise1pt\hbox{$\triangleleft$}} q$ whenever p intersects it. Trivially, any partition whose blocks are singletons is stable. *Unstable* partitions can be further stabilized (using R) by splitting their blocks.

Bisimulation problems can be stated as follows (see also [43] for more details and an extension of this point of view to *simulation*): given a partition $\pi^\star$ along with a relation R on $\bigcup \pi^\star$, find the coarsest $\pi_\star$ of all partitions that refine $\pi^\star$ and are R-stable. Formally the characterizing property of $\pi_\star$ is expressed as follows:

$$\pi_\star \sqsubseteq \pi^\star \wedge \forall \pi \sqsubseteq \pi^\star (\pi \mathrel{\widehat{=}} R \Leftrightarrow \pi \sqsubseteq \pi_\star),$$

and $\pi_\star$ can be seen as the equivalence relation with the largest possible classes that refines $\pi^\star$ and is coherent with R (i.e., R-stable).

Proceeding top-down, one can begin with $\pi = \pi^\star$ to then replace within π, as long as there are blocks p, q for which $p \cap R \mathbin{\raise1pt\hbox{$\triangleleft$}} q$ and $p \setminus R \mathbin{\raise1pt\hbox{$\triangleleft$}} q$ are both non-null, p by the latter two sets $p \cap R \mathbin{\raise1pt\hbox{$\triangleleft$}} q$ and $p \setminus R \mathbin{\raise1pt\hbox{$\triangleleft$}} q$. If $\bigcup \pi^\star$ is finite, one will at last attain the desired $\pi_\star$ as value of π, more or less rapidly, depending on the order in which blocks are processed and split. Within the stabilization process, the basic splitting action, namely, replacing p by $p \cap R \mathbin{\lceil} q$ and $p \setminus R \mathbin{\lceil} q$, can be packaged together with many other actions of the same kind. For example (as proposed in [82]), one can trace all p's which can be split by the same q and replace each of them by the resulting two blocks before seeking another q. Proceeding the other way around (as we will do), one can locate a p which is unstably relative to at least one q and then supersede p inside π, in a single shot, by all equivalence classes into which p gets partitioned by the equivalence relation

$$ x \sim_R y \leftrightarrow_{\mathrm{Def}} \forall q \in \pi \left(x \in R \mathbin{\lceil} q \Leftrightarrow y \in R \mathbin{\lceil} q \right) . $$

In the two cases which we will study, R will be $\ni$, while the initial partition $\pi^\star$ will first satisfy $\bigcup \pi^\star = \mathsf{HF}$ (Sect. 6.2.1) and then $\bigcup \pi^\star = \mathsf{HF}^{1/2}$ (Sect. 6.3.2). Despite $\bigcup \pi^\star$ being infinite in either case, infinite repetition of the basic splitting action will end into something valuable. To set the ground for this on a simple preliminary case, consider the following example.

Example 6.1 Suppose here that $\pi^\star = \{\mathsf{HF}\}$, let $\pi_0 = \pi^\star$, and then for $n = 0, 1, 2, \ldots$ inductively:

- prove that there is exactly one infinite block $p_n \in \pi_n$;
- prove that p_n is a culprit of the instability of π_n, as the sets

$$ \{x : x \in p_n \mid x \cap p_n \neq \emptyset\} \text{ and } \{x : x \in p_n \mid x \cap p_n = \emptyset\} $$

  are both non-null (actually, the former is infinite);
- put
$$ \pi_{n+1} = (\pi_n \setminus \{p_n\}) \cup \{\{x : x \in p_n \mid x \cap p_n \neq \emptyset\}, \{x : x \in p_n \mid x \cap p_n = \emptyset\}\}, $$
that is, split the class p_n by using p_n itself as a splitter.

  At the conclusion, $\{\{x : x \in p_n \mid x \cap p_n = \emptyset\} : n \in \mathbb{N}\}$ turns out to be the partition of HF whose blocks are the rank-equality classes. These blocks are all finite, but not singletons: an indication, since stable partitioning must give us the bisimilarity classes, that stability has not been attained as yet.

In what we are about to see, we resume work with the partition just outlined in the above example. We will sequence successive splitting actions fairly enough that the stable partition will result after denumerably many actions; along the way, we will impose an *order* on the singleton blocks.

6.2.1   Successive Partition Refinements

Processing the collection HF by means of the above splitting technique will amount to defining a countable sequence $(\mathscr{X}^n)_{n \in \mathbb{N}}$ of partitions $\mathscr{X}^n = \{X_i^n : i \in \mathbb{N}\}$, whose blocks are linearly ordered. Each partition $\mathscr{X}^{n+1}$ will turn out to be an *ordered refinement* of $\mathscr{X}^n$, namely (for all $i, j, h, k \in \mathbb{N}$):

$$\exists i' (X_i^{n+1} \subseteq X_{i'}^n), \tag{6.2}$$

$$X_i^{n+1} \subseteq X_k^n \ \wedge \ X_j^{n+1} \subseteq X_h^n \ \wedge \ k > h \Rightarrow i > j. \tag{6.3}$$

That is, $\mathscr{X}^{n+1} \sqsubseteq \mathscr{X}^n$, and the ordering of the sub-blocks into which the blocks of $\mathscr{X}^n$ get split in the formation of $\mathscr{X}^{n+1}$ will be consistent with the previous ordering.

For all n, we will maintain the invariant

$$\mathsf{Is\_finite}(X_i^n) \wedge \left(x \in X_h^n \ \wedge \ \mathsf{rank}(y) < \mathsf{rank}(x) \ \wedge y \in X_k^n \Rightarrow h > k \right), \tag{6.4}$$

implying that the blocks of $\mathscr{X}^n$ are all finite and they are ordered in a way complying with rank comparison among their elements—hence complying, in this well-founded case, with membership.[1] This is important because we want sets to be sorted *à la* Ackermann when, at the end of the process, the partition will be $\ni$-stable and blocks will be singletons. To meet (6.4) at the outset, we define $\mathscr{X}^0$ by putting

$$X_i^0 = \{x : x \in \mathsf{HF} \mid \mathsf{rank}(x) = i\} \quad \text{for all } i \in \mathbb{N}.$$

Preliminary to defining $\mathscr{X}^{n+1}$, we consider the smallest index h such that the block X_h^n *can be split* in the sense that there exist $x, y \in X_h^n$, and some k, such that x shares elements with X_k^n, whereas y does not. We also consider the equivalence relation $\sim_\ni$ on X_h^n given by

$$x \sim_\ni y \Leftrightarrow \forall k (X_k^n \cap x = \emptyset \leftrightarrow X_k^n \cap y = \emptyset).$$

Then we consider the partition induced by $\sim_\ni$ on X_h^n, ordered as follows: given two $\sim_\ni$-classes $Z', Z \subseteq X_h^n$, put Z' *before* Z if and only if, for $w \in Z'$ and $z \in Z$, the largest mismatch position k between w, z "favors" z, i.e.,

$$X_k^n \cap w = \emptyset \ \wedge \ X_k^n \cap z \neq \emptyset \ \wedge \ \forall j > k (X_j^n \cap w = \emptyset \leftrightarrow X_j^n \cap z = \emptyset).$$

It plainly ensues from the definition of $\sim_\ni$ that the mismatch position does not depend on the choice of w and z; hence this relationship imposes an order

---

[1] Notice that since our initial partition will separate sets at different ranks, we are guaranteed to have $h \neq k$ in the second conjunct of (6.4).

$Z_0, Z_1, \ldots, Z_m$ $(m \geqslant 1)$ on the $\sim_\ni$-equivalence classes of X_h^n. On this ground we can put

$$X_i^{n+1} = \begin{cases} X_i^n & \text{if } i < h, \\ Z_{i-h} & \text{if } h \leqslant i \leqslant h + m, \\ X_{i-m}^n & \text{if } h + m < i. \end{cases} \tag{6.5}$$

In the well-founded case at hand, an inductive argument on n shows that the smallest index h such that X_h^n can be split coincides with the smallest index h such that X_h^n is not a singleton; moreover, it turns out that the relation $\sim_\ni$ induces a partition of X_h^n into singleton blocks. These verifications are straightforward, and we leave them as exercises to the reader (see Exercise 6.13).

Properties (6.2), (6.3), and (6.4) hold throughout the construction and every element of HF will eventually belong to a singleton class. Given $n \in \mathbb{N}$ and $x \in$ HF, let $f(x, n) \in \mathbb{N}$ be such that

$$x \in X_{f(x,n)}^n.$$

Then one can easily prove that the full Ackermann order (as originally defined in Sect. 3.3) is the limit of the $\mathscr{X}^n$'s, that is:

$$x \prec y \Leftrightarrow \exists n \big(f(x, n) < f(y, n) \big).$$

The previous construction will be generalized in Sect. 6.3.2 to hypersets, by producing a sequence $(\mathscr{Y}^n)_{n \in \mathbb{N}}$ of ordered partitions, whose limit linearly orders $\mathsf{HF}^{1/2}$. For all $n \in \mathbb{N}$, the ordered partition $\mathscr{Y}^{n+1}$ will still be an ordered refinement of $\mathscr{Y}^n$, but we will not have the possibility to prove that $\mathscr{Y}^{n+1}$ results from splitting into *singleton* classes the first class of $\mathscr{Y}^n$ which is not a singleton: in spite of the close analogy between the constructions, the splitting process will behave differently in the non-well-founded case.

6.2.2 Computational Complexity

We conclude this section by showing how to specialize this splitting technique from the entire collection of HF sets to the vertices of a membership graph D. The goal here is to compute the Ackermann order on the vertices of D in time linear in the number of arcs of D. In the ongoing, we follow the exposition from [99], whose idea had appeared earlier, e.g., in [30].

Let n be the number of vertices of D and m be the number of its arcs. Observe that $m \geqslant n - 1$ holds, because the undirected graph underlying D is connected (recall Chap. 4).

At each step k of our procedure, we maintain an ordered partition π of $V(D)$. Initially, the blocks of π contain the vertices of D having the same rank, and the blocks of π are ordered by increasing ranks of their elements. Recall from Exercise 3.7 that for any two well-founded sets x and y, if $\mathsf{rank}(x) < \mathsf{rank}(y)$, then it holds that $x \prec y$. For each possible rank $k \in \{0, \ldots, n - 1\}$ in increasing order, we split the corresponding block of π into singleton blocks ordered in Ackermann order, exploiting the fact that the out-neighbors of the vertices of rank k have already been fully ordered. At the end of the procedure, each vertex of D will belong to a singleton block of π and the order of the blocks of π is the Ackermann order on the elements of x.

Expressed in algorithmic terms, at each iteration $k = 0, \ldots, n - 2$, we process the vertices of D of rank $i < k$. Moreover, recalling that our relation here is the arc relation, we consider vertices having in-neighbors among the vertices of rank k in decreasing Ackermann order. For every such vertex z and for every block P of π containing in-neighbors of z, we specialize the general strategy of pp. 181–182 as follows:

- we remove from P all elements which also belong to $N^-(z)$;
- these elements of $P \cap N^-(z)$ are collected in a new block P' which we insert in π, immediately after P.

It is an easy exercise for the reader (Exercise 6.11) to show that at the end of this procedure, each block of π is a singleton block and that these blocks are ordered inside π in Ackermann order. It remains to show in Lemma 6.1 that this procedure can be implemented in time $O(m)$. See also Example 6.2.

Example 6.2 Consider the membership graph D depicted below, whose vertices we labeled in increasing Ackermann order:

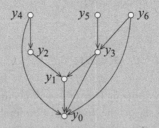

The algorithm starts with the ordered partition of $V(D)$ by rank:

$$[y_0]\ [y_1]\ [y_2, y_3]\ [y_4, y_5, y_6]$$

(continued)

Example 6.2 (continued)
The ranks 0 and 1 are already singletons. The block $[y_2, y_3]$ containing the vertices of rank 2 of D is split as follows. List $L = \langle y_1, y_0 \rangle$: vertex y_1 does not produce any splits, while vertex y_0 splits the block $[y_2, y_3]$ into singleton blocks. The resulting ordered partition is

$$[y_0]\ [y_1]\ [y_2]\ [y_3]\ [y_4, y_5, y_6]$$

The block $[y_4, y_5, y_6]$ containing the vertices of rank 3 of D is split as follows. List $L = \langle y_3, y_2, y_0 \rangle$: vertex y_3 produces the split.

$$[y_0]\ [y_1]\ [y_2]\ [y_3]\ [y_4]\ [y_5, y_6]$$

Vertex y_2 does not produce other splits, and vertex y_0 finally leads to a partition consisting of singleton blocks appearing inside it in Ackermann order:

$$[y_0]\ [y_1]\ [y_2]\ [y_3]\ [y_4]\ [y_5]\ [y_6]$$

Lemma 6.1 *The above procedure for computing the ordered partition π can be implemented in time $O(m)$.*

Proof We store π as a doubly linked list, with pointers to the first and last element of each block. The ranks of the vertices of D can be computed in time $O(m)$ by traversing the vertices of D in topological order and computing the rank value given by relation (6.1) for each vertex.

For every rank k, we collect the out-neighbors of the vertices of rank k in a temporary list L, in decreasing Ackermann order. We then scan L with a variable z. For each in-neighbor y of z belonging to a block P, we remove y from P and add it to the new block P' (in case P contains more in-neighbors of z, then all of them will belong to the same new block P') placed in π after P. We then continue updating π by advancing z to the next element of L.

Let V_k denote the vertices of D of rank k and let m_k denote the number of arcs outgoing from the vertices in V_k. The construction of L can be done in time $O(m_k)$ by scanning the out-neighbors of the vertices in V_k and placing them in an array of length m_k on the position given by their index in the Ackermann order on $V(D)$. Updating the ordered partition π for a vertex z in L takes time $O(|N^-(z) \cap V_k|)$, because π is stored as a doubly linked list. Therefore, updating π for each rank k takes time $O(m_k)$, and thus the entire procedure takes time $O(m)$. $\dashv$

6.3   Encoding Sets and Hypersets

We start this section by presenting an encoding of transitive well-founded sets
with n elements by tuples of subsets of $\{0, \ldots, n-2\}$. This was introduced
by Peddicord in 1962 [92] with the purpose of counting transitive sets with n
elements (recall Problem 6.1). Another value of it is that it provides a more compact
representation of such sets than the Ackermann encoding.[2]

Computing the Ackermann order (but not the actual encodings) is however a
subroutine of this encoding. The new look at the Ackermann order via partition
refinements, and in particular the algorithm from Sect. 6.2.2, will be used for
showing that this encoding can be computed in time linear in the number of arcs
of the membership graph of the transitive set.

In Sect. 6.3.2 we continue by encoding hypersets. We extend the idea of the
successive partition refinements to $\mathsf{HF}^{1/2}$ and discuss a natural encoding obtainable
from this order of $\mathsf{HF}^{1/2}$ hypersets by dyadic rational numbers.

6.3.1   Peddicord's Encoding of Transitive Sets

We start by defining an Ackermann-like, anti-lexicographic order on the family of
subsets of the set $\{0, \ldots, n-2\}$ of integers. For any two subsets $x, y \subseteq \{0, \ldots, n-2\}$,
we set

$$x \lhd y \leftrightarrow_{\mathrm{Def}} \max\{x \setminus y\} < \max\{y \setminus x\},$$

where we take by convention $\max \emptyset = -1$. For example, $\emptyset \lhd \{0\} \lhd \{0,1\} \lhd
\{0,2\}$. Notice that the semantics of $\lhd$ strongly recalls the one of $\prec$ and, ultimately,
the comparison between natural numbers written in binary notation. This is definitely no coincidence, since $\lhd$ and $\prec$ coincide when natural numbers are interpreted
as von Neumann's numerals (i.e., as finite von Neumann ordinals). Exercise 6.14
asks the reader to prove this fact.

The tuples of the following set will encode transitive sets with n elements:

$$\mathsf{P}_n = \left\{ \langle s_1, s_2, \ldots, s_{n-1} \rangle \mid s_1 = \{0\} \wedge \bigwedge_{i \in \{1, \ldots, n-2\}} (s_i \lhd s_{i+1} \wedge s_{i+1} \subseteq \{0, \ldots, i\}) \right\}.$$

We say that the elements of P_n are *Peddicord set systems*.

---

[2]One easily sees that the Ackermann encoding of a simple set such as $\{\cdots \{\emptyset\} \cdots\}$ requires an
exponential number of bits, as discussed also in [52].

Given a transitive set x with n elements, its Peddicord set system $p(x)$ is obtained by assigning to each $y \in x$ the set s_y of indices that the elements of y take in the Ackermann order restricted to the elements of x. More precisely:

1. Let the elements of x, listed in Ackermann order, be $y_0, y_1 \ldots, y_{n-1}$.
2. For each $i \in \{1, \ldots, n-1\}$, let s_i be such that, for all j, $j \in s_i$ if and only if $y_j \in y_i$.
3. The Peddicord set system of x is $p(x) = \langle s_1, s_2, \ldots, s_{n-1} \rangle$.

Example 6.3 Consider the transitive set x of cardinality 5, whose membership graph is drawn below.

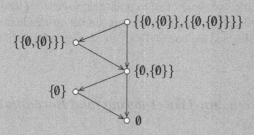

The labeling of the five elements of x by $y_0, \ldots, y_4$, shown below on the left, is according to their Ackermann order. The Peddicord set system of x is $p(x) = \langle \{0\}, \{0, 1\}, \{2\}, \{2, 3\} \rangle$, as drawn below on the right.

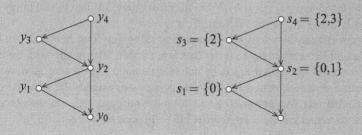

The range of the encoding p on sets of cardinality n is indeed P_n, since, first, for any transitive set, s_1 is always $\{0\}$; second, for all $i \in \{1, \ldots, n-2\}$, $s_i \lhd s_{i+1}$ holds because $y_i \prec y_{i+1}$; and, third, $s_{i+1} \subseteq \{0, \ldots, i\}$ holds because $u \in v$ implies $u \prec v$ for any well-founded sets u and v. The following theorem shows that p is indeed a one-to-one mapping.

Theorem 6.5 *For any $n \geqslant 2$ and any tuple $\langle s_1, \ldots, s_{n-1} \rangle \in \mathsf{P}_n$, there is a unique transitive set x of cardinality n such that $p(x) = \langle s_1, \ldots, s_{n-1} \rangle$.*

Proof The injectivity of p is clear by construction. For surjectivity, we construct the transitive set $x = \{y_0, \ldots, y_{n-1}\}$ by setting $y_0 = \emptyset$, $y_1 = \{y_0\} = \{\emptyset\}$ and, for every $i \in \{2, \ldots, n-1\}$, setting $y_i = \{y_j : j \in s_i\}$. All y_i's are pairwise distinct since $s_i \lhd s_{i+1}$ holds for all $i \in \{1, \ldots, n-2\}$. Thus the set x has cardinality n, and it is transitive since $y_i \subseteq x$ holds by construction for all $i \in \{0, \ldots, n-1\}$.   ⊣

Using the algorithm outlined in Sect. 6.2.2, and observing that, having the Ackermann order on the elements of x, the construction of the sets $s_1, \ldots, s_{n-1}$ can be easily carried out, we have that the encoding p takes $O(m)$ time to compute.

Corollary 6.1 *Given a transitive set x whose membership graph has m arcs, its Peddicord set system $p(x)$ can be computed in time $O(m)$.*

As a historic note, we would like to point the reader to [106] for an encoding of labeled directed acyclic graphs, which is similar to Peddicord's. In fact, both Peddicord's encoding and the one of [106] are similar to Prüfer's 1918 encoding of labeled trees [96].

6.3.2 An Ackermann-Like Ordering (and Encoding) of Hypersets

We can easily realize (cf. also [30, 57, 93]) that an extension of the Ackermann order to the entire collection $\mathsf{HF}^{1/2}$ cannot be carried out naively hoping to maintain the anti-lexicographic property (i.e., Proposition 3.1). To see this, consider the following distinct hypersets $h = \{h'\}$, $h' = \{h, \emptyset\}$. Since $\emptyset \prec h$, we have $\max_{\prec}\{x : x \in h' \setminus h\} = h$ and $\max_{\prec}\{x : x \in h \setminus h'\} = h'$. Proposition 3.1 then would imply $h \prec h'$ if and only if $h' \prec h$, a contradiction.

We now discuss how to apply the splitting technique from Sect. 6.2.1 to $\mathsf{HF}^{1/2}$. Even though we will obtain an order that extends the Ackermann order on HF, we should warn the reader that this is by no means unique. On the one hand, our order will depend on a rank notion for $\mathsf{HF}^{1/2}$ hypersets that extends the standard one for well-founded sets. We propose one such notion on page 192. On the other hand, different strategies can be used to order $\mathsf{HF}^{1/2}$ hypersets; see, e.g., [57].

6.3.2.1 The Order on Non-well-founded Hereditarily Finite Sets

Let us say that a linear order $\prec$ is *an Ackermann order* if it extends the Ackermann order of HF to a superset of HF. In order to get such an order on $\mathsf{HF}^{1/2}$, we will mimic the splitting process given for HF in Sect. 6.2.1. We will analogously build a sequence $(\mathscr{Y}^n)_{n \in \mathbb{N}}$ of ordered partitions $\mathscr{Y}^n = \{Y_i^n : i \in \mathbb{N}\}$ of $\mathsf{HF}^{1/2}$, where each partition $\mathscr{Y}^{n+1}$ is an ordered refinement of $\mathscr{Y}^n$. The $\mathscr{Y}^n$'s are constructed inductively again, starting with an $\mathscr{Y}^0$ which, by way of first approximation, is taken arbitrarily; as we will see, a linear order on $\mathsf{HF}^{1/2}$ will result as the limit of the sequence $(\mathscr{Y}^n)_{n \in \mathbb{N}}$ if all blocks in $\mathscr{Y}^0$ are finite. One further restraint must be met by $\mathscr{Y}^0$ so that this $\prec$ be an Ackermann order on $\mathsf{HF}^{1/2}$ (see page 194).

At step $n + 1$, the ordered partition $\mathscr{Y}^{n+1}$ is defined as a refinement of $\mathscr{Y}^n$, in complete analogy with the splitting action exploited in the well-founded case. We say that a block Y_i^n *can be split* if it contains two nonequivalent elements with respect to the relation $\sim_\ni$ defined (as above) by

$$x \sim_\ni y \Leftrightarrow \forall j \left(Y_j^n \cap x = \emptyset \leftrightarrow Y_j^n \cap y = \emptyset \right). \qquad (6.6)$$

By considering the smallest number h such that Y_h^n can be split, and the partition of the block Y_h^n induced by $\sim_\ni$, we proceed exactly as before to sort the $\sim_\ni$-equivalence classes of Y_h^n as $Z_0, Z_1, \ldots, Z_m$ $(m \geq 1)$. Then we put:

$$Y_i^{n+1} = \begin{cases} Y_i^n & \text{if } i < h, \\ Z_{i-h} & \text{if } h \leq i \leq h + m, \\ Y_{i-m}^n & \text{if } h + m < i. \end{cases} \qquad (6.7)$$

In sight of getting a linear order of $\mathsf{HF}^{1/2}$, we define as before the dyadic relation

$$x \prec y \Leftrightarrow \exists n \left(f(x, n) < f(y, n) \right) \qquad (6.8)$$

over $\mathsf{HF}^{1/2}$ in terms of the function $f : \mathsf{HF}^{1/2} \times \mathbb{N} \longrightarrow \mathbb{N}$ such that $x \in Y_{f(x,n)}^n$.

However, as we see in the following example, the relation $\prec$ is not necessarily a linear order.

> *Example 6.4* Suppose $\mathscr{Y}^0 = \{\mathsf{HF}^{1/2}\}$, $x = \Omega$, and $y = \{\emptyset, \Omega\}$ for the unique hyperset Ω introduced in (2.1). Then $f(x, 2) < f(y, 2)$ and hence $x \prec y$. As is easily proved by induction, for all n the class $Y_{f(x,n)}^n$ contains, besides x, the sequence $\emptyset^n, \emptyset^{n+1}, \emptyset^{n+2}, \ldots$ where $\emptyset^1 = \emptyset$ and $\emptyset^{n+1} = \{\emptyset^n\}$. It follows that $f(x, n)$ coincides with the smallest index h such that Y_h^n can be split. This implies that the non-singleton class $Y_{f(y,n)}^n$ is never split, and if $z \in Y_{f(y,n)}^n \setminus \{y\}$, then neither $z \prec y$ nor $y \prec z$ holds.

We next give a necessary and sufficient condition for the relation $\prec$ defined by (6.8) to be a linear order on $\mathsf{HF}^{1/2}$:

Lemma 6.2 *The relation $\prec$ is a linear order if and only if*

$$\forall x, y \in \mathsf{HF}^{1/2} \left(\forall n \left(f(x, n) = f(y, n) \right) \to \forall n \, \forall j \left(Y_j^n \cap x = \emptyset \leftrightarrow Y_j^n \cap y = \emptyset \right) \right)$$

$$(6.9)$$

(i.e., iff any sets x, y in $\mathsf{HF}^{1/2}$ that remain forever together in the same block never differ).

Proof Condition (6.9) is clearly necessary so that $\prec$ be a linear order.

Conversely, suppose (6.9) holds. Preliminary to proving that $\prec$ is a linear order, observe that $\prec$ is irreflexive and transitive; hence we must only prove that when $x \neq y$ holds, there exists $n \in \mathbb{N}$ such that $f(x, n) \neq f(y, n)$. This in turn follows from the fact that the relation $\flat \subseteq \mathsf{HF}^{1/2} \times \mathsf{HF}^{1/2}$ defined by

$$x \flat y \Leftrightarrow \forall n \left(f(x, n) = f(y, n) \right)$$

is a bisimulation. To see this, suppose $x \flat y$ and $x' \in x$; then $x' \in Y^n_{f(x',n)} \cap x$ for all $n \in \mathbb{N}$. By (6.9) we obtain that also $Y^n_{f(x',n)} \cap y \neq \emptyset$ for all $n \in \mathbb{N}$. Since y is a finite set, from $Y^n_{f(x',n)} \cap y \neq \emptyset$ for all $n \in \mathbb{N}$, we deduce the existence of an element $y' \in y$ belonging to all classes $Y^n_{f(x',n)}$. This implies that $x' \flat y'$.

Likewise, $x \flat y$ and $y' \in y$ implies the existence of an $x' \in x$ such that $x' \flat y'$. $\dashv$

One simple (and natural) choice to achieve condition (6.9) of Lemma 6.2 is to start the splitting process from a partition composed by finite sets, as the following Corollary shows.

Corollary 6.2 *If* $\mathscr{Y}^0 = \{Y^0_i : i \in \mathbb{N}\}$, *where every* Y^0_i *is finite, then* $\prec$ *linearly orders* $\mathsf{HF}^{1/2}$.

Proof Using Lemma 6.2 we can prove that $\prec$ is a linear order by proving that (6.9) holds. Assume x and y are such that there exist a stage n and a position j such that $Y^n_j \cap x = \emptyset$ and $Y^n_j \cap y \neq \emptyset$. If x and y belong to the same class Y^n_i, it follows from our hypothesis on $\mathscr{Y}^0$ that at stage n the number of elements belonging to classes preceding Y^n_i is finite. This is sufficient to guarantee that x and y will be eventually separated. $\dashv$

Corollary 6.2 ensures the existence of infinitely many linear orders on $\mathsf{HF}^{1/2}$ built up using the splitting procedure. Among them, as we will show below, we find an Ackermann order when the first partition $\mathscr{Y}^0$ is defined by resorting to a suitable notion of *rank*.

6.3.2.2 A Rank Notion for $\mathsf{HF}^{1/2}$

In order to be able to apply Corollary 6.2, we now want to define a suitable notion of *rank* for the hereditarily finite hypersets $\mathsf{HF}^{1/2}$.

We want our definition to have the following features:

- on the one hand, when specialized to the case of well-founded sets, it must coincide with the standard notion given as Definition 3.1 in Chap. 3;
- on the other hand, in order to work as initialization for our splitting procedure, it must guarantee that only a finite number of hypersets have any given finite rank.

In the well-founded case, the notion of rank is easily rendered graph theoretically: it corresponds to the maximum length of a path issuing from the point x in the pointed membership graph of x. Hence, sets of a given finite rank correspond to extensional acyclic graphs in which the longest path has a bounded length. In order to preserve, for hypersets, the finiteness-by-rank property—i.e., only a finite number of hyperextensional graphs can be built, for any finite given rank—we could try simply putting a bound on the length of a longest path in the membership graph of the hyperset.

Example 6.5 shows that this issue is delicate: even for rigid graphs—that is, graphs having the identity as unique automorphism, recall Sect. 3.4—there are infinitely many graphs in which the length of a longest (simple) path is bounded by the same number.

Example 6.5 Consider the graph below for arbitrarily large n.

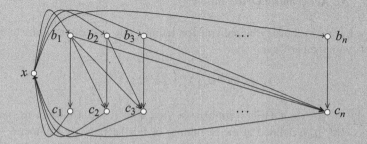

The maximum length of a path from x in the graph is two. Moreover, as one easily sees, the graph is rigid.

The above example suggests that if we want to limit the number of (pointed) membership graphs that we are able to build at any given rank, the notion of rank must not only limit paths in their *length*, but—for example—must also put a limit on the number of outgoing arcs from any given vertex.

A possible definition of rank satisfying all of our constraints is the one given below. Given $x \in \mathsf{HF}^{1/2}$ and $y \in \mathsf{trCl}(\{x\})$, we denote by $\overline{d}(x, y)$ the length of a longest path from x to y in the pointed membership graph of x.

Definition 6.1 For any $x \in \mathsf{HF}^{1/2}$, the *rank* $\mathsf{rank}(x)$ of x is[a]

$$\max \left\{ \overline{d}(x, y), \log^*(|y|) : y \in \mathsf{trCl}(\{x\}) \right\}.$$

The key properties relative to the above definition are stated as the two lemmas below, whose proofs are left to the reader (Exercises 6.16 and 6.17).

---

[a]The function $\log^*(\cdot)$ stands for the iterated (binary) logarithm.

Lemma 6.3 *The following bound holds:*

$$\left|\{x \in \mathsf{HF}^{1/2} \mid \mathrm{rank}(x) \leqslant n\}\right| \leqslant (\beth(n))^n.$$

Lemma 6.4 *For all $x \in \mathsf{HF}$, we have that*

$$\max\left\{\overline{d}(x,y), \log^*(|y|) : y \in \mathrm{trCl}\left(\{x\}\right)\right\} = \sup\left\{\mathrm{rank}(y) + 1 : y \in x\right\}.$$

Remark 6.1 Notice that the graph in Example 6.5 is *not* hyperextensional. The following question is as yet open: given $r \in \mathbb{N}$, is it true that there exist only finitely many *hyperextensional* graphs in which the longest path has length less than or equal to r?

6.3.2.3   An Ackermann Order on $\mathsf{HF}^{1/2}$

We now have all necessary ingredients for defining our Ackermann order on $\mathsf{HF}^{1/2}$. We start with the partition

$$\mathscr{Y}^0 = \{Y_i^0 : i \in \mathbb{N}\},$$

where $Y_i^0 = \{x : x \in \mathsf{HF}^{1/2} \mid \mathrm{rank}(x) = i\}$, for all $i \geqslant 0$. Consider the splitting sequence $(\mathscr{Y}^n)_{n \in \mathbb{N}}$ defined at the beginning of this section. By Lemma 6.3, each class Y_i^0 contains a finite number of hypersets, and from Corollary 6.2 it follows that the order $\prec$ defined by

$$x \prec y \Leftrightarrow \exists n \in \mathbb{N}\big(f(x,n) < f(y,n)\big)$$

is a linear order on $\mathsf{HF}^{1/2}$. Since the construction is a generalization of the splitting procedure on HF, and well-founded sets only have well-founded sets as elements, the order $\prec$ extends the Ackermann order on HF.

Example 6.6 Consider the hypersets $a = \{b\}$, $b = \{c,a\}$, $c = \{a,\emptyset\}$, $d = \{e\}$, $e = \{f\}$, $f = \{d,\emptyset\}$ depicted below

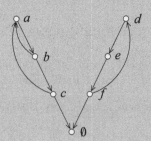

(continued)

Example 6.6 (continued)

The sets a, d have rank equal to 3, while b, c, e, f have rank equal to 2. Hence a, d belong to Y_3^0, while b, c, e, f belong to Y_2^0. The splitting procedure goes as follows:

$$[\emptyset, \ldots] \ldots [b, c, e, f \ldots][a, d \ldots] \ldots$$

$$[\emptyset] \ldots [e \ldots][c, f \ldots][b \ldots][a, d \ldots] \ldots$$

$$. [\emptyset] \ldots [e \ldots][c, f \ldots][b \ldots][d \ldots][a \ldots] \ldots$$

$$[\emptyset] \ldots [e \ldots][f \ldots][c \ldots][b \ldots][d \ldots][a \ldots] \ldots$$

Hence, the final order $\prec$ on a, b, c, d, e, f satisfies

$$\emptyset \prec e \prec f \prec c \prec b \prec d \prec a$$

Remark 6.2 Notice that the extended Ackermann order $\prec$ resulting from the above construction is by no means unique. Arguing as in the preceding section, in fact, we see that the splitting process could have started with any partition $\mathscr{Y}^0 = \{Y_i^0 : i \in \mathbb{N}\}$ composed of finite sets Y_i^0 with $Y_i^0 \supseteq \{x : x \in \mathsf{HF} \mid \mathrm{rank}(x) = i\}$: the limit of the sequence $(\mathscr{Y}^n)_{n \in \mathbb{N}}$ would then have been an Ackermann order as well.

The above remark suggests that in the presence of hypersets, the splitting procedure can be grounded on different notions of *rank*.

6.3.2.4   Hereditarily Finite Hypersets as Dyadic Numbers

Making use of the Ackermann order $\prec$ on $\mathsf{HF}^{1/2}$ obtained as explained above, we now propose an encoding of $\mathsf{HF}^{1/2}$ hypersets by numbers. Traditional sets, HF, will retain in our bijection the same images as before; together with those images, which span all natural numbers, the images of hereditarily finite hypersets will span the set of all *dyadic rationals* $\mathbb{Q}_2$:

$$\mathbb{Q}_2 = \left\{ \frac{n}{2^m} : n, m \in \mathbb{N} \right\},$$

namely, the set of all rational numbers whose binary expansion requires a finite number of digits.

The choice of this numeric domain stems from the rationale that we want the membership relation to be readable, as before, from the binary representation of numbers. In fact, the encoding $\mathbb{N}_A$ allows us to say that a 1 in jth position of the binary expansion of $\mathbb{N}_A(h)$ reflects the fact that if $\mathbb{N}_A(h') = j$, then $h' \in h$. The map $\mathbb{Q}_A$ of hypersets to dyadic rational numbers that we are going to introduce will generalize this fact by adding the property that a 1 in kth *decimal* position $\mathbb{Q}_A(h)$ reflects the fact that the kth non-well-founded set h' is an element of h.

Natural numbers have a twofold purpose in the Ackermann encoding, being used both as *positions* inside a code $\mathbb{N}_A(a)$ and as the code itself. Our strategy in defining $\mathbb{Q}_A$ consists in separating these two functions. This will be done by introducing *two* functions, one assigning a position to each hereditarily finite hyperset and the other assigning a code to it. A function $\mathbb{Z}_A$ will map hypersets into *integers*, using natural numbers for well-founded sets and negative integers for (proper) hypersets. As a by-product we naturally obtain a *(dyadic) rational number* $\mathbb{Q}_A(h)$ as code of h with the property that

there is a 1 in position $\mathbb{Z}_A(h')$ of the binary expansion of $\mathbb{Q}_A(h)$ if and only if $h' \in h$,

with both $\mathbb{Z}_A$ and $\mathbb{Q}_A$ having $\mathbb{N}_A$ as their restriction to HF.
     If $a \in \mathsf{HF}^{1/2}$, define:

$$\mathbb{Z}_A(a) = \begin{cases} |\{b : b \in \mathsf{HF} \mid b \prec a\}| & \text{if } a \in \mathsf{HF}, \\ -|\{b : b \in \mathsf{HF}^{1/2} \setminus \mathsf{HF} \mid b \prec a\}| - 1 & \text{if } a \in \mathsf{HF}^{1/2} \setminus \mathsf{HF}. \end{cases}$$

We now define the bijection $\mathbb{Q}_A$ from $\mathsf{HF}^{1/2}$ to dyadic numbers as follows:

$$\mathbb{Q}_A(a) = \Sigma_{b \in a} 2^{\mathbb{Z}_A(b)}.$$

We have thus obtained that:

- $\mathbb{Q}_A : \mathsf{HF}^{1/2} \to \mathbb{Q}_2$ extends the Ackermann function $\mathbb{N}_A : \mathsf{HF} \to \mathbb{N}$; that is, $\mathbb{Q}_A(x) = \mathbb{N}_A(x)$ holds when $x \in \mathsf{HF}$.
- A simple reading of the code $y \in \mathbb{Q}_2$ allows us to inductively determine the elements of the set a such that $y = \mathbb{Q}_A(a)$. This is because from the digits of y we determine the positions $\mathbb{Z}_A(b)$ of all $b \in a$, and since the bijection $\mathbb{Z}_A$ is effective, from $\mathbb{Z}_A(b)$ we are able to determine b.
- A simple recursive routine manipulating sets allows us to build the code $y = \mathbb{Q}_A(a)$ from any given hereditarily finite hyperset a, because if we know a we can compute $\mathbb{Z}_A(b)$ for all $b \in a$ and hence $\mathbb{Q}_A(a)$.

Panel 6.1 discusses another possible extension of Ackermann's encoding to hypersets, which takes *real* values.

Panel 6.1 A real-valued encoding of sets?
Consider the following definition of a map, which results from the definition of $\mathbb{N}_A$ when a minus sign is put in each exponent.

$$\mathbb{R}_A(x) = \Sigma_{y \in x} 2^{-\mathbb{R}_A(y)}$$

$\mathbb{R}_A$ bears a strong formal similarity with $\mathbb{N}_A$ but calls into play real numbers. This shift allows us to prove the existence of a unique solution to the following equation:

$$x = 2^{-x}, \tag{6.10}$$

putting us in a condition to view $\mathbb{R}_A$ as a real-valued map whose domain can be, for example, rational hypersets.

To see the existence and uniqueness of a solution for (6.10), it suffices to observe that the two curves $y = x$ and $y = 2^{-x}$ are increasing and decreasing, respectively, and they intersect only in the first quadrant of the Cartesian plane.

It is not known whether $\mathbb{R}_A$ is injective on either HF or $\text{HF}^{1/2}$. In [25] a few properties of $\mathbb{R}_A$ are proved (see also Exercises 6.19, 6.20, and 6.21) and the following two conjectures are put forward.

Conjecture 6.1 The function $\mathbb{R}_A$ is injective on HF.

Conjecture 6.2 The function $\mathbb{R}_A$ is injective on $\text{HF}^{1/2}$.

Exercises

6.1 Let x be a transitive well-founded set of cardinality n such that

$$|\{y \in x \,|\, \text{for all } z \in x, \; y \notin z\}| = s.$$

Show that there are exactly 2^{n-s} different well-founded sets whose transitive closure is x.

6.2 Recall Problem 6.4 in which $c_{n,s}$ denoted the number of non-isomorphic extensional acyclic graphs with n vertices, out of which s are sources. Express the number of well-founded sets whose transitive closure has cardinality n in terms of $c_{n,s}$.

6.3 Implement a program for computing the values from Theorem 6.1.

6.4 Implement a program for computing the values from Theorem 6.3.

6.5 Implement a program for computing the values from Theorem 6.4.

6.6 Having implemented the program for the values from Theorem 6.4, what can you say about $s_n/c_{n,s}$ as n grows? Can you observe other relations between the recurrence relations in this chapter? See [112] for proofs of some relations between these quantities.

6.7 Adapt the proof of Theorem 6.2 for writing a recurrence relation for the number of directed acyclic graphs on vertex set $\{1, \ldots, n\}$.

6.8 Given n and $s \leqslant n$, find a recurrence relation for the number of directed acyclic graphs on vertex set $\{1, \ldots, n\}$ having exactly n sources.

6.9 A directed acyclic graph is called *essential* [4, 107] if for every arc $\langle u, v \rangle \in E(G)$, it holds that $N^-(u) \neq N^-(v) \setminus u$. Derive a recurrence relation for the number of essential directed acyclic graphs on n vertices. *Hint.* Define the *depth* of a vertex v in a directed acyclic graph as the length of a longest directed path from a source to v. Count by vertices of maximum depth.

6.10 (*) Given n and m, find the number of extensional acyclic graphs with n vertices and m arcs. *Hint.* Try counting also by rank.

6.11 Show that the partition refinement algorithm described on p. 186 for computing the Ackermann order on the elements of a transitive set is correct.

6.12 By applying the technique outlined in Example 6.1, prove that the set $\{\{x : x \in p_n \mid x \cap p_n = \emptyset\} : n \in \mathbb{N}\}$ is the partition of HF whose blocks are the rank-equality classes.

6.13 Consider, for $n \in \mathbb{N}$, the partition $\{X_h^n : h \in \mathbb{N}\}$ inductively defined by (6.5) and prove, inductively on n, that

- the smallest index h such that X_h^n can be split coincides with the smallest index h such that X_h^n is not a singleton;
- the relation $\sim_\ni$ induces a partition of X_h^n into singleton blocks.

6.14 Prove that the relation $\lhd$ defined on p. 188 coincides with $\prec$ when natural numbers are interpreted as von Neumann's numerals.

6.15 (*) Derive a direct counting recurrence for the cardinality of the set

$$P_n = \left\{ \langle s_1, s_2, \ldots, s_{n-1} \rangle \mid s_1 = \{0\} \wedge \bigwedge_{i \in \{1, \ldots, n-2\}} (s_i \prec s_{i+1} \wedge s_{i+1} \subseteq \{0, \ldots, i\}) \right\},$$

without exploiting its one-to-one correspondence with transitive sets with n elements. *Hint.* See [99].

6.16 Prove Lemma 6.3.

6.17 Prove Lemma 6.4.

6.18 (*) Consider the problem of finding a sort of *inverse* of $\mathbb{Q}_A$, namely, a bijection $h : \mathbb{Q}_2 \to \mathsf{HF}^{1/2}$ that extends von Neumann's injection of $\mathbb{N}$ into HF so that

$$\forall a \in \mathbb{Q}_2 \ \forall x \in h(a) \ \exists b \in \mathbb{Q}_2 \ (b \leqslant a \wedge x = h(b)).$$

Prove that no such h exists.

6.19 Prove that the map $\mathbb{R}_A$ defined in Panel 6.1 assumes arbitrarily large values.

6.20 Prove that for all $i \in \mathbb{N}$:

- $\mathbb{R}_A(h_i) \neq \mathbb{R}_A(h_{i+1})$;
- $\mathbb{R}_A(h_i) \neq \mathbb{R}_A(h_{i+2})$.

6.21 (*) Prove that $\mathbb{R}_A$ is injective on the sets $\emptyset^i$ (as defined in Example 6.4), for all $i \in \mathbb{N}$.

Chapter 7
Random Generation of Sets

In this chapter we study how to generate a well-founded set of "size" n, uniformly at random. To wit, we will see algorithms that, given n, produce a well-founded set of size n at random, so that each set of size n has equal probability to occur. Procedures of this kind can be of use for testing the correctness of algorithm implementations or for testing conjectures about data that is reasonably represented by well-founded sets.

As has emerged in the previous chapter, we can measure the "size" of a well-founded set by different criteria: we can opt either for the cardinality of its transitive closure or for its rank. However, the latter notion of size poses no problems, as Exercise 7.7 asks the reader to show. We will thus consider well-founded sets x whose transitive closures have cardinality n. Furthermore, we will simplify our exposition in the same manner as we have done in the previous chapter, by assuming that x is transitive, namely, that $x = \mathrm{trCl}\,(x)$. (Exercise 7.8 asks the reader to solve the random generation problem without this transitivity assumption.)

This chapter also serves the purpose of describing three general methods of generating combinatorial objects uniformly at random. The first two of these, to be discussed in Sect. 7.1, are based on a so-called combinatorial decomposition of the objects. In Sect. 6.1 we have seen various such decompositions of transitive well-founded sets, and our running example will be the one by rank, requested by Problem 6.2. The third random generation method, to be presented in Sect. 7.2, is based on a Markov chain. Such technique has the advantage that it requires less insight into the structure of the objects to be generated. It suffers a major drawback, though, namely, that it is often hard to establish for how long the Markov chain stochastic process must be iterated in order to ensure that objects are generated with *uniform* probability distribution. In fact, this chapter falls short of providing such a bound.

© Springer International Publishing AG 2017

E.G. Omodeo et al., *On Sets and Graphs*, DOI 10.1007/978-3-319-54981-1_7

7.1 Approaches Based on Combinatorial Decompositions

In this section we will introduce two simple methods for generating uniformly at random a transitive well-founded set of cardinality n. They are based on a so-called combinatorial decomposition of such a set, namely, on a description of how a transitive well-founded set is made up of smaller transitive well-founded sets. In Sect. 6.1 we already used such decompositions for counting sets.

We start by giving a high-level overview of these two random generation methods, and in Sects. 7.1.1 and 7.1.2, we will adapt them to our task at hand.

Both methods are best understood if they are first explained on a tree. This tree will be a conceptual representation of the recursive description of a transitive well-founded set, very much similar to the recursion tree of evaluating the recurrence formulas from Chap. 6.1. Every leaf of this tree will correspond to a transitive well-founded set, thus generating one set at random is equivalent to randomly selecting one leaf of this conceptual tree.

For example, recall the recurrence

$$s_n^r = \sum_{k=1}^{n-r} s_{n-r}^k \binom{(2^k - 1)2^{n-r-k}}{r}, \qquad \text{where } s_1^1 = 1. \tag{7.1}$$

on page 177, about the number of transitive well-founded sets x with $|x| = n$ and with r elements of maximum rank $\mathsf{rank}(x) - 1$. Removing these r elements of maximum rank from x, we are left with a set x' that can have $k = 1, 2, \ldots, n-r$ elements of maximum rank $\mathsf{rank}(x') - 1$. The quantity $s_{n-r}^k \binom{(2^k-1)2^{n-r-k}}{r}$ tells us how many such sets x' exist. It also tells us, implicitly, how each set x is constructed from x'.

The tree T_n^r capturing this combinatorial decomposition (7.1) of s_n^r is defined recursively:

- The root of T_n^r is labeled with s_n^r, which is the number of leaves of T_n^r.
- The root of T_n^r has $\sum_{k=1}^{n-r} \binom{(2^k-1)2^{n-r-k}}{r}$ children: for every $k \in \{1, \ldots, n-r\}$, it has $\binom{(2^k-1)2^{n-r-k}}{r}$ children that are roots of a different copy of T_{n-r}^k (and thus labeled with s_{n-r}^k). The order of the children will be fixed, but ultimately irrelevant for our purpose.
- T_1^1 consists of a single node labeled with 1, because $s_1^1 = 1$, and if some s_n^r is equal to 0 then T_n^r is empty.

Each leaf of T_n^r thus corresponds to a transitive well-founded set x with $|x| = n$ and with r elements of maximum rank $\mathsf{rank}(x) - 1$. Example 7.1 illustrates such a tree.

Example 7.1 Let $n = 4$ and $r = 1$. Recall that s_n^r stands for the number of transitive well-founded sets x with $|x| = n$ and with r elements of maximum rank $\text{rank}(x) - 1$, and T_n^r is the conceptual tree of the combinatorial decomposition given by recurrence (7.1). We have:

$$s_4^1 = \binom{(2^1 - 1)2^{4-1-1}}{1} s_3^1 = \binom{(2^1 - 1)2^{4-1-1}}{1} \binom{(2^1 - 1)2^{3-1-1}}{1} s_2^1.$$

The tree T_4^1 capturing this recurrence is the following one.

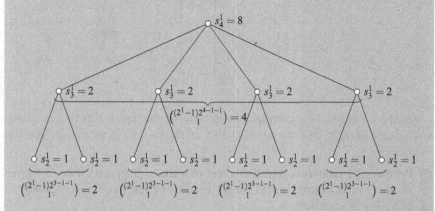

A transitive well-founded set x with n elements can have $1, 2, \ldots$ vertices of maximum rank $\text{rank}(x) - 1$. We can now construct a tree T_n with root r, whose children are the roots of the trees $T_n^1, T_n^2, \ldots$; thus, every transitive well-founded set with n elements now corresponds to a leaf of T_n. Selecting a random leaf of T_n is equivalent to selecting a transitive well-founded set x with n elements.

We depict below the tree T_4, where each node v is labeled by the number of leaves in the subtree rooted at v. Below each leaf we also draw the transitive well-founded set x to which that leaf corresponds. We draw in bold one arbitrary path from the root to a leaf.

(continued)

Example 7.1 (continued)

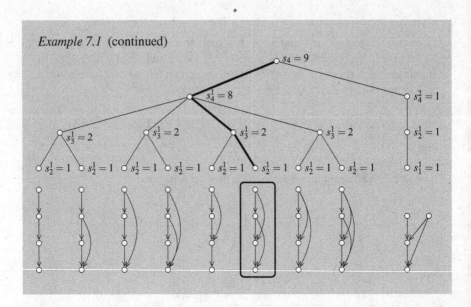

Having established the correspondence between sets and leaves of a combinatorial decomposition tree, we can now abstract the random generation problem as follows.

> **Problem 7.1 (Random selection a leaf of a tree)**   Suppose that we are given the root r of an ordered tree T. For any node v of T, the tree T supports only the following two queries:
>
> - list the children of v in the order in which they appear in T;
> - print the number $s(v)$ of leaves in the subtree of T rooted at v.
>
> Select a leaf v of T uniformly at random, by a traversal of T from r to v.

In the remaining part of this section, we describe, based on Problem 7.1, the two methods that we will use for generating a transitive well-founded set uniformly at random. The first one is called the *recursive method*, and we will use it in Sect. 7.1.1. The second one is called the *ranking and unranking* method, and we will use it in Sect. 7.1.2.

In the recursive method, we start at the root r of T and list the children of r, say $v_1, \dots, v_k$. For each v_i we get the number $s(v_i)$ of leaves in the subtree of T rooted at v_i. In order to guarantee the uniform distribution among the leaves, we choose one v_i with probability proportional to

$$\frac{s(v_i)}{s(v_1) + \cdots + s(v_k)} = \frac{s(v_i)}{s(r)}.$$

We then descend to v_i and iterate this process until reaching a leaf. See also Example 7.2.

Example 7.2 Consider the tree T depicted below, in which each node v is labeled by the number $s(v)$ of leaves in the subtree rooted at v. We must pick uniformly at random one among the six leaves of T.

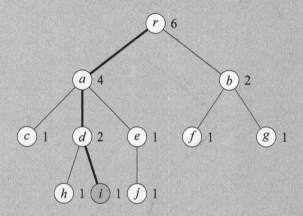

We start the random generation process from the root r, by picking at random between a, with probability $\frac{4}{6}$, and b, with probability $\frac{2}{6}$. Suppose we have chosen a; we now randomly choose among c, with probability $\frac{1}{4}$, d, with probability $\frac{2}{4}$, and e, with probability $\frac{1}{4}$. Supposing we have chosen d, then one of its two leaves must be picked with the same probability, $\frac{1}{2}$. Finally, we pick leaf i, which is indeed generated with uniform probability:

$$\frac{4}{6} \cdot \frac{2}{4} \cdot \frac{1}{2} = \frac{1}{6}.$$

The main idea of the ranking and unranking method is to construct a bijection, also called a *ranking*, between the $s(r)$ leaves of T and the numbers $\{0, \ldots, s(r)-1\}$. Then we can generate a random number z in $\{0, \ldots, s(r)-1\}$ and *unrank* z to obtain the leaf being ranked as z.

Let, as before, $v_1, \ldots, v_k$ be the children of r, and let $s(v_i)$ be the number of leaves in the subtree of T rooted at v_i. We first check what is the index $i \in \{1, \ldots, k\}$ such that

$$z \in \left[\sum_{t=1}^{i-1} s(v_t), \sum_{t=1}^{i} s(v_t) \right).$$

Then we descend to v_i and update $z := z - \sum_{t=1}^{i-1} s(v_t)$. We iterate this step until reaching a leaf of T (or, equivalently, until $z = 0$). See also Example 7.3.

Example 7.3 Consider the tree T depicted below, in which each node v is also labeled by the number $s(v)$ of leaves in the subtree rooted at v. We must pick uniformly at random one among the seven leaves of T.

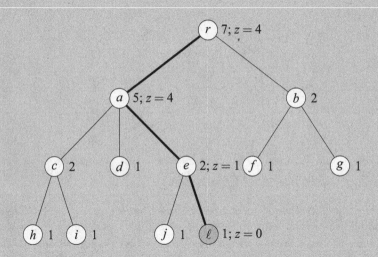

We generate uniformly at random a number $z \in \{0, \ldots, 6\}$, say $z = 4$, and start traversing T from r. Since $z \in [0, 5)$, we need to descend to a and update z as $z := z - 0 = 4$. We then have that $z \in [2+1, 2+1+2)$; thus, we must descend to e and update $z := z - (2+1) = 1$. Finally, we have $z \in [1, 1+1)$, we descend to ℓ and update $z := z - 1 = 0$. Leaf ℓ is generated with uniform probability because z was generated with uniform probability.

Let us now give a formal statement of the main task of this chapter.

> **Problem 7.2 (Random generation of transitive sets)**   Given an integer n, produce a transitive well-founded set with n elements uniformly at random.

We are now ready to apply these two methods directly to Problem 7.2, without explicitly building a decomposition tree.

7.1.1   The Recursive Method

We now apply the recursive method to Problem 7.2, using the decomposition of a transitive set by elements of maximum rank.

Given n, we recursively generate a transitive set X with n elements as in Algorithm 7.1. We first choose the number r of elements of X of maximum rank. This is done on line 2 of Algorithm 7.1, by choosing $r \in \{1, \ldots, n\}$ at random with probability

$$\frac{s_n^r}{s_n^1 + \cdots + s_n^n}.$$

We then recursively call the random generation function to obtain a transitive well-founded set Y with $n - r$ elements, whose elements of maximum rank are in stored in $Z \subseteq Y$. We need to add to X its r elements $\{v_1, \ldots, v_k\}$ of maximum rank. Each v_i thus consists of a *nonempty* subset of Z and an arbitrary subset of $Y \setminus Z$. See also Fig. 7.1 for an illustration of this process.

The key in implementing Algorithm 7.1 and analyzing its complexity is in lines 2 and 4. In Exercises 7.3 and 7.4, we ask the reader for efficient algorithms for implementing these two steps. We thus obtain the following result:

Algorithm 7.1: RANDOMTRANSITIVESETRECURSIVE(n)

Input: An integer n and values s_i^r precomputed by recurrence (7.1), for all $1 \leqslant i \leqslant n$ and $1 \leqslant r \leqslant i$.

Output: A pair (X, R) consisting of a transitive set X with n elements and its subset R of elements of maximum rank.

---

1 **if** $n = 0$ **then return** $(\emptyset, \emptyset)$;

2 choose $r \in \{1, \ldots, n\}$ at random with probability $\dfrac{s_n^r}{\sum_{j=1}^{n} s_n^j}$;

3 $(Y, Z) :=$ RANDOMTRANSITIVESETRECURSIVE($n - r$);

4 generate uniformly at random pairwise distinct sets $v_1, \ldots, v_r \subseteq Y$ such that $v_i \cap Z \neq \emptyset$
   holds for every $i \in \{1, \ldots, r\}$;

5 $X := Y \cup \{v_1, \ldots, v_r\}$;

6 **return** $(X, \{v_1, \ldots, v_r\})$.

Fig. 7.1 An illustration of
the recursive method for
randomly generating a
transitive well-founded set X
with n elements; elements
$v_1, \ldots, v_r$ have maximum
rank in X. Notice the
similarity with Fig. 6.1

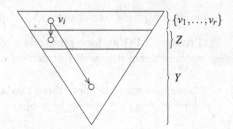

Algorithm 7.2: RANDOMTRANSITIVESETUNRANKING(n, r, z)

Input: Integers n, r, integer $z \in \{0, \ldots, s_n^r\}$, and values s_i^r precomputed by recurrence (7.1), for all $1 \leqslant i \leqslant n$ and $1 \leqslant r \leqslant n$.

Output: A pair (Z, R) consisting of a transitive set Z with n elements and its subset R of r elements of maximum rank.

1 **if** $n = 0$ **then return** $(\emptyset, \emptyset)$;

2 let $i \in \{1, \ldots, n\}$ be such that $z \in \left[\sum\limits_{k=1}^{i-1} s_n^k \binom{(2^k-1)2^{n-r-k}}{r}, \sum\limits_{k=1}^{i} s_n^k \binom{(2^k-1)2^{n-r-k}}{r} \right)$;

3 $z := z - \sum\limits_{k=1}^{i-1} s_n^r \binom{(2^k-1)2^{n-r-k}}{r}$;
4 $y := z$ **div** s_{n-r}^i;
5 $z := z$ **mod** s_{n-r}^i;
6 $(X, Y) := $ RANDOMTRANSITIVESETUNRANKING$(n-r, i, z)$;
7 $R := $ SUBSETLEX(X, Y, y);
8 **return** $(X \cup R, R)$.

Theorem 7.1 *Algorithm 7.1 generates uniformly at random a transitive well-founded set with n elements in polynomial time.*

7.1.2 Ranking and Unranking

We now apply the ranking and unranking method to Problem 7.2, using the same decomposition of a transitive set by elements of maximum rank (Algorithm 7.2). This method was first applied to transitive sets in [99], via a counting recurrence for the number of Peddicord set systems that encode them (recall Sect. 6.3 and Exercise 6.15).

Before starting Algorithm 7.2, we must decide how many elements of maximum rank does our random set have. We can do this as in the previous section, by choosing a number $r \in \{1, \ldots, n\}$ with probability proportional to

$$\frac{s_n^r}{s_n^1 + \cdots + s_n^n} = \frac{s_n^r}{s_n}.$$

We then generate uniformly at random a number $z \in \{0, \ldots, s_n^r - 1\}$ and call RANDOMTRANSITIVESETUNRANKING(n, r, z). This is equivalent to descending from the root of the decomposition tree to one of its children. This function will then generate the desired well-founded transitive set $X \cup R$ (with elements of maximum rank stored in R, $|R| = r$) which is the zth leaf in this decomposition subtree. Given z, it first decodes the number i elements of maximum rank in X, by choosing i depending on the interval to which z belongs. (Recall that for each such possible i, there are $s_n^i \binom{(2^i-1)2^{n-r-i}}{r}$ possible well-founded sets.) Afterward, it updates z as $z := z - \sum_{k=1}^{i-1} s_n^r \binom{(2^k-1)2^{n-r-k}}{r}$ and computes $y := z \operatorname{div} s_{n-r}^i$. Value y is the rank of one of the $\binom{(2^i-1)2^{n-r-i}}{r}$ possible ways of adding the r elements of maximum rank, given the well-founded transitive set X of cardinality $n - r$, with $|Y| = i$ elements of maximum rank. Then, it updates $z := z \bmod s_{n-r}^i$. See Example 7.4 for an example of this algorithm, simulated on a decomposition tree.

By SUBSETLEX(X, Y, y) we denote a function returning the yth set $\{v_1, \ldots, v_i\}$ with the property that $v_1, \ldots, v_i \subseteq X$ are all pairwise distinct and $v_j \cap Y \neq \emptyset$ holds for every $j \in \{1, \ldots, i\}$. The order among such sets $\{v_1, \ldots, v_i\}$ can be any arbitrary order, for example, the lexicographic one, when one assumes the order on the elements of X to be the Ackermann order. Exercise 7.6 asks the reader for an efficient algorithm implementing this function. We thus obtain:

Theorem 7.2 *A well-founded transitive set with n elements can be generated uniformly at random in polynomial time, using Algorithm 7.2.*

This *unranking* algorithm can be inverted into a *ranking* algorithm. Namely, given a well-founded transitive set X with n elements out of which r have maximum rank, we can write a polynomial algorithm RANDOMTRANSITIVESETRANKING(X) returning the number $z \in \{0, \ldots, s_n^r - 1\}$, such that RANDOMTRANSITIVESETUNRANKING$(n, r, z) = X$ (Exercise 7.7).

Example 7.4 We depict below the tree T_n, for $n = 4$, from Example 7.1. Each node now also has a name and is again labeled by the number of leaves in the subtree rooted at it.

The random generation method described in this section starts by choosing $r \in \{1, 2\}$ with probability s_4^r/s_4. Suppose it has chosen $r = 1$. It then chooses $z \in \{0, \ldots, s_4^1 - 1\}$ uniformly at random, say $z = 5$, and calls RANDOMTRANSITIVESETUNRANKING$(n = 4, r = 1, z = 5)$. The only possible value for i is $i = 1$. We update z as $z := z - 0 = 5$ and compute $y := z \operatorname{div} s_3^1 = 2$. We then update $z := z \bmod s_3^1 = 1$. We have to descend to the yth child of a (namely, d since $y = 2$), using the value $z = 1$.

At node d, the only possible value for i is $i = 1$. We update $z := z - 0 = 1$ and compute $y := z \operatorname{div} s_2^1 = 1$. We then update $z := z \bmod s_2^1 = 0$. Finally, we descend to the yth child of d (namely, node k since $y = 1$), using $z = 0$.

(continued)

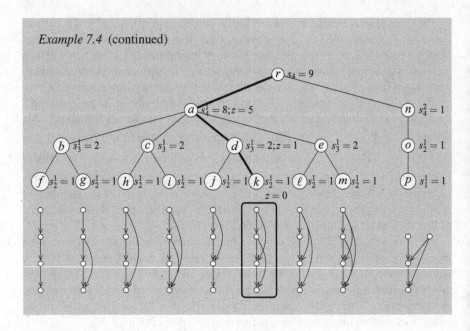

Example 7.4 (continued)

7.2 A Markov Chain Monte Carlo Approach

The idea of the Markov chain Monte Carlo random generation method is a very simple one: start with an arbitrary object X from the set $\mathscr{S}$ of objects to be randomly sampled; iteratively make small random changes to X, as long as the resulting object X belongs to $\mathscr{S}$. Such an approach is easily implementable, as e.g., it does not require discovering and evaluating recurrence formulas. However, in order for all objects from $\mathscr{S}$ to be sampled with the same uniform probability, this process must satisfy a few properties. As we will see, this part will be easy; the difficult part is deriving an upper bound (preferably prolynomial) on the number of steps that the process must be iterated in order to achieve the uniform distribution. This method was first applied to transitive sets in [95].

In fact, we do not give such a proof for the Markov chain that we will present in this section. Therefore, the two algorithms from the previous section remain the only ones for which we can guarantee that each well-founded transitive set can be generated with uniform probability, in polynomial time.

Despite these drawbacks, the simplicity of such a Markov chain Monte Carlo method makes it a more likely candidate to tackle the problem of generating hypersets uniformly at random (see also Panel 7.1). In fact, there is no known combinatorial decomposition to date of a hyperset, or a hyperextensional graph, let alone counting recurrences as studied in Sect. 6.1 for well-founded sets.

Panel 7.1   Eliminating arcs and vertices without collisions in hypersets.
As a consequence of Theorem 7.3, we can conclude that Algorithm 7.3 can explore the entire space of extensional acyclic graphs with n vertices by simply adding and removing graph elements.

Can we move among hypersets using similar strategies? For example, can we guarantee that, given a *cyclic* membership graph, there exists a vertex whose elimination does not cause any collision—i.e., bisimulation equivalence classes containing more than one vertex? What about arc elimination?

The problem has been tackled in [22] where the following membership graph is produced as an example of graph *not* admitting any vertex elimination without collisions. This means that there are hypersets in whose transitive closure not a single element can be removed without triggering a sequence of collisions. This is different from the well-founded case, where the membership graphs have at least one source, and this can be removed without creating collisions.

Arc elimination without collisions in hypersets is much more promising (see [22]) and, as a consequence, also an adaptation of the Markov chain presented in Algorithm 7.3 to hypersets, seems to viable. Exercise 7.12 asks the reader to find an arc in the membership graph membership graph from Fig. 7.2.

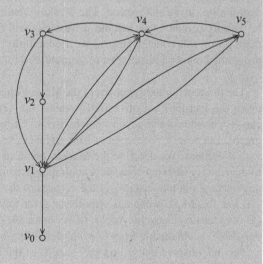

Fig. 7.2 A membership graph such that the elimination of any vertex causes (at least) a collision. For example, the elimination of v_1 causes the collision of v_2 and v_0 and the elimination of v_0 causes the collision of all other vertices (see Exercise 7.11).

We now continue with a few basic definitions.

> **Definition 7.1 (Markov chain)** A *discrete time finite stochastic process* is a sequence $X = (X_t : t \in \mathbb{N})$, where X_t are $\mathscr{S}$-valued random variables and $\mathscr{S}$ is a finite set, called the *state space* of X. We say that X is a *Markov chain* if
>
> $$\forall t \in \mathbb{N}, \ \Pr(X_{t+1} = s_{t+1} \mid X_t = s_t, \dots, X_0 = s_0) = \Pr(X_{t+1} = s_{t+1} \mid X_t = s_t).$$
>
> Moreover, a Markov chain X is said to be *time homogeneous* if
>
> $$\forall s, s' \in \mathscr{S}, \ \exists p_{ss'}, \ \forall t \in \mathbb{N}, \ \Pr(X_{t+1} = s \mid X_t = s') = p_{ss'}.$$

> **Definition 7.2 (Markov chain properties)** A time-homogeneous Markov chain over the state space $\mathscr{S}$ is said to be:
>
> - *irreducible* if $\forall s, s' \in \mathscr{S}, \ \exists t \in \mathbb{N}, \ \Pr(X_t = s' \mid X_0 = s) > 0$;
> - *aperiodic* if $\forall s \in \mathscr{S}, \ \gcd(\{t \in \mathbb{N} \mid \Pr(X_t = s \mid X_0 = s) > 0\}) = 1$;
> - *symmetric* if $\forall s, s' \in \mathscr{S}, \ p_{ss'} = p_{s's}$.
>
> In the above, $\gcd(A)$ denotes the greatest common divisor of the elements of the set A.

A well-known result (see, e.g., [55]) states that any finite, irreducible, aperiodic, and symmetric time-homogeneous Markov chain converges toward the uniform distribution on its state space. The Markov chain to be introduced below will be by definition symmetric, and the aperiodicity will follow from its irreducibility. Our main proof will concern instead its irreducibility; as mentioned above, we do not prove any bound on the number of steps needed for reaching the uniform distribution over $\mathscr{S}$.

The random iterative changes of the Markov chain are graph theoretic ones. Thus, given the fact that transitive well-founded sets are in bijection with isomorphism classes of extensional acyclic graphs (recall Sect. 3.4), we describe this chain for graphs. Let now $\mathscr{S}_n$ be the set of isomorphism classes of extensional acyclic graphs with n vertices. We describe the Markov chain in Algorithm 7.3. Note that thanks to the rigidity of extensional acyclic graphs (Lemma 3.6) and Lemma 2.1, we need not worry about the "names" of the vertices of the graph.

First, this Markov chain is symmetric: given any two isomorphism classes G' and G'' of extensional acyclic graphs, G' is obtainable from G'' with probability p if and only if G'' is obtainable from G' with probability p.

Second, observe that from every state G of the Markov chain, there is a nonzero probability of it remaining in the same state G (e.g., by choosing $u, v \in V(G)$ with $u = v$). Thus, its aperiodicity will follow from its irreducibility.

Theorem 7.3 *The Markov chain described in Algorithm 7.3 is irreducible.*

Proof Let G be an arbitrary extensional acyclic graph. We show that there is a sequence of transitions $(G = G_0, G_1, \dots, G_t)$, where G_t is isomorphic to an acyclic tournament. Since the Markov chain is symmetric, this will imply that there is a

---

Algorithm 7.3: A Markov chain for the random generation of isomorphism classes of extensional acyclic graphs with n vertices (equivalently, of transitive well-founded sets of cardinality n).

$i := 0$;
let G_i be an arbitrary extensional acyclic graph with n vertices;
repeat
    choose uniformly at random $u, v \in V(G_i)$;
    if $\langle u, v \rangle$ *is an arc of* G_i **then**
    | $G' := (V(G_i), E(G_i) \setminus \{\langle u, v \rangle\})$;
    else
    | $G' := (V(G_i), E(G_i) \cup \{\langle u, v \rangle\})$;
    end
    if G' *is an extensional acyclic graph* **then**
    | $G_{i+1} := G'$;
    else
    | $G_{i+1} := G_i$;
    end
    $i := i + 1$;
until *reaching the uniform distribution*;
return G_i.

---

nonzero probability that the Markov chain, starting in G, arrives to the isomorphism class of any other extensional acyclic graph G' with n vertices.

For any vertex $v \in V(G)$, we define

$$R(v) =_{\mathrm{Def}} \{u \in V(G) \setminus \{v\} \mid \mathrm{rank}(u) \leqslant \mathrm{rank}(v)\}$$

as the set of vertices of G different from v and of rank at most the rank v.

We order the vertices of G in decreasing order on rank, with vertices of the same rank being ordered arbitrarily. For every vertex v in this order, we add arcs to all vertices $u \in R(V)$ in decreasing order on their rank. See also Example 7.5.

Note that this is possible, first of all, because the addition of an arc $\langle v, u \rangle$ does not create a cycle in the resulting graph. Second, observe that the subgraph of G induced by the vertices $V(G) \setminus R(v)$ is an acyclic tournament. Therefore, an arc addition would create a collision only between v and a vertex $w \in R(v)$. This is however not the case, since after the first addition of such an arc, $\mathrm{rank}(v)$ becomes strictly greater than $\mathrm{rank}(w)$, for all $w \in R(v)$, which guarantees the absence of collisions. ⊣

Example 7.5 We draw below an extensional acyclic graph G_0 with six vertices. The sequence of graphs $G_0, G_1, \ldots, G_8$ is the one constructed in the proof Theorem 7.3. We highlight the vertex from which we add arcs, together

(continued)

Example 7.5 (continued)
with the newly added arc, in each G_i. The extensional acyclic graph G_8 is
isomorphic to an acyclic tournament.

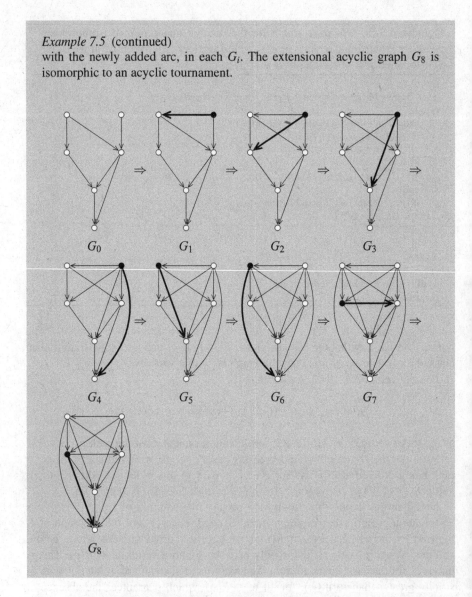

As mentioned, the symmetry and aperiodicity of the Markov chain imply the
following corollary:

Corollary 7.1 *The Markov chain described in Algorithm 7.3 converges to the
uniform distribution.*

Exercises

7.1 Give an algorithm for generating uniformly at random:

(a) a well-founded set of rank at most n;
(b) a well-founded set of rank exactly n.

Hint. Use Ackermann's encoding.

7.2 Simulate a run of the function RANDOMTRANSITIVESETRECURSIVE(n) (Algorithm 7.1) for $n = 6$.

7.3 Give an efficient algorithm for implementing line 2 of Algorithm 7.1. That is, given numbers $a_1, \ldots, a_n$, write an efficient algorithm for randomly selecting an index $i \in \{1, \ldots, n\}$, such that i is chosen with probability

$$\frac{a_i}{a_1 + \cdots + a_n}.$$

7.4 Give an efficient algorithm for implementing line 4 of Algorithm 7.1. That is, given nonnull sets Y and $Z \subseteq Y$, write an efficient algorithm for randomly generating r pairwise distinct sets $v_1, \ldots, v_r \subseteq Y$ such that $v_i \cap Z \neq \emptyset$ holds for all $i \in \{1, \ldots, r\}$.

7.5 Simulate a run of the function RANDOMTRANSITIVESETUNRANKING(n, r, z) $= X$ (Algorithm 7.2) for $n = 6$, $r = 2$, $z = 237$. You may use the values s_i^r from Table 7.1.

7.6 Write an efficient algorithm implementing function SUBSETLEX(X, Y, y) from Algorithm 7.2.

7.7 Given a well-founded transitive set X with n elements out of which r have maximum rank, write a polynomial-time algorithm RANDOMTRANSITIVESETUNRANKING(X) returning the number $z \in \{0, \ldots, s_n^r - 1\}$, such that RANDOMTRANSITIVESETUNRANKING(n, r, z) $= X$.

Table 7.1 The values s_i^r, for $i \in \{1, \ldots, 6\}$ and $r \in \{1, 2, 3\}$

n \ r	1	2	3
1	1	0	0
2	1	0	0
3	2	0	0
4	8	1	0
5	76	12	0
6	1504	290	8

7.8 Recall Exercise 6.2, namely, that the number of well-founded sets whose transitive closure has cardinality n can be expressed in terms of $c_{n,s}$. Describe how to apply the recursive method to $c_{n,s}$ (by e.g., modifying Algorithm 7.1).

7.9 Modify the Markov chain from Algorithm 7.3 so that it generates a weakly extensional acyclic graphs with vertex set $\{1, \ldots, n\}$ uniformly at random.

7.10 Propose a symmetric, aperiodic, and irreducible Markov chain that generates weakly extensional acyclic graphs with n vertices $\{1, \ldots, n\}$ and m arcs. *Hint:* Construct a Markov chain that only reverses arcs.

7.11 Prove that the elimination of any vertex in the hyperset depicted in Panel 7.1 causes a collision.

7.12 Determine an arc whose elimination does not cause any collision in the hyperset depicted in Panel 7.1.

Chapter 8
Infinite Sets and Finite Combinatorics

In tackling the set-satisfiability problem, in Sect. 3.5, we have not gone beyond the analysis of formulae with a single prefixed universal quantifier: we have seen how to determine whether or not a formula of the form $\forall y\, \mu$ is satisfiable, where μ stands for a propositional combination of membership and equality literals. When the response is affirmative, the formula-testing algorithm produces a satisfying set assignment $x \mapsto \boldsymbol{x}$, whose domain consists of the variables x other than y appearing in μ; it also ensures that all sets used in the construction of the images $\boldsymbol{x}$ are *(hereditarily) finite*.

What happens if we allow multiple universal quantifiers to appear in the prefix of the formula subject to the set-satisfiability analysis? A legal input, now, is any formula $\forall y_1 \cdots \forall y_m\, \mu$, with μ devoid of quantifiers; to state it briefly, it is a $\forall^*$-formula which we want to make true via a substitution $x \mapsto \boldsymbol{x}$ of sets to its free variables. A decision algorithm can still be devised—at the current state of the art, under the assumption that $\in$ is a well-founded relationship, see [74, 75]—but it is much more complicated than the one encountered in Sect. 3.5.2. A feature of the more general satisfiability problem which we are addressing now, causing the decision algorithm to become utterly challenging, is that sometimes *infinite* sets forcibly enter into the construction of the satisfying assignment.

This book is focused on finite combinatorics; hence, we must refrain from talking about all intricacies of the known decision algorithm for $\forall^*$-formulae; but, luckily, regardless of whether they are finite, the sets which enter into play in the modeling of a $\forall^*$-formula can always be associated with a finite graph which patterns their construction. We will, hence, devote this chapter to discussing the infinitudes—rather "regular," in the sense just mentioned—which enter into the modeling of $\forall^*$-formulae, or equivalently into proving prenex $\exists^*\forall^*$-sentences within the framework of one of various competing axiomatic elementary set theories. Our main motivation for undertaking this discussion is that the study of those infinitudes will provide a revealing insight on Ramsey's celebrated theorem, fundamental in finite combinatorics. This also provides a good reason for not restraining our discussion to well-founded sets: by leaving out of the game the well-foundedness of membership,

© Springer International Publishing AG 2017   217
E.G. Omodeo et al., *On Sets and Graphs*, DOI 10.1007/978-3-319-54981-1_8

we will find a straighter bridge between the satisfiability problem for pure first-order logic and the set-satisfiability problem.

8.1   The Bernays-Schönfinkel-Ramsey Class of Formulae

Since this chapter revolves around the satisfiability of prenex $\forall^*$-formulae, we will have to consider also $\exists^*\forall^*$-sentences. The study of such formulae and sentences was undertaken over one century ago by B̲ernays and S̲chönfinkel, and later deepened by R̲amsey; this is why we will refer to them by the acronym "BSR."

In this section we will show how to reduce the satisfiability problem for BSR formulae over a first-order signature consisting of an arbitrary number of relators (in any number of arguments) to the analogous problem referring to formulae over a signature in a single dyadic relator. Then we address the issue of translating the BSR formulae over a first-order signature consisting of a single dyadic relator into set-theoretic BSR formulae (wherein the dyadic relator designates membership).

Definition 8.1 A formula $\Psi(x_1,\ldots,x_n) = \forall y_1 \cdots \forall y_m\, \mu(x_1,\ldots,x_n)$ is said to be in the BSR class if $\mu(x_1,\ldots,x_n)$ is unquantified. In case $\mu(x_1,\ldots,x_n)$ is written in $\mathscr{L}_\in$, the formula is said to be in the *set-theoretic* BSR class.

A satisfying assignment for $\Psi(x_1,\ldots,x_n) = \forall y_1 \cdots \forall y_m\, \mu(x_1,\ldots,x_n)$ in the set-theoretic BSR must be given by sets. This is in contrast with the case in which the basic language of $\Psi(x_1,\ldots,x_n)$ is going to be interpreted in merely logical terms. We begin by showing that the set-theoretic BSR class turns out to be much more expressive than the purely logical BSR class. This is proved by giving a satisfiability-preserving set-theoretic rewriting for BSR-formulae into set-theoretic ones.

The higher expressive power of the set-theoretic BSR class is a consequence of the underlying (axiomatic) assumptions we are making on sets and can be illustrated by the following simple examples. The formula

$$\Psi_1(x_1, x_2, x_3) \equiv \forall y\big(x_1 \in x_3 \wedge x_2 \notin x_3 \wedge (y \in x_1 \leftrightarrow y \in x_2)\big),$$

is set-theoretically unsatisfiable, due to the axiom of extensionality.[1] Again extensionality is responsible to declare

---

[1] We will feel free to ascribe to the BSR-class also equisatisfiable versions of Ψ_1 such as

$$\Psi_1'(x_1, x_2, x_3) \equiv x_1 \in x_3 \wedge x_2 \notin x_3 \wedge \forall y(y \in x_1 \leftrightarrow y \in x_2),$$

or as

$$\Psi_1'' \equiv \exists x_1, x_2, x_3 \forall y\big(x_1 \in x_3 \wedge x_2 \notin x_3 \wedge (y \in x_1 \leftrightarrow y \in x_2)\big).$$

$$\Psi_2(x_1,\ldots,x_n) \equiv \forall y_1 \cdots \forall y_n \left(\bigvee_{i=0}^{n-1} \bigvee_{j=i+1}^{n} \bigwedge_{k=1}^{n} \left(y_k \in x_i \leftrightarrow y_k \in x_j \right) \right),$$

an *injectively* unsatisfiable BSR or $\forall^*$ scheme.

Our satisfiability-preserving set-theoretic rewriting will be presented as a translation of BSR sentences from an uninterpreted, purely logical context into one referring to a model of the standard Zermelo-Fraenkel theory of sets. We can in fact assume that the language $\mathscr{L}$ of pure logic is nothing but the language $\mathscr{L}_E$ of graph theories plus equality. To see that this assumption is, in fact, inessential, it is sufficient to observe that any occurrence of an *n*-ary relation symbol $R(z_1,\ldots,z_n)$ other than E can be replaced by the following conjunction of *n* atomic formulae:

$$\mathrm{E}\left(z_1, x_1^R\right) \wedge \ldots \wedge \mathrm{E}\left(z_n, x_n^R\right),$$

where $x_1^R,\ldots,x_n^R$ are (fresh) existentially quantified variables, to be used to eliminate R only. Roughly speaking, $\mathrm{E}(\cdot, x_j^R)$ captures the *j*-th *projection* $R_j = \{z_j \mid R(z_1,\ldots,z_n)\}$ of R. It is an easy exercise to prove that any interpretation satisfying a formula before the replacement can be turned into a graph satisfying the same formula after the replacement.

Let us stress again that in set theory—as opposed to the case of logic—and in connection to the satisfiability problem at hand, it is immaterial whether or not we regard equality as a *primitive* relator in the signature of the language. Anyhow, since we know that we can eliminate "=" from set-theoretic BSR sentences without leaving the BSR class, by substituting any literal $z = w$ for $\forall y(y \in z \leftrightarrow y \in w)$, we will feel free to use it when this can improve readability.

The simplest possible way to produce our translation is a simple replacement of E by $\ni$. This takes us to the first problem we must tackle, that is, E might be cyclic, while $\in$, by axiomatic assumption, cannot. We overcome this problem by representing each logical variable z in *split* form, by means of a source-target pair, z_s, z_t, of set variables. This transformation reflects a common way of proceeding in graph theory, for example, to reduce cycle cover problems to matching problems in bipartite graphs (cf. [58]).

Theorem 8.1 *For every Φ in the BSR class written in $\mathscr{L}_E$, there exists $\Phi^\in$ in the set-theoretic BSR class such that Φ is satisfiable if and only if $\Phi^\in$ is satisfiable in (well-founded) Set Theory.*

Proof We assume Φ to be a BSR *sentence* of the following form:

$$\Phi \equiv \exists x_1 \cdots \exists x_n \, \forall y_1 \cdots \forall y_m \, \varphi(x_1,\ldots,x_n,y_1,\ldots,y_m),$$

where φ is an unquantified matrix in the language $\mathscr{L}_E$. Put

$$\Phi^\in \equiv (\exists d)\,(\exists x_{s,1}, x_{t,1}, \ldots, x_{s,n}, x_{t,n} \in d)\,(\forall y_{s,1}, y_{t,1}, \ldots, y_{s,m}, y_{t,m} \in d)$$

$$\varphi^\in(x_{s,1}, x_{t,1}, \ldots, x_{s,n}, x_{t,n}, y_{s,1}, y_{t,1}, \ldots, y_{s,m}, y_{t,m}),$$

where $\varphi^\in$ results from φ through replacement of each literal of the form $z_i \mathrel{E} w_j$, with $z, w \in \{x, y\}$, by

$$z_{s,i} \ni w_{t,j},$$

and of each literal of the form $z_i = w_j$, with $z, w \in \{x, y\}$, by the conjunction

$$z_{s,i} = w_{s,j} \wedge z_{t,i} = w_{t,j}.$$

To prove the implication from left to right in our claim, assume that Φ is satisfiable and that (V, E) is a finite directed graph satisfying it, and perform the following cycle-untying transformation: replacement of each vertex v by distinct vertices v_s and v_t and of each arc $\langle u, w \rangle \in E$ by an arc leaving u_s and entering w_t. Bearing this transformation in mind, consider functions f, g from $V_{s,t} = \{v_s, v_t \colon v \in V\}$ into sets subject to the following constraints:

- $f(u_s) = \{f(w_t) \colon \langle u, w \rangle \in E\} \cup \{g(u_s)\}$;
- $f(u_t) = \{g(u_t)\}$;
- the function g is injective and $|g(v)| > 1$ for every v.

Once fixed the function g, the function f is determined uniquely in view of the acyclicity of the graph $(V_{s,t}, \{\langle u_s, w_t \rangle \colon \langle u, w \rangle \in E\})$.

The functions f and g associate two sets with each vertex in $V_{s,t}$, mimicking E by the acyclic relation $\ni$ even in case E has cycles. The function f is injective on $\{v_t \colon v \in V\}$, by the injectivity of g. Moreover, for every u and w, $f(w_t) \neq g(u_s)$, since $|f(w_t)| = 1$ while $|g(u_s)| > 1$. Therefore, f is injective on the whole $V_{s,t}$, since if $u_s \neq u'_s$ then $g(u_s) \in f(u_s) \setminus f(u'_s)$ (and, symmetrically, $g(u'_s) \in f(u'_s) \setminus f(u_s)$).

Based on the injectivity—just proved—of the Mostowski-like collapsing function f, equality and membership literals are properly modeled: in fact, if one interprets d as $\{f(v) \colon v \in V_{s,t}\}$ and $x_{s,i}, x_{t,i}$ as $f(v_{s,i}), f(v_{t,i})$, where v_i is the vertex assigned to x_i by the satisfying assignment for Φ, then the resulting set assignment will satisfy $\Phi^\in$.

Conversely, assuming $\Phi^\in$ is satisfied by a set-theoretic interpretation, define (V, E) to be the graph with vertices $V = \{v_1, \ldots, v_n\}$ such that $v_i = v_j$ holds precisely when the interpretations $x_{t,i}, x_{s,i}$ of $x_{t,i}$ and $x_{s,i}$ equal the corresponding interpretations, $x_{t,j}, x_{s,j}$, of $x_{t,j}$ and $x_{s,j}$, and with arcs $E = \{\langle v_i, v_j \rangle \colon i, j = 1, \ldots, n \mid x_{t,j} \in x_{s,i}\}$. On these grounds we have that

$$v_i = v_j \text{ if and only if } x_{s,i} = x_{s,j} \wedge x_{t,i} = x_{t,j},$$

$$\langle v_i, v_j \rangle \in E \text{ if and only if } x_{t,j} \in x_{s,i},$$

from which it plainly follows that Φ is satisfiable in (V, E). $\dashv$

8.2 Taming Infinite Set Models

ZF is a standard acronym for the Zermelo-Fraenkel theory. As already mentioned in Panel 3.1, its axioms are essentially the ones introduced in this book (see Sect. 2.1), with foundation and infinity axioms, without AFA, and with the axiom of choice left aside. Throughout this section, we also indicate by:

FA: von Neumann's Foundation axiom;
AFA: Aczel's Anti-Foundation Axiom (cf. [2]), antithetic to FA, which:

- on the one hand requires that a graph whatsoever can be decorated, thereby enriching the universe of all sets—the entities forming the enlarged universe are often called *hypersets* (cf. [8]);
- on the other hand strengthens the equality criterion enforced by the extensionality axiom EA, to avoid excessive proliferation of hypersets;

ZF^-: the subtheory of ZF resulting from withdrawal of the axiom of infinity (**I**);
$ZF^- - FA$: the subtheory of ZF^- resulting from withdrawal of FA.

8.2.1 Stating Infinity by Means of a BSR Sentence

In fairly conventional terms, the specification

$$\iota(a) \equiv \emptyset \in a \land (\forall x \in a)(\exists y \in a)(x \in y)$$

helps in asserting that infinite sets exist. It is, in fact, plain that any well-founded set a satisfying $\iota(a)$ must be infinite. Unless we assume FA, though, $\iota(a)$ turns out to be satisfiable by means of a hereditarily finite hyperset as simple as $\Omega' = \{\Omega', \emptyset\}$ (Fig. 8.1).

A formulation of the axiom of infinity for whose working FA is immaterial, and which clings to the statement of this axiom as proposed originally by Zermelo in 1908, can be based on the following slightly strengthened variant of $\iota(a)$:

$$\tilde{\iota}(a) \equiv \emptyset \in a \land (\forall x \in a)(\exists y \in a)(y = \{x\}) \,.$$

Fig. 8.1 The hyperset
$\Omega' = \{\Omega', \emptyset\}$

In terms of the latter, one will assert: $(\exists a)\bar{\iota}(a)$. Such an infinite set can be so easily summoned up thanks to the alternation "$(\forall x \in a)(\exists y \in a)$" of universal and existential quantification, which imposes a condition "local" to each element of a. The class of $\forall^* \exists^*$-formulae hence has the ability to express infinity; unfortunately, though, it constitutes an undecidable fragment of the elementary set-theoretic language, even when this language refers only to well-founded sets [85, 90].

Let us stick, for a while, to the well-founded setting of FA. As we will see, infinity is then expressible by means of a BSR-formula $\iota\iota(a, b)$ involving, besides the two free variables a and b, only two universally quantified variables [86, 87]. To figure out how to perform this "strength reduction" of infinitude from $\forall^* \exists^*$ to $\exists^* \forall^*$, while also explaining the machinery that pushes any two sets satisfying such a formula to be infinite, we take a step back and consider one of the most basic infinite sets, namely, the set ω of natural numbers conceived in the way suggested by Fig. 8.2.

In order to devise an infinitely satisfiable formula having ω as a model, we can start by observing that ω satisfies the two conditions[2]

$$\bigcup \omega \subseteq \omega,$$
$$(\forall x \in \omega)(\forall y \in \omega)(x = y \vee x \in y \vee y \in x).$$

The conjunction of these captures much of the structure of ω, without yet enforcing the infinity of ω. For, it is satisfied by any finite transitive set over which membership induces a strict linear ordering (such sets are, precisely, the natural

Fig. 8.2 The set of natural numbers, seen as the least $\omega \supseteq \{\emptyset\}$ closed under the successor function

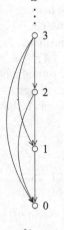

$$\omega = \{i : i = \{0, \dots, i-1\}\}$$

---

[2]We are making use of the union-set operator to increase readability: $\bigcup a \subseteq b$ in fact abbreviates the BSR-formula $(\forall x \in a)(\forall y \in x)(y \in b)$.

numbers as conceived *à la* von Neumann). The missing ingredient for infinitude can be provided by the formula $\iota(a)$. In fact, if we define $\iota_\omega(a)$ to be the conjunction of

(i) $\iota(a)$

(ii) $\bigcup a \subseteq a$

(iii) $(\forall x \in a)(\forall y \in a)(x = y \lor x \in y \lor y \in x)$,

still we have that $\iota_\omega(\omega)$ holds; moreover, as desired, any well-founded set $\boldsymbol{a}$ satisfying $\iota_\omega(a)$ must be infinite.

Trying to replace the $\forall\exists$-subformula $\iota(a)$ by some purpose-built BSR formula is a good example of the difficulties arising when it comes to describing an infinite (hyper)set by means of a set-theoretic formula and even more so by a BSR-formula: one must play a subtle game on keeping the formula satisfiable, while avoiding finite satisfiability. What makes this even more interesting is that in order to exclude finite satisfiability, we reason only about *finite* sets—since we proceed by contradiction—turning our proofs into a finitary combinatorial game.

Since this game cannot succeed in capturing ω, we will get a set akin to ω as the union of two parts, to be represented by the free variables a, b of the sought BSR formula. The intended model that we want to capture consists of the disjoint sets ω_1, ω_2 described in Fig. 8.3. First, notice that these satisfy analogs of the two conditions about ω shown above:

$$\bigcup \omega_1 \subseteq \omega_2 \land \bigcup \omega_2 \subseteq \omega_1,$$
$$(\forall x \in \omega_1)(\forall y \in \omega_2)(x \in y \lor y \in x).$$

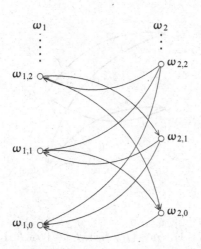

$$\begin{array}{l} \omega_1 = \{\omega_{1,i} : i \in \omega\} \text{ and } \omega_{1,j} = \{\omega_{2k} : 0 \leqslant k < j\} \\ \omega_2 = \{\omega_{2,i} : i \in \omega\} \text{ and } \omega_{2,j} = \{\omega_{1,k} : 0 \leqslant k \leqslant j\} \end{array} \Bigg\} \text{ for all } j \in \omega.$$

Fig. 8.3 The elements of $\omega_1 \cup \omega_2$ form a double-stranded well-founded spiral, as required by ι

Second, the conditions $\omega_1 \neq \omega_2$, $\omega_1 \notin \omega_2$, and $\omega_2 \notin \omega_1$ also hold. As was first observed in [87] (albeit already implicit in [86]), these $\forall\forall$-formulae provide a satisfactory formulation of infinity. Indeed, let $\iota\iota(a, b)$ be the conjunction of:

(i) $a \neq b \land a \notin b \land b \notin a$
(ii) $\bigcup a \subseteq b \land \bigcup b \subseteq a$
(iii) $(\forall x \in a)(\forall y \in b)(x \in y \lor y \in x)$.

Even though the above condition (i) might seem to bring little information, it will turn out to be the right substitute for $\iota(a)$. On the one hand, requiring $a \neq b$ is tantamount to imposing that the two sets cannot both be empty, in analogy with the first conjunct of $\iota(a)$. On the other hand, the requirement $a \notin b \land b \notin a$ proves to be a substitute for the second conjunct of $\iota(a)$. Indeed, it can be shown that conditions (ii) and (iii) enact such a rigid structure that if either one of the sets a, b—jointly satisfying $\iota\iota$—turned out to be finite, then the other one would belong to it, contravening (i). In Sect. 8.2.3 we will formalize this argument in detail.

Let us now move on to a setting deprived of FA; are there (hereditarily) finite ill-founded sets satisfying $\iota\iota(a, b)$? If we simply retract the requirement that sets must be well founded (by carrying out our considerations within $\mathsf{ZF}^- - \mathsf{FA}$), the answer is *yes*: unconventional sets doing to the case are in fact described in Fig. 8.4. Those are not hypersets, though, since their membership graphs are not hyperextensional (see Exercise 8.3). The following questions then arise naturally:

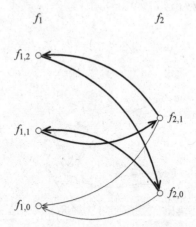

$$f_1 = \{f_{1,j} : 0 \leqslant j \leqslant 2\}, \; f_{1,0} = \emptyset, \qquad f_{1,1} = \{f_{2,1}\}, \qquad f_{1,2} = \{f_{2,0}\},$$
$$f_2 = \{f_{2,j} : 0 \leqslant j \leqslant 1\}, \; f_{2,0} = \{f_{1,0}, f_{1,1}\}, \; f_{2,1} = \{f_{1,0}, f_{1,2}\}.$$

Fig. 8.4 Hereditarily finite sets f_1 and f_2 such that $\iota\iota(f_1, f_2)$ holds [89]. The directed cycle drawn with *thick lines* in the membership graph of f_1, f_2 reveals that these are neither well-founded sets nor hypersets: vertices $f_{2,0}$ and $f_{2,1}$—and, likewise, $f_{1,1}$ and $f_{1,2}$—should in fact be decorated by the same hyperset

1. Is there any BSR formula capturing infinity under $ZF^- - FA$?
2. Does $\iota\iota(a, b)$ still express infinity if, instead of simply dropping FA, we supersede it, along with the Extensionality Axiom EA, by AFA?

Question 1 was answered positively in [89], where the driving engine forcing a and b to be infinite was carefully distilled, by the insertion into $\iota\iota(a, b)$ of the new conjunct

(iv) $(\forall x_1, x_2 \in a)(\forall y_1, y_2 \in b)(x_2 \in y_2 \in x_1 \in y_1 \rightarrow x_2 \in y_1)$,

clearly derivable from (iii) when FA is assumed. Even though well foundedness is no longer assumed, the resulting $\forall\forall\forall\forall$ formula is again satisfied by the well-founded sets ω_0 and ω_1, and all sets satisfying it are infinite.

As we are about to see, an even more succinct distillate from FA is

$(\star)$ $(\forall y_1, y_2 \in b)(y_1 \subseteq y_2 \vee y_2 \subseteq y_1)$,

which in fact follows from $\iota\iota(a, b)$ if the above (iv) is assumed. Although a recasting of this formula into purely universal form seems to require four universal quantifiers, in Sect. 8.2.6 we succeed in making the necessary tweaks to $\iota\iota(a, b)$ to produce a $\forall\forall\forall$ formula expressing infinity independent of FA, hence lowering the number of universal quantifiers from four to three. This result appeared in [77].

Regarding Question 2, we will show that, surprisingly, under AFA we can derive $(\star)$ from conditions (i)–(iii) and $a \cap b = \emptyset$. Moreover, even in the non-well-founded setting of AFA, it will turn out that *all* sets satisfying $\iota\iota(a, b) \wedge (a \cap b = \emptyset)$ are *well-founded* This was also shown in [77]. We also give a $\forall\forall\forall$-formula that, under AFA, is satisfied only by infinite hypersets having all peculiarities of non-well-foundedness: membership cycles, as well as infinite descending membership chains devoid of repeated elements. This is based on [76].

8.2.2 Stating Infinity, Under Different Set-Theoretic Axioms

What do we mean by saying that a sentence such as the one,

$$\mathsf{Inf} \equiv (\exists a)\, \tilde{\iota}(a),$$

introduced above, *expresses infinity*? Setting aside concerns about the possibility that an inconsistency arises between such a sentence and $ZF^- - FA$, there are two readings of the concept in question: one *internal* to $ZF^- - FA$, and one *external*, i.e., referring to the models of this theory.[3]

Internal version: This discussion presupposes that complementary predicates Finite $(\cdot)$ and Infinite $(\cdot)$ have been defined in a way reflecting the usual meaning

---

[3]We are focusing on $ZF^- - FA$ and Inf only momentarily and for the sake of definiteness: our considerations easily carry over to other axiomatic theories and to different statements of infinity.

Fig. 8.5 Laws regarding
finiteness and infinitude

1.	$\mathsf{Finite}(F) \wedge G \subseteq F \;\rightarrow\; \mathsf{Finite}(G)$
2.	$\mathsf{Finite}(\emptyset)$
3.	$\mathsf{Finite}(F) \;\rightarrow\; \mathsf{Finite}(F \cup \{X\})$
4.	$\mathsf{Finite}(F) \wedge \mathsf{Finite}(G) \;\rightarrow\; \mathsf{Finite}(F \cup G)$
5.	$\mathsf{Finite}(F) \;\rightarrow\; \mathsf{Finite}(\{t(x) : x \in F\})$
6.	$\mathsf{Finite}(A) \wedge A \neq \emptyset \;\rightarrow\; (\exists m \in A)(\forall y \in A \setminus \{m\})(m \nsubseteq y)$
7.	$\mathsf{Infinite}(I) \;\leftrightarrow\; \neg\mathsf{Finite}(I)$

of their names and so that all laws displayed in Fig. 8.5 are met in $\mathsf{ZF}^- - \mathsf{FA}$;
i.e., one can derive these laws without resorting to Inf or to FA. Concerning the
finiteness predicate, a definition drawn, essentially, from [109] and well-suited for
our purposes goes as follows:

$$\mathsf{Finite}(X) \;\leftrightarrow_{\mathrm{Def}}\; (\forall y \in \mathcal{P}(\mathcal{P}(X)) \setminus \{\emptyset\})\, (\exists m)\, (y \cap \mathcal{P}(m) = \{m\}).$$

This states that a set X is finite if and only if every family y of subsets of X, unless
null, has an element m which is minimal with respect to $\subseteq$.

Thus, Inf is an internal expression of infinity because $\neg\mathsf{Finite}(a)$, and hence
$\mathsf{Infinite}(a)$, can be derived from $\tilde{\imath}(a)$ in $\mathsf{ZF}^- - \mathsf{FA}$. A strategy to achieve this is
to show that a certain set a'' of the form $\{t(x) : x \in a'\}$, with $a' \subseteq a$, owns no
maximal element relative to $\subseteq$, but then $\neg\mathsf{Finite}(a'')$ can be derived, thanks to law
6. of Fig. 8.5, and so $\neg\mathsf{Finite}(a')$ by law 5., and therefore $\neg\mathsf{Finite}(a)$ by law 1.
Referring to the above definition of $\mathsf{Finite}(\cdot)$, and taking advantage of the transitive
embedding axiom (see Exercise 2.2), we can implement this strategy by instantiating
$t(x)$ as the transitive closure of x (see Exercise 2.6).

External version: before shifting to the semantic level, notice that the scheme
$(\neg\mathsf{Finite}(a)) \rightarrow a \nsubseteq \{X_1, \ldots, X_n\}$, with any number n of distinct variables X_i, is
derivable from the laws 1., 2., and 3.; accordingly, all statements $a \nsubseteq \{X_1, \ldots, X_n\}$
are derivable from $\tilde{\imath}(a)$. It plainly follows, for every model $\mathscr{U} = (\mathscr{U}, \in)$ of $\mathsf{ZF}^- -$
FA, that if $\mathscr{U} \models \tilde{\imath}(a)$ for some a in $\mathscr{U}$, then such an a will satisfy infinitely many
literals $x \in a$ with x in $\mathscr{U}$.

Repeatedly, in what follows, we will show that some sentence

$$(\exists x_1) \cdots (\exists x_n)\, (\exists y_1) \cdots (\exists y_m)\, \varphi(x_1, \ldots, x_n, y_1, \ldots, y_m)$$

(where, in order to fulfill our aims, φ will be a BSR formula involving at least $n > 1$
distinct free variables $x_1, \ldots, x_n$) *expresses infinity*. In analogy with the paradigmatic
case just sketched, we will show this by indicating how to derive from the postulates
of an axiomatic set theory, say ZF^- (but we will also play inside $\mathsf{ZF}^- - \mathsf{FA}$ or inside
$\mathsf{ZF}^- - \mathsf{FA} + \mathsf{AFA}$), that no x_i can have an inclusion-maximal element, but then, by
laws 6. and 7., we can conclude

$$\mathsf{ZF}^- \vdash (\forall x_1) \cdots (\forall x_n)\, (\forall y_1) \cdots (\forall y_m)$$

$$\left(\varphi(x_1, \ldots, x_n, y_1, \ldots, y_m) \rightarrow \bigwedge_{i=1}^{n} \mathsf{Infinite}(x_i) \right).$$

As is then plain, for any model $\mathscr{U} = (\mathscr{U}, \in)$ of ZF$^-$ and for each x_1 in $\mathscr{U}$, if

$$\mathscr{U} \models (\exists x_2) \cdots (\exists x_n)\, (\exists y_1) \cdots (\exists y_m)\, \varphi(x_1,\, x_2, \ldots, x_n,\, y_1, \ldots, y_m)$$

holds, then infinitely many literals $y \in x_1$ with y in $\mathscr{U}$ must be true.

Thereby, $(\exists x_1) \cdots (\exists x_n)\, (\exists y_1) \cdots (\exists y_m)\, \varphi(x_1, \ldots, x_n,\, y_1, \ldots, y_m)$ will turn out to *express infinity* according to both readings of this locution: internal and external.

8.2.3 Basic Pattern for an Infinite Well-Founded Spiral

Having so far outlined the rules of our game and the main results that we will achieve, we now enter the more general context in which we aim at expressing infinity via a BSR formula involving an arbitrarily big number n of free variables.

We refrain, momentarily, from assuming either FA or AFA, and present a sequence

$$u^{[2]}(x_1, x_2),\ u^{[3]}(x_1, x_2, x_3),\ \ldots,\ u^{[n]}(x_1, \ldots, x_n),\ \ldots$$

of "amplifications" of $u(a, b)$, each of which can be satisfied by means of infinite well-founded sets. For each $n \geq 2$, the formula $u^{[n]}(x_1, \ldots, x_n)$ results from conjoining the clauses

(i) $x_1 \neq \emptyset \wedge \bigwedge_{i=1}^{n} \left(x_i \notin x_{(i \bmod n)+1} \right)$
(ii) $\bigwedge_{i=1}^{n} \left(\bigcup x_{(i \bmod n)+1} \subseteq x_i \right)$
(iii) $(\forall y_1 \in x_1) \cdots (\forall y_n \in x_n) \left(\bigvee_{i=1}^{n} y_i \in y_{(i \bmod n)+1} \right)$

together, where $i \bmod n$ indicates the integer remainder after division of i by n.

To readily see how to meet such conditions, note that

$$u^{[n]} \left(\{\, \omega_{1,0},\, \omega_{1,1},\, \omega_{1,2} \ldots \},\, \ldots,\, \{\, \omega_{n,0},\, \omega_{n,1},\, \omega_{n,2} \ldots \} \right)$$

is true if, for every nonnegative integer j, the equalities

$$\omega_{1,j} = \{\, \omega_{n,0}, \ldots, \omega_{n,j-1} \} \quad \text{and}$$

$$\omega_{i,j} = \{\, \omega_{i-1,0}, \ldots, \omega_{i-1,j} \},\ \text{for } i = 2, \ldots, n,$$

hold—so that, in particular, $\omega_{1,0} = \emptyset$. (Figures 8.6 and 8.7 portray the case when $n = 3$.)

We will show that $u^{[n]}$ is satisfied only by infinite sets under ZF$^-$. Our first lemma shows that $u^{[n]}$ is reminiscent of Zermelo's infinity axiom, as formulated via either $\iota(a)$ or $\bar{\iota}(a)$.

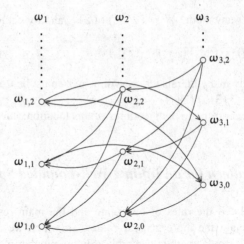

$$\begin{aligned}
\omega_1 &= \{\omega_{1,j} : j \in \omega\} \text{ and } \omega_{1,j} = \{\omega_{3,k} : 0 \leqslant k < j\} \\
\omega_2 &= \{\omega_{2,j} : j \in \omega\} \text{ and } \omega_{2,j} = \{\omega_{1,k} : 0 \leqslant k \leqslant j\} \\
\omega_3 &= \{\omega_{3,j} : j \in \omega\} \text{ and } \omega_{3,j} = \{\omega_{2,k} : 0 \leqslant k \leqslant j\}
\end{aligned} \right\} \text{ for all } j \in \omega.$$

Fig. 8.6 Well-founded sets $\omega_1, \omega_2, \omega_3$ satisfying $\iota\iota^{[3]}$

Fig. 8.7 A more suggestive, 3-dimensional representation of the membership graph of Fig. 8.6

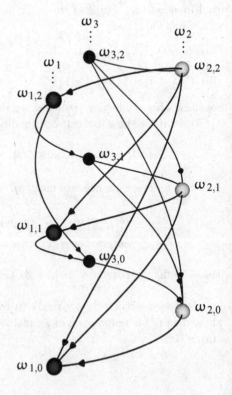

Lemma 8.1 *Independently   of* **FA**, *for   any   sets* $\omega_1, \ldots, \omega_n$ *such   that* $\iota\iota^{[n]}(\omega_1, \ldots, \omega_n)$ *is true, the following conditions hold, for* $i = 1, \ldots, n$:

- $\omega_i \neq \emptyset$;
- $(\forall y \in \omega_i)(\exists z \in \omega_{(i \bmod n)+1})(y \in z)$.

Proof As a preliminary remark, observe that

$$x \in \omega_{(i \bmod n)+1} \text{ always implies } \omega_i \not\subseteq x \, ;$$

for, assuming the contrary, the equality $\omega_i = x$ would hold because

$$x \subseteq \bigcup \omega_{(i \bmod n)+1} \subseteq \omega_i \, ,$$

but then $\omega_i \in \omega_{(i \bmod n)+1}$ would hold, contradicting clause (i) of $\iota\iota^{[n]}(\omega_1, \ldots, \omega_n)$.

Under the temporary assumption that some ω_i can be null, given that $\omega_1 \neq \emptyset$, consider a k for which $\omega_k = \emptyset$ and $\omega_{(k \bmod n)+1} \neq \emptyset$. We can pick an element $x \in \omega_{(k \bmod n)+1}$ but then get into a contradiction, because we know from the previous remark that $\omega_k \not\subseteq x$.

Suppose next that $y \in \omega_i$. Put $y_i = y$ and $h_0 = i$. Then, for $k = 1, \ldots, n - 1$, choose $h_k \in \{1, \ldots, n\}$ so that $(h_k \bmod n) + 1 = h_{k-1}$ and—taking the initial remark into account—pick an element $y_{h_k} \in \omega_{h_k} \setminus y_{h_{k-1}}$. At this point, since $y_1 \in \omega_1, \ldots, y_n \in \omega_n$, and since $y_h \notin y_{(h \bmod n)+1}$ save for $h = i$, we get $y_i \in y_{(i \bmod n)+1}$ from clause (iii) of $\iota\iota^{[n]}(\omega_1, \ldots, \omega_n)$. ⊣

Our next lemma shows that if any of the free variables of $\iota\iota^{[n]}$ is substituted by an infinite set in order to make $\iota\iota^{[n]}$ true, then all of them must be substituted by infinite sets.

Lemma 8.2 *Independently   of* **FA**, *for   any   sets* $\omega_1, \ldots, \omega_n$ *such   that* $\iota\iota^{[n]}(\omega_1, \ldots, \omega_n)$ *is true, if any of* $\omega_1, \ldots, \omega_n$ *is infinite, then each of* $\omega_1, \ldots, \omega_n$ *is infinite.*

Proof Suppose that $\omega_{(i \bmod n)+1}$ is infinite for some $i \in \{1, \ldots, n\}$. Then ω_i must be infinite as well, because $\bigcup \omega_{(i \bmod n)+1} \subseteq \omega_i$ as by condition (ii) of $\iota\iota^{[n]}$. Trivially, in fact, $\mathcal{P}(x)$ is finite whenever x is a finite set; moreover, $y \subseteq \mathcal{P}(\bigcup y)$ holds for any set y; accordingly, if ω_i were finite, then $\bigcup \omega_{(i \bmod n)+1}$ would be finite, $\mathcal{P}(\bigcup \omega_{(i \bmod n)+1})$ would also be finite, and $\omega_{(i \bmod n)+1}$ would be finite.

This argument can be iterated to show that each of $\omega_1, \ldots, \omega_n$ is infinite. ⊣

Next we will see that $\iota\iota^{[n]}$ gets satisfied by means of infinite sets once we manage to ensure, for at least one of the sets ω_i assigned to the free variables of $\iota\iota^{[n]}$, that the elements of ω_i are pairwise comparable by inclusion according to the condition $(\star)$ seen on p. 225.

Theorem 8.2 *Independently of* FA, *for any sets* $\omega_1, \ldots, \omega_n$ *such that the conditions*

- $\iota\iota^{[n]}(\omega_1, \ldots, \omega_n)$,
- $(\forall y_1, y_2 \in \omega_n)(y_1 \subseteq y_2 \vee y_2 \subseteq y_1)$,

both hold, we have that $\omega_1, \ldots, \omega_n$ *are infinite.*

Proof Assume for a contradiction that ω_n is finite. Thanks to the second condition of the hypothesis, ω_n has an $\subseteq$-maximum element y_0. From Lemma 8.1 we get that $\omega_{n-1} \subseteq y_0$. Since $y_0 \subseteq \bigcup \omega_n \subseteq \omega_{n-1}$, we get $y_0 = \omega_{n-1}$, contradicting condition $\omega_{n-1} \notin \omega_n$ of $\iota\iota^{[n]}$.

Since ω_n is infinite, Lemma 8.2 gives us that each of $\omega_1, \ldots, \omega_n$ is infinite. $\dashv$

8.2.4   Well-Founded Infinities, Under (Anti-)Foundation or Without It

In this section we take a stance on adopting one of FA, AFA rather than neither one. When FA is taken as an axiom, we will see the claim of Theorem 8.2 can be refined in terms of rank; more importantly, we will see that the second condition of that claim is not necessary to ensure the infinity of each set involved in any solution of $\iota\iota^{[n]}$. When the antithetic axiom AFA is adopted, the hypothesis of Theorem 8.2 will be met, for $n = 2$, provided ω_1 and ω_2 are disjoint. No matter whether FA or AFA is assumed, basically the same parsimonious expression of infinitude, $\iota\iota$, hence works.

Theorem 8.3 *Under* ZF$^-$, *sets* ω_i *for which* $\iota\iota^{[n]}(\omega_1, \ldots, \omega_n)$ *is true always share the same rank, which is a limit ordinal. (Consequently each* ω_i *has infinitely many elements.)*

Proof Lemma 8.1 ensures that no ω_i is empty. Suppose next that some ω_k has a successor rank, so that there is a $y \in \omega_k$ whose rank satisfies $\mathsf{rank}(y) + 1 = \mathsf{rank}(\omega_k)$. By exploiting n consecutive times the second claim of Lemma 8.1, we can construct a membership chain starting in y and ending in a y' belonging to the same ω_k with which we have started. Since we are assuming that the membership relation is well founded, $y \neq y'$ holds, and $\mathsf{rank}(y) < \mathsf{rank}(y')$ also holds, which contradicts the rank maximality of y within ω_k. Each ω_i hence has a limit rank.

The task of verifying that $\mathsf{rank}(\omega_1) = \cdots = \mathsf{rank}(\omega_n)$ is left as Exercise 8.4. $\dashv$

The theorem just seen, instantiated with $n = 2$, yields that the $\forall\forall$ formula $\iota\iota(a, b)$ introduced on p. 224 is satisfied, under ZF$^-$, exclusively by infinite sets. A direct reduction to primitive symbols of the condition $a \neq \emptyset$ of $\iota\iota^{[2]}(a, b)$ would have required the introduction of a new free variable for eliminating $\emptyset$; but we have resorted to $a \neq b$ which, in light of the clauses (ii) and (i) of $\iota\iota(a, b)$, does imply $a \neq \emptyset$ (see Exercise 8.1).

$$\mathbf{X}_0 = \{Y_0\} = \{\mathbf{X}_1 \cup \mathbf{A}_0\} \qquad \mathbf{X}_1 = \{Y_1\} = \{\mathbf{X}_0 \cup \mathbf{A}_1\}$$

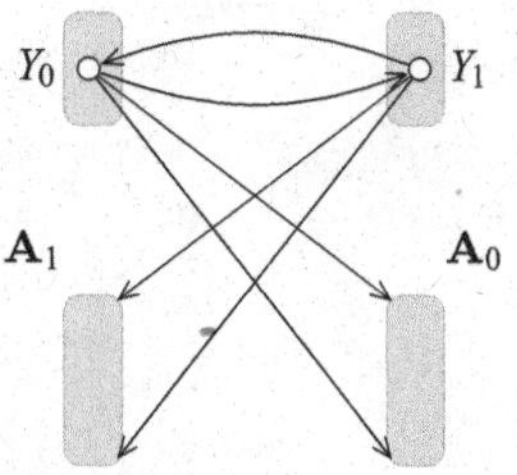

If $\mathbf{A}_0, \mathbf{A}_1, \mathbf{X}_0, \mathbf{X}_1$ are sets fulfilling the conditions

$$\mathbf{X}_i \neq \emptyset, \ \mathbf{A}_i \notin \mathbf{X}_i, \text{ and}$$

$$\text{for all } y \in \mathbf{X}_i, \ \mathbf{A}_i \subseteq y \wedge y \setminus \mathbf{A}_i \subseteq \mathbf{X}_{i-1} \wedge y \cap \mathbf{X}_i = \emptyset$$

for $i = 0, 1$, then $\mathbf{X}_0, \mathbf{X}_1$ are those unique sets which solve the system

$$X_0 = \{Y_0\}, \ X_1 = \{Y_1\}, \quad Y_0 = X_1 \cup \mathbf{A}_0, \ Y_1 = X_0 \cup \mathbf{A}_1$$

of equations.

Fig. 8.8 A consequence of AFA

Making use of FA, we could have established directly from Theorem 8.2 that $\iota\iota(a, b)$ is not satisfied when a, b are finite sets: $(\forall y_1, y_2 \in b)(y_1 \subseteq y_2 \vee y_2 \subseteq y_1)$ in fact follows from $\iota\iota(a, b)$, thanks to its clause (iii) (see Exercise 8.2).

We show next that, surprisingly, under $\mathsf{ZF}^- - \mathsf{FA} + \mathsf{AFA}$, essentially the same formulation of infinity as in the well-founded context is satisfied exclusively by *infinite well-founded* sets. Let $\widetilde{\iota\iota}(a, b)$ be the conjunction of $\iota\iota(a, b)$ with $a \cap b = \emptyset$. In Fig. 8.8 we point out a plain consequence of AFA, which we use in our first lemma (Fig. 8.9).

Lemma 8.3 *Under* $\mathsf{ZF}^- - \mathsf{FA} + \mathsf{AFA}$, *if* ω_1, ω_2 *are sets such that* $\widetilde{\iota\iota}(\omega_1, \omega_2)$ *is true, then there are no infinite descending membership chains in* $\omega_1 \cup \omega_2$.

Proof First of all, note that $x \notin x$ holds for every $x \in \omega_1 \cup \omega_2$, since otherwise we would have $x \in \omega_1 \cap \omega_2$, contradicting the clause $\omega_1 \cap \omega_2 = \emptyset$ of $\widetilde{\iota\iota}(\omega_1, \omega_2)$.

Arguing by contradiction, suppose that the elements of $C = \{c_0, c_1, \dots\} \subseteq \omega_1 \cup \omega_2$ form an infinite descending membership chain $c_0 \ni c_1 \ni c_2 \ni \cdots$ (so that $c_0 \neq c_1$). Consider then the set

$$\mathbf{X} = \{ y : y \in \omega_1 \cup \omega_2 \wedge \text{ there exist } m \in \omega \text{ and } y_1, y_2, \dots, y_m \in \omega_1 \cup \omega_2$$

$$\text{such that } y \ni y_1 \ni y_2 \ni \cdots \ni y_m \text{ and } y_m \in C \}.$$

Fig. 8.9 A graphical representation of the proof of Lemma 8.3

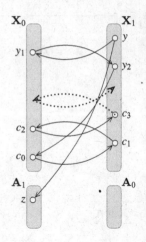

Now we can refer the observation in Fig. 8.8 to the sets $\mathbf{X}_0 = \mathbf{X} \cap \omega_1$, $\mathbf{X}_1 = \mathbf{X} \cap \omega_2$, and $\mathbf{A}_i = (\bigcup \mathbf{X}_i) \setminus \mathbf{X}$ for $i = 0, 1$.

Indeed, the $\mathbf{X}_i$'s are nonnull, as $c_0 \in \mathbf{X}_0$ and $c_1 \in \mathbf{X}_1$ (or vice versa). Since for any $y \in \mathbf{X}$, we have $y \cap \mathbf{X} \neq \emptyset$, we deduce $\mathbf{A}_i \notin \mathbf{X}_i$, for $i = 0, 1$. Let now $y \in \mathbf{X}_i$. To see that $\mathbf{A}_i \subseteq y$, consider a $z \in \mathbf{A}_i$ such that $z \notin y$. From (iii), we get $y \in z$, which implies $z \in \mathbf{X}$, against the choice of z. Requirement $y \setminus \mathbf{A}_i \subseteq \mathbf{X}_{1-i}$ follows from (ii), while $y \cap \mathbf{X}_i = \emptyset$ follows from $\omega_1 \cap \omega_2 = \emptyset$. Moreover, note that $\mathbf{X}_0 \cup \mathbf{A}_1 = \omega_1$ and that $\mathbf{X}_1 \cup \mathbf{A}_0 = \omega_2$.

Therefore, $\mathbf{X}_0 = \{\mathbf{Y}_0\}$, $\mathbf{X}_1 = \{\mathbf{Y}_1\}$, where $\mathbf{Y}_0 = \mathbf{X}_1 \cup \mathbf{A}_0$, $\mathbf{Y}_1 = \mathbf{X}_0 \cup \mathbf{A}_1$. Hence $\mathbf{Y}_0 = \omega_2$ and $\mathbf{Y}_1 = \omega_1$, which, given that $\mathbf{X}_0 \subseteq \omega_1$ and $\mathbf{X}_1 \subseteq \omega_2$, entails $\omega_2 \in \omega_1$ and $\omega_1 \in \omega_2$, contradicting (i). ⊣

Lemma 8.4 *Under* $\mathsf{ZF}^- - \mathsf{FA} + \mathsf{AFA}$, *if* ω_1, ω_2 *are sets such that* $\widetilde{u}(\omega_1, \omega_2)$ *is true, then* $(\forall y_1, y_2 \in \omega_2)(y_1 \subseteq y_2 \vee y_2 \subseteq y_1)$ *holds.*

Proof Arguing by contradiction, assume that $y_1, y_2 \in \omega_2$ and x_1, x_2 are such that $x_1 \in y_1 \setminus y_2$ and $x_2 \in y_2 \setminus y_1$. Conditions (ii) and (iii) of $\widetilde{u}$ imply that $x_1, x_2 \in \omega_1$ and that there is a membership cycle $x_1 \in y_1 \in x_2 \in y_2 \in x_1$ in $\omega_1 \cup \omega_2$, and hence also an infinite descending membership chain (with repeated elements), contradicting Lemma 8.3. ⊣

Theorem 8.4 *Under* $\mathsf{ZF}^- - \mathsf{FA} + \mathsf{AFA}$, *if* ω_1, ω_2 *are sets such that* $\widetilde{u}(\omega_1, \omega_2)$ *is true, then* ω_1 *and* ω_2 *are infinite.*

Proof The claim follows from Theorem 8.2 and Lemma 8.4. ⊣

Passing now to $\mathsf{ZF}^- - \mathsf{FA}$, let $\overline{u}(a, b)$ be the conjunction of $u(a, b)$ with the following:

(iv) $(\forall x_1, x_2 \in a)(\forall y_1, y_2 \in b)(x_2 \in y_2 \in x_1 \in y_1 \rightarrow x_2 \in y_1)$.

Within the framework proposed in Sect. 8.2.3, we can easily deduce the following result of [89] about $\overline{u}$.

Theorem 8.5 *Under* ZF⁻ − FA, *if* ω_1, ω_2 *are sets such that* $\overline{u}(\omega_1, \omega_2)$ *is true, then* ω_1 *and* ω_2 *are infinite.*

Proof By Theorem 8.2 it suffices to show that $(\forall y_1, y_2 \in \omega_2)(y_1 \subseteq y_2 \vee y_2 \subseteq y_1)$ holds. Assuming the contrary, we can pick $y_1, y_2 \in \omega_2$, and $x_1 \in y_1 \setminus y_2$ and $x_2 \in y_2 \setminus y_1$. Conditions (ii) and (iii) of $\overline{u}$ imply that $y_2 \in x_1$. Condition (iv) of $\overline{u}$ then entails $x_2 \in y_1$, leading to a contradiction.   ⊣

8.2.5 Basic Pattern for an Infinite Non-well-founded Spiral

> *Bottomless wonders spring from simple rules · · · repeated without end.*
>
> Benoit Mandelbrot

All formulations of infinity seen up to now are satisfied by (basically the same) well-founded sets. This was just an isolated case, since a plethora of infinite non-well-founded sets can be expressed by BSR formulae, even in the presence of **AFA**. We start with a formula involving an arbitrary number $n \geq 2$ of free variables and then analyze the case $n = 2$ in greater detail (Fig. 8.10).

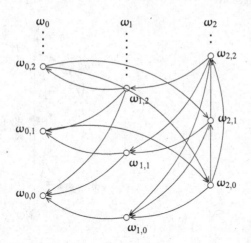

$$\omega_0 = \{\omega_{0,i} : i \in \omega\}, \ \omega_{0,j} = \{\omega_{2,k} : 0 \leqslant k < j\},$$
$$\omega_1 = \{\omega_{1,i} : i \in \omega\}, \ \omega_{1,j} = \{\omega_{0,k} : 0 \leqslant k \leqslant j\},$$
$$\omega_2 = \{\omega_{2,i} : i \in \omega\}, \ \omega_{2,j} = \{\omega_{1,k} : 0 \leqslant k \leqslant j\} \cup \{\omega_{2,k} : k > j\}, \ \forall j \in \omega.$$

Fig. 8.10 Hypersets $\omega_0, \omega_1, \omega_2$ satisfying $\underline{u}^3$

Let $\underline{u}^{[n]}(x_0, \ldots, x_{n-1})$ result from conjunction of the following clauses[4]:

(i) $x_0 \neq \emptyset \wedge \bigwedge_{i=0}^{n-1} \left(x_{(i-1) \bmod n} \notin x_i \right)$

(ii') $\bigwedge_{i=0}^{n-2} \left(\bigcup x_i \subseteq x_{(i-1) \bmod n} \right) \wedge \bigcup x_{n-1} \subseteq x_{n-2} \cup x_{n-1}$

(iii) $(\forall y_0 \in x_0, \ldots, \forall y_{n-1} \in x_{n-1}) \left(\bigvee_{i=0}^{n-1} y_i \in y_{(i+1) \bmod n} \right)$

(iv') $(\forall y_1, y_2 \in x_{n-1})(y_1 \in y_2 \rightarrow y_2 \subseteq y_1)$

To see that $\underline{u}^{[n]}$ is satisfiable by means of hypersets, observe that $\underline{u}^{[n]}(\omega_0, \ldots, \omega_{n-1})$ holds, where each $\omega_i = \{\omega_{i,j} \mid j \in \omega\}$, and

$$\omega_{0,j} = \{\omega_{n-1,k} : k < j\},$$
$$\omega_{i,j} = \{\omega_{i-1,k} : k \leq j\}, \text{ if } i \in \{1, \ldots, n-2\},$$
$$\omega_{n-1,j} = \{\omega_{n-2,k} : k \leq j\} \cup \{\omega_{n-1,k} : k > j\}.$$

so that, in particular, $\omega_{0,0} = \emptyset$. We omit a proof that sets ω_i are hypersets, that is, that their membership graphs are hyperextensional; we will prove this fact for $n = 2$ in Sect. 8.2.6. A graphical representation of sets ω_i, for $n = 3$, is depicted in Fig. 8.11.

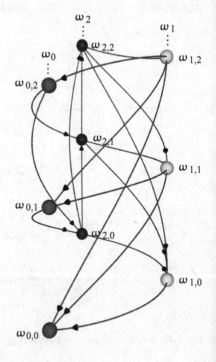

Fig. 8.11 A more suggestive three-dimensional representation of the membership graph of Fig. 8.10

---

[4]We use the shorthand $\subseteq$ in the context $a \in b \rightarrow b \subseteq a$ with the meaning $(\forall x)(a \in b \rightarrow (x \in b \rightarrow x \in a))$.

Fig. 8.12 Sketch of the
derivation of condition (iv)
of $\overline{u}$

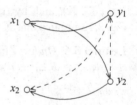

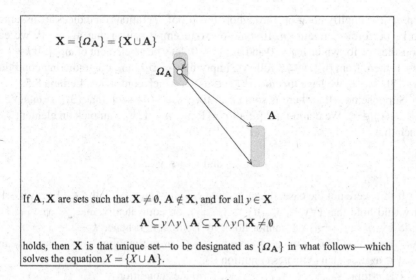

$\mathbf{X} = \{\Omega_{\mathbf{A}}\} = \{\mathbf{X} \cup \mathbf{A}\}$

$\Omega_{\mathbf{A}}$

$\mathbf{A}$

If $\mathbf{A}, \mathbf{X}$ are sets such that $\mathbf{X} \neq \emptyset$, $\mathbf{A} \notin \mathbf{X}$, and for all $y \in \mathbf{X}$

$$\mathbf{A} \subseteq y \wedge y \setminus \mathbf{A} \subseteq \mathbf{X} \wedge y \cap \mathbf{X} \neq \emptyset$$

holds, then $\mathbf{X}$ is that unique set—to be designated as $\{\Omega_{\mathbf{A}}\}$ in what follows—which
solves the equation $X = \{X \cup \mathbf{A}\}$.

Fig. 8.13 A plain consequence of AFA

The main difference between $\underline{u}^{[n]}$ and $u^{[n]}$ lies in the relaxation of condition
$\bigcup x_{n-1} \subseteq x_{n-2}$ into $\bigcup x_{n-1} \subseteq x_{n-2} \cup x_{n-1}$. Once we allow membership relations
to hold between elements of x_{n-1}, we can use them to "propagate" membership
relations across x_{n-2} and x_{n-1}. For example, when $n = 2$, condition $(\forall x_1, x_2 \in$
$a)(\forall y_1, y_2 \in b)(x_2 \in y_2 \in x_1 \in y_1 \rightarrow x_2 \in y_1)$ of $\overline{u}(a, b)$ is implied by condition
(iv') if $y_1 \in y_2$ holds (see Fig. 8.12).

We start by laying the groundwork for a proof that $\underline{u}^{[n]}$ is infinitely satisfiable—
closely following the proof method employed in Sect. 8.2.3—and then show two
ways in which the conditions governing the membership relations among elements
of x_{n-1} can be instantiated. In doing this, we will repeatedly exploit the plain
consequence of **AFA** given in Fig. 8.13.

Lemma 8.5 *Under* ZF$^-$ − FA + AFA, *for any sets* $\omega_0, \ldots, \omega_{n-1}$ *such that*
$\underline{u}^{[n]}(\omega_0, \ldots, \omega_{n-1})$ *is true,* $(\forall y \in \omega_{n-1})(\omega_{n-2} \not\subseteq y)$ *holds.*

Proof Arguing by contradiction, suppose this is not the case, so that the set $\mathbf{X} =$
$\{y \in \omega_{n-1} \mid \omega_{n-2} \subseteq y\}$ is nonnull. Then, by (ii') and (iv'), we have $(\forall y \in \mathbf{X})(y \setminus$

$\omega_{n-2} \subseteq \mathbf{X}$), and, by (i), that $(\forall y \in \mathbf{X})(y \cap \mathbf{X} \neq \emptyset)$. The observation made in Fig. 8.13 implies that $\mathbf{X} = \{\Omega_{\omega_{n-2}}\}$, contradicting the last conjunct of (ii').   ⊣

Lemma 8.6 *Under* $\mathsf{ZF}^- - \mathsf{FA} + \mathsf{AFA}$, *for any sets* $\omega_0, \ldots, \omega_{n-1}$ *such that* $\underline{\mathfrak{u}}^{[n]}(\omega_0, \ldots, \omega_{n-1})$ *is true, the following conditions hold, for any* $i \in \{0, \ldots, n-1\}$:

- $\omega_i \neq \emptyset$;
- $(\forall y \in \omega_i)(\exists z \in \omega_{(i+1) \bmod n})(y \in z)$.

Proof To simplify notation, throughout this proof operations on indices are assumed to be performed modulo n. If some ω_i were empty, given that $\omega_0 \neq \emptyset$, we can consider a k for which $\omega_k = \emptyset$ and $\omega_{k+1} \neq \emptyset$; let y be an element of ω_{k+1}. If $k+1 \neq n-1$, then, from (ii'), $y \subseteq \emptyset$ follows, implying $y = \emptyset = \omega_k$, contradicting condition (i). Otherwise, we have that $\omega_{n-2} \subseteq y \in \omega_{n-1}$, which contradicts Lemma 8.5.

Suppose now that there is some $k \in \{0, \ldots, n-1\}$ such that $(\exists y_k \in \omega_k)(\forall z \in \omega_{k+1})(y_k \notin z)$. We claim that, for all $i = 1, \ldots, n-1$, we can pick an element y_{k-i} such that

$$y_{k-i} \in \omega_{k-i} \text{ and } y_{k-i} \notin y_{k-i+1}.$$

If this were not the case, then for some $y_{k-j+1} \in \omega_{k-j+1}$, with $j \in \{1, \ldots, n-1\}$, it would hold that $(\forall y \in \omega_{k-j})(y \in y_{k-j+1})$, or, equivalently, $\omega_{k-j} \subseteq y_{k-j+1}$. The case $k - j + 1 = n - 1$ cannot hold, by Lemma 8.5; hence $k - j + 1 \neq n - 1$. As $\bigcup \omega_{k-j+1} \subseteq \omega_{k-j}$, we also have $y_{k-j+1} \subseteq \omega_{k-j}$, implying $y_{k-j+1} = \omega_{k-j}$ and $\omega_{k-j} \in \omega_{k-j+1}$. This violates condition (i).

The n-tuple $y_0, \ldots, y_k, y_{k+1}, \ldots, y_{n-1}$ violates condition

$$(\forall y_0 \in \omega_0, \ldots, \forall y_{n-1} \in \omega_{n-1})\left(\bigvee_{i=0}^{n-1} y_i \in y_{i+1}\right),$$

of $\underline{\mathfrak{u}}^{[n]}$, as $y_k \notin y_{k+1}$ follows from the initial assumption, in consequence of $y_{k+1} \in \omega_{k+1}$.   ⊣

Theorem 8.6 *Under* $\mathsf{ZF}^- - \mathsf{FA} + \mathsf{AFA}$, *for any sets* $\omega_0, \ldots, \omega_{n-1}$ *such that the following two conditions hold:*

- $\underline{\mathfrak{u}}^{[n]}(\omega_0, \ldots, \omega_{n-1})$,
- $(\forall y_1, y_2 \in \omega_{n-1})(y_1 \subseteq y_2 \vee y_2 \subseteq y_1)$,

we have that $\omega_0, \ldots, \omega_{n-1}$ *are infinite.*

Proof If ω_{n-2} is finite, let $Y \subseteq \omega_{n-1}$ be a finite set such that $(\forall y \in \omega_n)(\exists z \in Y)(y \in z)$. From Lemma 8.6 it follows that Y is nonempty, while the second condition of the hypothesis guarantees that we can take $y_0 \in Y$ to be $\subseteq$-maximal in Y. This entails that $\omega_{n-2} \subseteq y_0$, which contradicts Lemma 8.5.

Let now k, $0 \leqslant k \leqslant n - 1$, be the greatest index such that ω_k is infinite, but $\omega_{(k-1) \bmod n}$ is finite. Since $\bigcup \omega_k \subseteq \omega_{(k-1) \bmod n}$ ensues from condition (ii) of $\underline{\iota\iota}^{[n]}$, the claim readily follows, as in the proof of Lemma 8.2. ⊣

Corollary 8.1 *Under* ZF⁻ − FA + AFA, *for any sets* $\omega_0, \ldots, \omega_{n-1}$ *such that the following two conditions hold*

$$\underline{\iota\iota}^{[n]}(\omega_0, \ldots, \omega_{n-1}),$$

(v) $(\forall y_1, y_2 \in \omega_{n-1})(y_1 = y_2 \vee y_1 \in y_2 \vee y_2 \in y_1)$

we have that $\omega_0, \ldots, \omega_{n-1}$ *are infinite.*

Proof The above condition (v) and condition (iv′) of $\underline{\iota\iota}^{[n]}$ imply $(\forall y_1, y_2 \in \omega_{n-1})(y_1 \subseteq y_2 \vee y_2 \subseteq y_1)$; the claim then follows, by Theorem 8.6. ⊣

8.2.6 Non-well-founded, or Just Ill-Founded, Infinities

We focus now on capturing non-well-founded infinity with the least number of universally quantified variables or free variables and propose three such ∀∀ formulae.

The only ∀∀ formulation of infinity in a context deprived of FA is $\widetilde{\iota\iota}$ given in Sect. 8.2.4, which however is satisfied exclusively by well-founded sets. This raises the question of whether an infinite and "genuinely" ill-founded set can be captured with only two universal quantifiers; should a negative answer emerge, it would suggest the likelihood that the decision algorithm of [12], devised for ∀∀ formulae about ordinary well-founded sets, can be recast to cope with Aczel's sets.

As just done in Corollary 8.1, our first formula requires that a membership relation be present between any two distinct elements of x_{n-1}. Let thus $\underline{\iota\iota}_1(a, b)$ be the conjunction of the following sub-formulae:

(i) $a \neq b \wedge a \notin b \wedge b \notin a$
(ii′) $\bigcup a \subseteq b \wedge \bigcup b \subseteq a \cup b \wedge (\forall y \in b)(y \notin y)$
(iii) $(\forall x \in a)(\forall y \in b)(x \in y \vee y \in x)$
(iv′) $(\forall x \in a)(\forall y_1, y_2 \in b)(y_1 \in y_2 \to y_2 \subseteq y_1)$
(v) $(\forall y_1, y_2 \in b)(y_1 = y_2 \vee y_1 \in y_2 \vee y_2 \in y_1)$

Consider now three elements $x \in a$ and $y_1, y_2 \in b$, such that $y_2 \in x \in y_1$. Condition (v) of $\underline{\iota\iota}_1(a, b)$ imposes a membership arc to connect y_1 with y_2, but tells us nothing about its orientation. As our goal is to find formulae which have no well-founded models, it comes natural to impose $y_1 \in y_2$, in order to obtain the membership cycle $y_2 \in x \in y_1 \in y_2$. Consequently, we introduce the formula $\underline{\iota\iota}_2(a, b)$ be obtained from $\overline{\iota\iota}_1(a, b)$ by replacing condition (v) with:

(v′) $(\forall x \in a)(\forall y_1, y_2 \in b)(y_1 \neq y_2 \wedge y_2 \in x \in y_1 \to y_1 \in y_2)$

Notice that, analogously to the well-founded case, $\omega_0 \cap \omega_1 = \emptyset$ holds by (iii) and the last conjunct of (ii′), for any sets ω_0, ω_1 satisfying $\underline{\iota\iota}_1$ or $\underline{\iota\iota}_2$ (Fig. 8.14).

Fig. 8.14 Condition (iv′) of $\underline{u}_1(a,b)$ and of $\underline{u}_2(a,b)$ (*left*). Condition (v′) of $\underline{u}_2(a,b)$ (*right*)

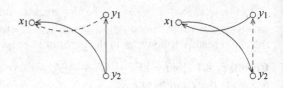

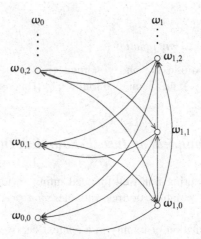

$\omega_0 = \{\omega_{0,j} : j \in \omega\}$, $\omega_{0,j} = \{\omega_{1,k} : 0 \leqslant k < j\}$,
$\omega_1 = \{\omega_{1,j} : j \in \omega\}$, $\omega_{1,j} = \{\omega_{0,k} : 0 \leqslant k \leqslant j\} \cup \{\omega_{1,k} : k > j\}$, $\forall j \in \omega$.

Fig. 8.15 A model of $\underline{u}_1 \wedge \underline{u}_2$

Theorem 8.7 *Under* ZF$^-$ − FA + AFA, $\underline{u}_1$ *and* $\underline{u}_2$ *are satisfiable.*

Proof A shared model ω_0, ω_1 for $\underline{u}_1, \underline{u}_2$ is shown in Fig. 8.15. We claim that $\underline{u}_1(\omega_0, \omega_1)$ and $\underline{u}_2(\omega_0, \omega_1)$ are true. At the outset, we will prove by induction on n that $\omega_{0,i} \neq \omega_{1,j}$ for all $i,j \in \{0, \ldots, n\}$ and that $\omega_{t,i} \neq \omega_{t,j}$ for all $i,j \in \{0, \ldots, n\}, i \neq j$, and $t \in \{0, 1\}$.

For $n = 0$, we have that $\omega_{0,0} \neq \omega_{1,0}$, as $\omega_{0,0} = \emptyset$, and $\emptyset \in \omega_{1,0}$. Supposing that the claim is true for n, we will show that it is also true for $n+1$. Since $\omega_{1,n} \in \omega_{0,n+1}$, and since $\omega_{1,n} \notin \omega_{0,i}$ holds for any $i \in \{0, \ldots, n\}$, we have that $\omega_{0,n+1} \neq \omega_{0,i}$ for any $i \in \{0, \ldots, n\}$. Moreover, for any $i \in \{0, \ldots, n\}$, $\omega_{0,n+1}$ differs from $\omega_{1,i}$, since $\omega_{1,i} \in \omega_{0,n+1}$, but $\omega_{0,n+1} \notin \omega_{1,i}$. In a similar manner, one can check that $\omega_{1,n+1} \neq \omega_{0,i}$ for any $i \in \{0, \ldots, n+1\}$ and that $\omega_{1,n+1} \neq \omega_{1,i}$ for any $i \in \{0, \ldots, n\}$.

The above statement guarantees that $\omega_0 \cap \omega_1 = \emptyset$ and that $\omega_0 \neq \omega_1$. Conditions (ii′)–(v), (v′) are also satisfied by the way ω_0 and ω_1 were constructed. Suppose now that there exists $i \in \omega$ such that $\omega_{0,i} = \omega_1$. Since $\omega_{1,i} \in \omega_1$ but $\omega_{1,i} \notin \omega_{0,i}$, we obtain a contradiction. Hence $\omega_1 \notin \omega_0$. Similarly, supposing that there exists $i \in \omega$ such that $\omega_{1,i} = \omega_0$, we observe that $\omega_{0,i+1} \in \omega_0$ but $\omega_{0,i+1} \notin \omega_{1,i}$, another contradiction, leading us to the conclusion $\omega_0 \notin \omega_1$. ⊣

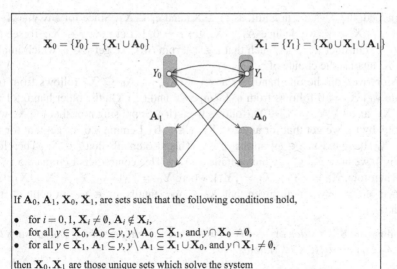

$$\mathbf{X}_0 = \{Y_0\} = \{\mathbf{X}_1 \cup \mathbf{A}_0\} \qquad\qquad \mathbf{X}_1 = \{Y_1\} = \{\mathbf{X}_0 \cup \mathbf{X}_1 \cup \mathbf{A}_1\}$$

If $\mathbf{A}_0$, $\mathbf{A}_1$, $\mathbf{X}_0$, $\mathbf{X}_1$, are sets such that the following conditions hold,

- for $i = 0, 1$, $\mathbf{X}_i \neq \emptyset$, $\mathbf{A}_i \notin \mathbf{X}_i$,
- for all $y \in \mathbf{X}_0$, $\mathbf{A}_0 \subseteq y$, $y \setminus \mathbf{A}_0 \subseteq \mathbf{X}_1$, and $y \cap \mathbf{X}_0 = \emptyset$,
- for all $y \in \mathbf{X}_1$, $\mathbf{A}_1 \subseteq y$, $y \setminus \mathbf{A}_1 \subseteq \mathbf{X}_1 \cup \mathbf{X}_0$, and $y \cap \mathbf{X}_1 \neq \emptyset$,

then $\mathbf{X}_0, \mathbf{X}_1$ are those unique sets which solve the system

$$X_0 = \{Y_0\},\ X_1 = \{Y_1\}, \quad Y_0 = X_1 \cup \mathbf{A}_0,\ Y_1 = X_0 \cup X_1 \cup \mathbf{A}_1$$

of equations.

Fig. 8.16 Another consequence of AFA

Even though it may seem that $\underline{\iota\iota}_2$ is just a particular case of $\underline{\iota\iota}_1$, in the following proposition we show that, in fact, under $\mathsf{ZF}^- - \mathsf{FA} + \mathsf{AFA}$, $\underline{\iota\iota}_2(a, b)$ holds whenever $\underline{\iota\iota}_1(a, b)$ holds. However, the converse is not true, as testified by the two hypersets ω_0', ω_1'—given in Proposition 8.1—which satisfy $\underline{\iota\iota}_2$, but without satisfying $\underline{\iota\iota}_1$. For this, let us point out in Fig. 8.16 another plain consequence of AFA, similar to the one given for $\widetilde{\iota\iota}$.

Lemma 8.7 *Under* $\mathsf{ZF}^- - \mathsf{FA} + \mathsf{AFA}$, *if* ω_0, ω_1 *are sets such that* $\underline{\iota\iota}_1(\omega_0, \omega_1)$ *is true, then* $\underline{\iota\iota}_2(\omega_0, \omega_1)$ *also holds.*

Proof Let $c_1, c_3 \in \omega_1$, $c_1 \neq c_3$, and $c_2 \in \omega_0$ such that $c_1 \in c_2 \in c_3$, but $c_3 \notin c_1$. From condition (v) of $\underline{\iota\iota}_1$ we have $c_1 \in c_3$, which implies that $c_2 \in c_1$ holds as well, from (iv').

Consider now the subset of elements of $\omega_0 \cup \omega_1$ from which there is an alternating membership chain between ω_0 to ω_1 to one of c_1 or c_2, that is,

$$\mathbf{X} = \{ y \colon y \in \omega_0 \cup \omega_1 \wedge \text{there exist } m \in \omega \text{ and } y_1, y_2, \ldots, y_m \in \omega_0 \cup \omega_1$$

such that $y = y_0 \ni y_1 \ni y_2 \ni \cdots \ni y_m$ and $y_m \in \{c_1, c_2\}$ and

$y_{2k} \in \omega_0$ and $y_{2k+1} \in \omega_1$, or $y_{2k} \in \omega_1$ and $y_{2k+1} \in \omega_0$, $\forall k$, $0 \leqslant k \leqslant (m-1)/2 \}$.

Then, we can apply the observation in Fig. 8.16, by taking $\mathbf{X}_0 = \mathbf{X} \cap \omega_0$, $\mathbf{X}_1 = \mathbf{X} \cap \omega_1$, and $\mathbf{A}_i = (\bigcup \mathbf{X}_i) \setminus \mathbf{X}$ for $i = 0, 1$ (thus, $\mathbf{X}_0 \cup \mathbf{A}_1 = \omega_0$ and $\mathbf{X}_1 \cup \mathbf{A}_0 = \omega_1$).

Indeed, the $\mathbf{X}_i$'s are nonnull, as $c_2 \in \mathbf{X}_0$ and $c_1 \in \mathbf{X}_1$. Since for any $y \in \mathbf{X}$, we have $y \cap \mathbf{X} \neq \emptyset$, we deduce $\mathbf{A}_i \notin \mathbf{X}_i$, for $i = 0, 1$. Let now $y \in \mathbf{X}_i$. To see that $\mathbf{A}_i \subseteq y$, consider a $z \in \mathbf{A}_i$ such that $z \notin y$. From (iii), we get $y \in z$, which implies $z \in \mathbf{X}$, against the choice of z.

Moreover, on the one hand, for any $y \in \mathbf{X}_0$, $y \setminus \mathbf{A}_0 \subseteq \mathbf{X}_1$ follows from (ii'), while $y \cap \mathbf{X}_0 = \emptyset$ follows from $\omega_0 \cap \omega_1 = \emptyset$ and (ii'). On the other hand, for any $y \in \mathbf{X}_1$, also $y \setminus \mathbf{A}_1 \subseteq \mathbf{X}_0 \cup \mathbf{X}_1$ follows from (ii'). Supposing now that $y \cap \mathbf{X}_1$ were empty, by (v) we get that for any $z \in \mathbf{X}_1$, $y \in z$. By Lemma 8.6 we get that for any $v \in \mathbf{X}_0$, there exists $z \in \omega_1$ so that $v \in z$. This also entails that $z \in \mathbf{X}_1$. Therefore, by (iv'), we have $\mathbf{X}_0 \subseteq y$, which entails $\omega_0 \subseteq y$. This contradicts Lemma 8.5.

Therefore, $\mathbf{X}_0 = \{\mathbf{Y}_0\}, \mathbf{X}_1 = \{\mathbf{Y}_1\}$, where $\mathbf{Y}_0 = \mathbf{X}_1 \cup \mathbf{A}_0$, $\mathbf{Y}_1 = \mathbf{X}_0 \cup \mathbf{X}_1 \cup \mathbf{A}_1$. Hence $\mathbf{Y}_0 = \omega_1$, which, given that $\mathbf{X}_0 \subseteq \omega_0$, entails $\omega_1 \in \omega_0$, in contradiction with (i). $\qquad\dashv$

Proposition 8.1 *Under* $\mathsf{ZF^-}-\mathsf{FA}+\mathsf{AFA}$, *there exist hypersets* ω_0' *and* ω_1' *such that* $\underline{\mathit{ll}}_2(\omega_0', \omega_1') \wedge \neg\underline{\mathit{ll}}_1(\omega_0', \omega_1'))$ *holds.*

Proof To obtain such a model ω_0', ω_1' of $\underline{\mathit{ll}}_2 \wedge \neg\underline{\mathit{ll}}_1$, it will suffice to replace, in the model ω_0, ω_1 of $\underline{\mathit{ll}}_2$, vertex $\omega_{1,0}$ by five new vertices connected to one another as shown in Fig. 8.17 (and connecting each of them with the rest of the vertices in the same way $\omega_{1,0}$ was connected with). The proof of Theorem 8.7 can be closely followed to show that these sets are indeed hypersets.

It can easily be seen that $\underline{\mathit{ll}}_2(\omega_0', \omega_1')$ is true. However, $\underline{\mathit{ll}}_1(\omega_0', \omega_1'')$ does not hold, since $\omega_{1,0}^3, \omega_{1,0}^4 \in \omega_0''$ but $\omega_{1,0}^3 \neq \omega_{1,0}^4$, $\omega_{1,0}^3 \notin \omega_{1,0}^4$, and $\omega_{1,0}^4 \notin \omega_{1,0}^3$, in conflict with condition (v) of $\underline{\mathit{ll}}_1(\omega_0', \omega_1')$. $\qquad\dashv$

Theorem 8.8 *Under* $\mathsf{ZF^-}-\mathsf{FA}+\mathsf{AFA}$, *if* ω_0, ω_1 *are sets such that either* $\underline{\mathit{ll}}_1(\omega_0, \omega_1)$ *or* $\underline{\mathit{ll}}_2(\omega_0, \omega_1)$ *is true, then* ω_0 *and* ω_1 *are infinite.*

Proof We will show that the second condition of the hypothesis of Theorem 8.6 holds in both cases. If $\underline{\mathit{ll}}_2(\omega_0, \omega_1)$ is true, then assume for a contradiction that $y_1, y_2 \in \omega_1$ and $x_1, x_2 \in \omega_0$ are such that $x_1 \in y_1 \setminus y_2$ and $x_2 \in y_2 \setminus y_1$. By (iii), we get $y_2 \in x_1$, which by (v') implies $y_1 \in y_2$, and thus $y_2 \subseteq y_1$, by (iv'). If $\underline{\mathit{ll}}_1(\omega_0, \omega_1)$ is true, the claim readily follows from conditions (iv') and (v) (and also from the above argument, in light of Lemma 8.7). $\qquad\dashv$

Fig. 8.17 Five new vertices replacing $\omega_{1,0}$ in the model of $\underline{\mathit{ll}}_2 \wedge \neg\underline{\mathit{ll}}_1$

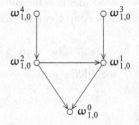

Having established that any sets ω_0, ω_1 satisfying $\underline{u}_1$ or $\underline{u}_2$ are infinite, we are concerned with characterizing their models. The following proposition, similar to a statement in the proof of [89, Proposition 6], characterizes in more detail the structure of $\underline{u}_1$ and $\underline{u}_2$.

Proposition 8.2 *Assuming that ω_0 and ω_1 satisfy $\underline{u}_2$, if X is a finite nonnull subset of $\omega_0 \cup \omega_1$, then there is an element $c_X \in X$ such that one of the following two properties hold:*

- $c_X \in \omega_0$ *and* $X \cap \omega_1 \subseteq c_X$,
- $c_X \in \omega_1$ *and* $X \cap \omega_0 \subseteq c_X$.

Proof We will prove the claim by induction on the cardinality of X. If X is a singleton the claim is clear, since $\omega_0 \cap \omega_1 = \emptyset$. Otherwise, if $X \cap \omega_0 = \emptyset$, then every element in X can be taken as c_X. Otherwise, pick $a_* \in X \cap \omega_0$ and let $X' = X \setminus \{a_*\}$.

By the induction hypothesis applied to X', there is a $c_{X'} \in X'$ satisfying our claim. If $c_{X'} \in \omega_0$ and $X' \cap \omega_1 \subseteq c_{X'}$, then, since $\omega_0 \cap \omega_1 = \emptyset$, we have $X \cap \omega_1 = X' \cap \omega_1$. Hence $X \cap \omega_1 \subseteq c_{X'}$ and we can take c_X to be $c_{X'}$. On the other hand, if $c_{X'} \in \omega_1$ and $X' \cap \omega_0 \subseteq c_{X'}$, we have two cases: $a_* \in c_{X'}$ and $a_* \notin c_{X'}$.

In the former case, it suffices to take $c_X = c_{X'}$. In the latter, since $(\forall x \in \omega_0)(\forall y \in \omega_1)(x \in y \vee y \in x)$, $c_{X'} \in a_*$. If $\omega_1 \cap X \subseteq a_*$, then it suffices to let $c_X = a_*$. Otherwise there must be a $b_* \in \omega_1 \cap X$ such that $b_* \notin a_*$. Hence, as before, we have $a_* \in b_*$. Since $b_* \neq c_{X'}$, from condition (v') we have $b_* \in c_{X'}$, and from condition (iv') we get $X' \cap \omega_0 \subseteq b_*$; so we can take $c_X = b_*$.   ⊣

Actually, this rich information about the structure of any model of $\underline{u}_2$ is enough to provide a second proof of the fact that $\underline{u}_2$ (and hence, by Lemma 8.7, that also $\underline{u}_1$) is infinitely satisfiable.

Proof (Theorem 8.8) Assume that ω_0, ω_1 satisfy $\underline{u}_2$ and that $\omega_0 \cap \omega_1$ is finite. From Proposition 8.2 applied to $X = \omega_0 \cap \omega_1$, we can find $c_X \in X$ satisfying one of the two claims of that proposition. In the first case, due to (ii'), we have $c_X = \omega_1$, contradicting (i). In the second case, $\omega_0 \subseteq c_X$ contradicts Lemma 8.5.   ⊣

To see that that all models of $\underline{u}_2$ are non-well-founded, let ω_0 and ω_1 satisfy $\underline{u}_1$ or $\underline{u}_2$. By Lemma 8.6, we can take $y_1, y_2 \in \omega_1$ and $x \in \omega_0$ such that $y_2 \in x \in y_1$. If $\underline{u}_2(\omega_0, \omega_1)$ holds, then $y_1 \in y_2$ follows, which produces a membership cycle in both $\mathsf{trCl}\,(\omega_0)$ and $\mathsf{trCl}\,(\omega_1)$ (transitive closures which, incidentally, due to (ii) and to Lemma 8.6, coincide).

Moreover, in both models ω_0, ω_1 proposed for $\underline{u}_1$ or for $\underline{u}_2$, there is an infinite descending membership chain in ω_1 with no repeated elements. In Proposition 8.3, we show that this property holds for any model of either $\underline{u}_1$ or $\underline{u}_2$, in blatant violation of **FA**. The following is a preparatory lemma.

Lemma 8.8 *If ω_0 and ω_1 satisfy $\underline{u}_2$, then for all $y \in \omega_1$, $y \cap \omega_1 \neq \emptyset$.*

Proof Let $y_1 \in \omega_1$. By Lemma 8.6, we can find $y_2 \in \omega_0$ and $y_3 \in \omega_1$ such that $y_1 \in y_2 \in y_3$. This implies, by condition (v'), that $y_3 \in y_1$, and hence $y_1 \cap \omega_1 \neq \emptyset$.   ⊣

In the following proposition, we resort to a useful variant of the transitive closure operation: the *relativized transitive closure operation*, depending on a parameter b, which sends every set s into the set $\mathsf{trCl}_b(s)$ formed by those s' which can reach s through membership without ever leaving b. The following definition readjusts the earlier definition of transitive closure to our current needs:

Definition 8.2 Given sets b and s, we define $\mathsf{trCl}_b(s)$ to be the set formed by those s' for which there is a finite-length path

$$s = s_0 \ni \cdots \ni s_{n+1} = s'\,.$$
$$\overset{\cap}{b} \quad \cdots \quad \overset{\cap}{b}$$

Proposition 8.3 *If ω_0 and ω_1 satisfy $\underline{u}_1$ or $\underline{u}_2$, then ω_1 contains an infinite descending membership chain with no repeated elements.*

Proof As ω_1 is infinite (from Theorem 8.8), Lemma 8.8 implies that the set $Z = \{z \in \omega_1 \mid z \in \mathsf{trCl}_{\omega_1}(\{z\})\}$ is nonnull. Moreover, we can find a $z_0 \in Z$ such that for all $z \in Z \setminus \{z_0\}$, $\mathsf{trCl}_{\omega_1}(\{z_0\}) \not\subseteq \mathsf{trCl}_{\omega_1}(\{z\})$. From condition (iv′) and the choice of z_0, we have that $(\forall y \in \mathsf{trCl}_{\omega_1}(\{z_0\}))(y \cap \omega_0 = z_0 \cap \omega_0)$. Additionally, $(\forall y \in \mathsf{trCl}_{\omega_1}(\{z_0\}))(\emptyset \neq y \cap \omega_1 \subseteq \mathsf{trCl}_{\omega_1}(\{z_0\}))$, and hence $z_0 = \Omega_{z_0 \cap \omega_0}$ entailing $z_0 \in z_0$. This violates the condition $(\forall y \in \omega_1)(y \notin y)$. ⊣

For the remainder of this section, we refrain again from assuming **FA** or **AFA**. The following formula $\overline{oi}$—the conjunction of the subsequent four conditions— is a variation of $\underline{u}_2$, so that the inclusion appearing in condition (iv′) is reversed; accordingly, (ii′) has to be changed into (ii″). These changes also guarantee that no further condition governing the membership relations among elements of b has to be required. Just like for $\overline{u}$, the working of $\overline{oi}$ is immaterial of **FA** or of **AFA**; however, this is now obtained with only three universal quantifiers, instead of the four of $\overline{u}$. Moreover, $\overline{oi}$ is satisfied only by non-well-founded sets, since $(\forall y \in b)(y \in y)$ follows from the last conjunct of (ii″) and from (iii).

(i) $a \neq b \wedge a \notin b \wedge b \notin a$
(ii″) $\bigcup(a \setminus b) \subseteq b \wedge \bigcup b \subseteq a \wedge b \subseteq a$
(iii) $(\forall x \in a)(\forall y \in b)(x \in y \vee y \in x)$
(iv″) $(\forall y_1, y_2 \in b)(y_1 \in y_2 \rightarrow y_1 \subseteq y_2)$

A hyperset model of $\overline{oi}$ is depicted in Fig. 8.18; the proof of Theorem 8.7 can be closely followed to show that these sets are indeed hypersets. Our first lemma is closely analogous to Lemma 8.1.

Lemma 8.9 *If ω_0, ω_1 are sets such that $\overline{oi}(\omega_0, \omega_1)$ is true, then $(\forall x \in \omega_0 \setminus \omega_1)(\exists y \in \omega_1)(x \in y)$.*

Proof Assume for a contradiction that $(\forall y \in \omega_1)(x_0 \notin y)$ holds for some $x_0 \in \omega_0 \setminus \omega_1$. By (iii), we have that $(\forall y \in \omega_1)(y \in x_0)$. Thus, since $\omega_1 \subseteq x_0 \subseteq \bigcup(\omega_0 \setminus \omega_1) \subseteq \omega_1$ by (ii′), we have $x_0 = \omega_1$, contradicting $\omega_1 \notin \omega_0$. ⊣

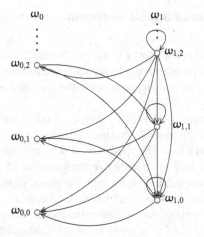

$$\omega_0 = \{\omega_{1,j} : i \in \omega\}, \qquad \omega_{0,j} = \{\omega_{1,k} : 0 \leqslant k < j\},$$
$$\omega_1 = \{\omega_{0,j} : i \in \omega\} \cup \omega_0,\ \omega_{1,j} = \{\omega_{0,k} : 0 \leqslant k \leqslant j\} \cup \{\omega_{1,k} : k \leqslant j\},\ \forall j \in \omega.$$

Fig. 8.18 Hypersets ω_0 and ω_1 such that $\overline{oi}(\omega_0, \omega_1)$ holds

Lemma 8.10 *If* ω_0, ω_1 *are sets such that* $\overline{oi}(\omega_0, \omega_1)$ *is true, then* $(\forall y_1, y_2 \in \omega_1)(y_1 \subseteq y_2 \vee y_2 \subseteq y_1)$.

Proof Immediate from the conjunct $\omega_1 \subseteq \omega_0$ of (ii″), (iii) and (iv″). $\dashv$

Lemma 8.11 *If* ω_0, ω_1 *are sets such that* $\overline{oi}(\omega_0, \omega_1)$ *is true, and* ω_1 *is finite, then* $(\exists y \in \omega_1)(\omega_1 \subseteq y)$.

Proof Suppose that this were not the case, and consider $y_0 \in \omega_1$ such that $y_0 \cap \omega_1$ is $\subseteq$-maximal. Let $y_1 \in \omega_1 \setminus y_0$. From (ii″) and (iii) we get $y_0 \in y_1$, which implies, by (iv″), that $y_0 \subseteq y_1$. Likewise we have $y_1 \in y_1$, which contradicts the maximality of y_0. $\dashv$

Theorem 8.9 *Under* $\mathsf{ZF}^- - \mathsf{FA}$, *if* ω_0, ω_1 *are sets such that* $\overline{oi}(\omega_0, \omega_1)$ *is true, then* ω_0 *and* ω_1 *are infinite.*

Proof Notice first that $\omega_1 \neq \emptyset$, since otherwise, by (ii″), either $\omega_0 = \emptyset$ or $\omega_0 = \{\emptyset\}$ would hold. This contradicts (i). Now, assume for a contradiction that ω_1 is finite. By Lemma 8.10, let y_0 stand for the $\subseteq$-maximum element of ω_1. From Lemma 8.9, we have that $\omega_0 \setminus \omega_1 \subseteq y_0$. From Lemma 8.11, let $y_1 \in \omega_1$ such that $\omega_1 \subseteq y_1$. From the maximality of y_0, we have $\omega_1 \subseteq y_1 \subseteq y_0$, and hence $\omega_0 \subseteq y_0$. Moreover, $y_0 \subseteq \bigcup \omega_1 \subseteq \omega_0$, entailing $y_0 = \omega_0$, contradicting $\omega_0 \notin \omega_1$. Note that the infinitude of ω_1 also implies the infinitude of ω_0, since $\bigcup \omega_1 \subseteq \omega_0$. $\dashv$

8.3   The *Spectrum* of a BSR Formula

In this final section, we prove that by using the "regular" form of infinity discussed in the previous section, together with the set-theoretic translation of logical formulae introduced in Sect. 8.1, we can elegantly solve the so-called *spectrum* problem for a BSR-formula.

In his celebrated paper [98], F. P. Ramsey studied the Bernays-Schönfinkel class with equality, motivated by the goal of characterizing the *spectrum* of a generic formula in that class. The spectrum of Φ is the collection of cardinalities of models of Φ; being able to characterize the spectrum of a formula goes a long way toward a deep understanding of its expressiveness. Ramsey's result—obtained by applying what today everybody dubs *Ramsey theorem*—is that given Φ in the BSR class, a threshold $\mathfrak{r}(\Phi)$ can be determined such that either all models have cardinality smaller than $\mathfrak{r}(\Phi)$, or models of *all* cardinalities greater than $\mathfrak{r}(\Phi)$ are available for Φ: a sort of "sophisticated" pumping lemma on models holds. Incidentally, Ramsey's construction of large models is illustrated by carrying out a syntactic analysis indicating—when possible—*where* to "pump" new elements into the model. The considerations condensed in his combinatorial preparatory result turn out to generalize the so-called pigeonhole principle.

In this section we illustrate a result—first proved in [78]—stating that the spectrum problem for Φ can be fully and compactly "rendered" in set-theoretic terms, thinking of tuples of sets as graphs. We will show that Φ has models of arbitrarily large cardinalities if and only if its set-theoretic "rendering" Φ^σ is set-theoretically satisfiable.

The reader can check (see Exercise 8.6) that an encoding functionally equivalent to the one presented here could have been carried out in pure logic as well. This, however, would have called into play a pretty large logical formula. The small size of Φ^σ makes it evident that much is gained, in terms of expressiveness, when stepping from $\mathscr{L}_E$ to $\mathscr{L}_\in$.

Let us begin by recalling the basics of Ramsey's combinatorial result: it is a proof of the fact that given a sufficiently large domain and a partition of all its k-tuples, an m-element *homogeneous* subset of the domain can always be found. The key notion is that of *homogeneity*: a set is homogeneous if every k-tuple of its elements belongs to the *same* class of the partition.

Since we are going to represent tuples of sets as (directed) graphs, our first task is to adapt the notion of homogeneity to the realm of graphs. To this end we begin by introducing the notion of *n-type*.

> **Definition 8.3**   Given a graph $G = (\{1, \ldots, n, n+1, \ldots, n+\ell\}, E)$, the *n-type* of $w \in \{n+1, \ldots, n+\ell\}$ is the graph $G' = (\{0, 1, \ldots, n\}, E')$, where
>
> $$E' = E \cup \{\langle 0, v \rangle \colon \langle w, v \rangle \in E\}.$$

In other words, the *n*-type of w is obtained focusing on the relationship of w (which is rewritten as 0 in G') with the special vertices $\{1, \ldots, n\}$. Since an *n*-type of a vertex is nothing but a graph with vertex set $\{0, 1, \ldots, n\}$, comparing two n-types means checking whether they have the same set of arcs.

By $G \upharpoonright \tau$, for a generic n-type τ of a vertex of G as above, we indicate the subgraph of G induced by all vertices of type τ. That is, $G \upharpoonright \tau$ is the graph that results from G when every vertex of type other than τ is eliminated.

Definition 8.4 We say that a graph $G = (\{1, \ldots, n, n + 1, \ldots, n + \ell\}, E)$ is *n-homogeneous* if:

- all $w \in \{n + 1, \ldots, n + \ell\}$ have the same n-type;
- $\{n + 1, \ldots, n + \ell\}$ *induces* either a clique or an independent set in the underlying undirected graph of G.

Thus, in an n-homogeneous graph G, all vertices not in $\{1, \ldots, n\}$ "behave" in the same way and, moreover, their mutual relationships are "easy."

The following result is a consequence of Ramsey's (combinatorial) theorem. As a matter of fact, the only instance of Ramsey's theorem that we are going to use in our proof is the following one: for every j there exists an i_j such that any undirected graph G of size at least i_j, contains either a clique or an independent set of size at least j.

Lemma 8.12 *Let $G_1, G_2, G_3, \ldots$ be an infinite sequence of (directed) graphs $G_j = (\{1, \ldots, n, n + 1, \ldots, n + \ell_j\}, E_j)$, where $0 < \ell_1 < \ell_2 < \cdots$. Then, there exist an n-type τ and infinitely many indexes $i_0, i_1, i_2, \ldots$, such that for all $j \in \mathbb{N}$:*

- *$G_{i_j} \upharpoonright \tau$ has an n-homogeneous induced subgraph Γ_j,*
- *Γ_j has size $n + j$.*

Proof For $j \in \mathbb{N}$, let $\tau_{j,1}, \ldots, \tau_{j,\ell_j}$ be the types of $n + 1, \ldots, n + \ell_j$ in G_j. Since an n-type is nothing but a graph with vertex set $\{0, 1, \ldots, n\}$, altogether, the number of distinct n-types is bounded by $2^{(n+1)^2}$. Let $t_j(\tau)$ be the number of times τ occurs within each sequence $\tau_{j,1}, \ldots, \tau_{j,\ell_j}$. Since the sizes of the G_j's are strictly increasing, there must exist an n-type τ such that the set $\{t_1(\tau), t_2(\tau), t_3(\tau), \ldots\}$ has no maximum. Indeed, arguing by contradiction and indicating by t_τ the maximum corresponding to each τ, we would have $\ell_k \leq \sum_\tau t_\tau$ for any k, contradicting $0 < \ell_1 < \ell_2 < \cdots$.

This plainly implies that we can extract an infinite subsequence $G_{i'_0}, G_{i'_1}, G_{i'_2}, \ldots$ of $G_1, G_2, G_3, \ldots$ so that the graphs $G_{i'_0} \upharpoonright \tau, G_{i'_1} \upharpoonright \tau, G_{i'_2} \upharpoonright \tau, \ldots$ have increasing sizes. By the finite Ramsey theorem (specifically, Theorem C of [98]), we can extract from the sequence of the $G_{i'_j} \upharpoonright \tau$'s a subsequence $G_{i_0}, G_{i_1}, G_{i_2}, \ldots$ so that the undirected graph of every $G_{i_j} \setminus \{1, \ldots, n\}$ contains either a clique, or an independent set, of size greater than or equal to j.

For $j \in \mathbb{N}$, can now easily get an Γ_j as an n-homogeneous induced subgraph of G_{i_j}, of size $n + j$. ⊣

So far we have sieved a sequence of graphs based on some properties of their undirected underlying graphs related to independent sets or cliques. In our next result, we do a final sieving based on their directed structure. This makes use of Lemma 8.12 above and of following instance of Ramsey's theorem, this time applied to (directed) graphs (see [34]): for every j there exists an i_j such that any tournament G of size at least i_j contains a *subgraph isomorphic to an* acyclic tournament of size at least j.

Corollary 8.2 *Let $G_1, G_2, G_3, \ldots$ be an infinite sequence of graphs such that each $G_j = (\{1, \ldots, n, n+1, \ldots, n+\ell_j\}, E_j)$ and $0 < \ell_1 < \ell_2 < \cdots$. Then, there exist an n-type τ and infinitely many indexes $i_0, i_1, i_2, \ldots$, such that for all $j \in \mathbb{N}$, $G_{i_j} \upharpoonright \tau$ has an n-homogeneous induced subgraph Γ_j and either:*

- *for all $j \in \mathbb{N}$, the vertices of $\Gamma_j \setminus \{1, \ldots, n\}$ form an independent set of cardinality j, or*
- *for all $j \in \mathbb{N}$, $\Gamma_j \setminus \{1, \ldots, n\}$ is an acyclic tournament of cardinality j.*

Proof By applying Lemma 8.12 we obtain an n-type τ and a subsequence $G_{i'_0}$, $G_{i'_1}, G_{i'_2}, \ldots$, such that every $G_{i'_j}$ contains either an independent set or a tournament of size j.

In case there are infinitely many independent sets, it is easy to build $G_{i_0}, G_{i_1}, G_{i_2}, \ldots$ as a subsequence. Otherwise we must have tournaments of arbitrarily large sizes, and hence applying Ramsey theorem for directed graphs, we can again build a subsequence satisfying our claim.   ⊣

The n distinguished vertices $\{1, \ldots, n\}$ mentioned in the above corollary constitute a (fixed) "scenario" for the satisfiability of a BSR formula: one among the (many) possible networks of membership relations for the interpretations of the existentially quantified variables. The above result can then be seen as a recasting of Ramsey's result on the existence of (arbitrarily) large homogeneous sets, to the case of arbitrarily large graphs with n distinguished/fixed vertices—i.e., possible "scenarios." Notice that the membership network among vertices different from $\{1, \ldots, n\}$ in Γ_j is, in any case, extremely simple either no memberships or an acyclic tournament.

Theorem 8.10 *To each BSR sentence Φ in $\mathscr{L}_E$, there corresponds a BSR sentence Φ^σ in $\mathscr{L}_\in$, of size $|\Phi^\sigma| = O\left(|\Phi|^2\right)$, such that Φ is satisfiable by arbitrarily large models if and only if Φ^σ is satisfiable in well-founded Set Theory.*

Proof Consider a BSR sentence

$$\Phi \equiv \exists x_1, \ldots, x_n \forall y_1, \ldots, y_m \; \varphi(x_1, \ldots, x_n, y_1, \ldots, y_m),$$

where φ is an unquantified matrix in the language $\mathscr{L}_E$.

In preparation for the definition of Φ^σ, consider a hypothetical sequence $G_i = (D_i, E_i)$, with $i \in \mathbb{N}$, of (directed) graphs of increasing sizes that satisfy Φ; each G_i will hence have n distinguished vertices $v_1^i, \ldots, v_n^i \in D_i$ used to interpret $x_1, \ldots, x_n$, and we are supposing that the D_i's have strictly increasing cardinalities (if not, we can achieve this by sieving out a subsequence of the G_i's before moving on).

We will amalgamate (copies of) all G_i's together *inside* $d_0 \cup d_1$, where d_0, d_1 are infinite sets satisfying the formula $\iota\iota(d_0, d_1)$ seen in Sect. 8.2.1.

The embedding of each G_i in $d_0 \cup d_1$ is a modification of the one employed in Theorem 8.1 and can be described as follows: for every vertex v_k^i we introduce a set $x_{s,k} \in d_0$, acting as its representative when v_k^i is considered as source; moreover, we introduce n vertices $x_{t,k,1}, \ldots, x_{t,k,n} \in d_1$ each acting as a potential target (of an arc outgoing from $x_{s,1}, \ldots, x_{s,n}$, respectively) when v_k^i is playing the role of target. The matrix φ^σ of Φ^σ will be designed so as to impose the constraints needed to tie $\in$ with E_i, while $\iota\iota(d_0, d_1)$ will ensure that sufficiently many targets—respecting the corresponding membership conditions—can always be found in d_1.

All the subformulae to be used must be intended (and verified) to be shortcuts for set-theoretic pure BSR formulae.

For any $i \in \mathbb{N}$, consider now a generic element $w^i \in D_i \setminus \{v_1^i, \ldots, v_n^i\}$ and consider the subgraph $G_i[\{w^i, v_1^i, \ldots, v_n^i\}]$ of G_i induced by $\{w^i, v_1^i, \ldots, v_n^i\}$. For any $i, j \in \mathbb{N}$ we say that $G_i[\{w^i, v_1^i, \ldots, v_n^i\}]$ is *isomorphic* to $G_j[\{w^j, v_1^j, \ldots, v_n^j\}]$ if the correspondence sending $v_1^i, \ldots, v_n^i$ to $v_1^j, \ldots, v_n^j$ and w^i to w^j, respectively, is an isomorphism with respect to the arc relation. In formulae:

$$G_i[\{w^i, v_1^i, \ldots, v_n^i\}] \cong G_j[\{w^j, v_1^j, \ldots, v_n^j\}].$$

We will assume that the sequence of the G_i's enjoys the following properties:

(i) for all $i, j \in \mathbb{N}$, given any $w^i \in D_i \setminus \{v_1^i, \ldots, v_n^i\}$ and any $w^j \in D_j \setminus \{v_1^j, \ldots, v_n^j\}$,

$$G_i[\{w^i, v_1^i, \ldots, v_n^i\}] \cong G_j[\{w^j, v_1^j, \ldots, v_n^j\}];$$

(ii) for all $i, j \in \mathbb{N}$, if $i < j$, then $|D_i| < |D_j|$;

(iii) either $D_i \setminus \{v_1^i, \ldots, v_n^i\}$ is an independent set in G_i, for all $i \in \mathbb{N}$,
   or $D_i \setminus \{v_1^i, \ldots, v_n^i\}$ is an acyclic tournament in G_i, for all $i \in \mathbb{N}$.

Should these conditions not be met, Corollary 8.2 tells us how we can enforce them by replacing the G_i's by a suitably related sequence of Γ_j's.

We are now in the position to define Φ^σ and to prove our main claim. Let Φ^σ be the following formula:

$$(\exists d_0, d_1)(\exists x_{s,1}, \ldots, x_{s,n+1} \in d_0)(\exists X_{t,1}, \ldots, X_{t,n+1} \subseteq d_0 \cup d_1)(\exists Y_s)(\exists \ell \in d_1)\Big(\iota\iota(d_0, d_1)$$

$$\wedge \bigwedge_{k=1}^{n+1} (X_{t,k} = \{x_{t,k,1}, \ldots, x_{t,k,n+1}\} \wedge X_{t,k} \cap d_1 \subseteq x_{s,k}) \wedge Y_s = (d_0 \setminus \ell) \cup \{x_{s,1}, \ldots, x_{s,n+1}\}$$

$$\wedge (\forall y_{s,1}, \ldots, y_{s,m} \in Y_s)\varphi^\sigma(x_{s,1}, \ldots, x_{s,n+1}, x_{t,1,1}, \ldots, x_{t,n+1,n+1}, y_{s,1}, \ldots, y_{s,m})\Big)$$

where φ^σ is obtained from φ by replacing every literal of the form $z_h \in w_j$, with $z, w \in \{x, y\}$, by $z_{s,h} \ni w_{t,h,j}$ with $w_{t,h,j} \equiv x_{t,h,n+1}$ when $w_j \equiv y_j$, and every literal of the form $z_h = w_j$ by $z_{s,h} = w_{s,j}$.

It is plain (see Sect. 2.2 and Exercise 8.7) that this can be formulated in $\mathscr{L}_{\in}$. Moreover, the fact that $|\varPhi^\sigma| \in O(|\varPhi|^2)$ immediately follows from the fact that the number of (set) variables introduced in $\varPhi^\sigma$ is quadratic and the size of the matrix φ^σ is equal to the size of φ.

We begin by proving that if $\varPhi$ is satisfiable by models of arbitrarily large cardinalities, then $\varPhi^\sigma$ is satisfiable in well-founded Set Theory.

Under our hypothesis, as observed above, we can assume we have a sequence of models G_i, for $i \in \mathbb{N}$ such that *(i)*, *(ii)*, and *(iii)* hold.

We claim that we can determine:

(a) $n + 1$ elements $\alpha_1, \ldots, \alpha_{n+1} \in d_0$,
(b) $(n + 1)(n + 2)$ elements $\beta_{k,1}, \ldots, \beta_{k,n+2} \in d_0 \cup d_1$, with $k = 1, \ldots, n + 1$,

so that, for any G_i, there is an arc from:

(1) v_j^i to v_k^i if and only if $\alpha_j \ni \beta_{k,j}$,
(2) w^i to v_k^i if and only if $\alpha_{n+1} \ni \beta_{k,n+1}$,
(3) v_j^i to w^i if and only if $\alpha_j \ni \beta_{j,n+1}$,
(4) w^i to w^i if and only if $\alpha_{n+1} \ni \beta_{n+1,n+1}$,
(5) two distinct vertices in $D_i \setminus \{v_1, \ldots, v_n\}$ if and only if $\alpha_{n+1} \ni \beta_{n+1,n+2}$.

The α's satisfying a) are used to interpret $x_{s,1}, \ldots, x_{s,n+1}$, respectively, while the β's satisfying b) are used to interpret $x_{t,k,1}, \ldots, x_{t,k,n+1}$, respectively.

See Fig. 8.19, where we depicted a scenario in which the various choices described in (a) and (b) have been made on *stripes* to be seen as associated with

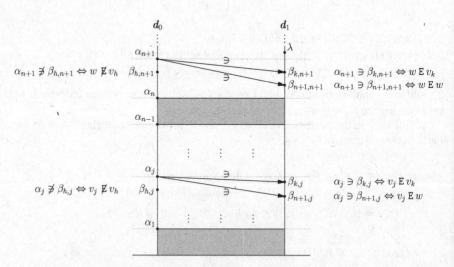

Fig. 8.19 A possible scenario for the choice of $\alpha_1, \ldots, \alpha_n$ (corresponding to $v_1, \ldots, v_n$ and hence to $x_1, \ldots, x_n$). This illustrates, among other things, the encodings of: presence of an arc between v_j and v_k and between v_j and w; absence of the arc between v_j and v_h. The elements have been chosen in *stripes*, the last stripe being associated with α_{n+1} (which corresponds to w and hence to a generic universal variable)

$\alpha_1, \ldots, \alpha_n$, followed by a stripe associated with $\alpha_{n+1} \in Y_s$. Let λ be the element in d_1 of minimum rank greater than the rank of α_{n+1}. The elements above α_{n+1} in d_0 are exactly the elements in $d_0 \setminus \lambda$ and are meant to constitute, along with $\alpha_1, \ldots, \alpha_n$, the *infinite* interpretation of Y_s.

In Y_s, in fact, all the domains of the G_i's are "glued" together, and the validity of $\alpha_{n+1} \ni \beta_{n+1,n+2}$ tells whether the $D_i \setminus \{v_1, \ldots, v_n\}$'s are tournaments or independent sets.

To see that our claim holds, it is sufficient to recall that each of d_0 and d_1 has infinitely many elements and that for any pair of elements $a \in d_0, b \in d_1$, either $a \in b$ or $b \in a$ holds and that for no $a, b \in d_0$ it can be the case that $a \in b$.

At this point we can complete our set-theoretic interpretation as follows:

- interpret ℓ as the element $\lambda \in d_1$ (hence a subset of d_0) consisting of elements of rank smaller than the rank of α_{n+1},
- interpret Y_s as $(d_0 \setminus \lambda) \cup \{\alpha_1, \ldots, \alpha_{n+1}\}$.

This concludes the proof that the satisfiability of Φ by models of arbitrarily large cardinalities implies the satisfiability of Φ^σ in well-founded Set Theory.

For the opposite direction of our main claim, namely, that if Φ^σ is satisfiable in well-founded Set Theory, then Φ is satisfiable by models of arbitrarily large cardinalities, recall that d_0 and d_1 satisfying $\iota\iota(d_0, d_1)$ must both be infinite. Therefore, since ℓ must be interpreted as an element of d_1—namely, as a subset of d_0 which has infinitely many majorants in d_0—it follows that Y_s must be interpreted by an infinite set. At this point an infinite graph satisfying Φ can be built as follows: introduce n vertices $v_1, \ldots, v_n$, to be associated with the set interpretations of the x-variables; introduce infinitely many vertices w^i, to be associated with the set interpretations of the y-variables, i.e., elements of Y_s; introduce an arc between two vertex if and only if the set interpretation of the s-indexed associated variable has the set interpretation of the t-indexed associated variable as an element. The very definition of φ^σ ensures that all literals made true by the set-theoretic interpretation force the corresponding literals to be true under the graph-theoretic interpretation. Hence, since the set-theoretic interpretation guarantees the truth of φ^σ, the matrix φ will be true when x-variables are interpreted by $v_1, \ldots, v_n$ and y-variables vary among (the v's and) the w^i's.

To conclude, observe that the existence of an *infinite* model for Φ implies our claim that there are finite models of arbitrarily large cardinalities for Φ. $\dashv$

The result just proved is an example of using the set-theoretic language to describe and analyze an entire collection of models of a given logical formula. More precisely, the models analyzed are nothing but graphs, and the natural correspondence between membership and arc relation is all we need for the analysis. The extra ingredient used above is the ability to encode—at a low syntactic complexity—the existence of infinite sets. The union of the two infinite sets d_0 and d_1 provides the background needed to represent infinitely many finite graphs by elements of a single set.

Exercises

8.1 Show that the clauses (i) and (ii) of $u(a,b)$ imply that $a \neq \emptyset$.

8.2 Show that the three formulae

$$(\forall x_1, x_2 \in a)(\forall y_1, y_2 \in b)(x_2 \in y_2 \in x_1 \in y_1 \rightarrow x_2 \in y_1),$$
$$(\forall y_1, y_2 \in b)(y_1 \subseteq y_2 \vee y_2 \subseteq y_1),$$
$$a \cap b = \emptyset,$$

follow from $u(a,b)$ under **FA**.

8.3 Explain why $u(f_1, f_2)$ is false when f_1, f_2 are hypersets which solve the equations appearing in front of the caption of Fig. 8.4.

8.4 Prove that for any n-tuple $\langle \omega_1, \ldots, \omega_n \rangle$ such that $u^{[n]}(\omega_1, \ldots, \omega_n)$ is true, all of the sets $\omega_1, \ldots, \omega_n$ have the same (limit) rank.

8.5 Prove the claim appearing in Fig. 8.8.

8.6 Prove that, given a BSR sentence Φ, a sentence Φ' in the same language can be written, such that Φ' is satisfiable if and only if Φ is satisfiable in models of arbitrarily large cardinalities.

8.7 Prove that Φ^σ can be written in $\mathscr{L}_\in$.

8.8 (*) Prove that Φ^σ can be written in $\mathscr{L}_\in$ *without* equality.

Appendix A
Excerpts from a **Referee**-Checked Proof-Script

This appendix examines and carefully explains parts of the Ref scenario, available in its entirety at the URL aetnanova.units.it/scenarios/GraphsAsTransitiveSets/, which has been introduced in Chap. 5. As described there, the scenario treats various issues regarding

- finite graphs, in particular *connectedness*, and regarding
- finite digraphs, in particular *extensionality*, *weak extensionality*, and *acyclicity*.

Moreover it relates graphs to digraphs via orientations.

Such issues have been thoroughly discussed all over this book. In particular, as just said, this appendix parallels Chap. 5; but here we will keep a much closer eye on technical details which we have skipped over elsewhere. This will convey a feel to the reader of which level of accuracy formalizations require in order to benefit from present-day proof technology.

We will prove that every connected undirected graph endowed with at least three vertices has a non-cut vertex, namely, a vertex whose removal does not disrupt connectedness. This proof is rather low-level; consequently it presupposes very little, and this is why we have chosen it as our detailed case study. We will not afford this part of the experiment in its entirety, though, but will concentrate on the Ref-based proofs of two major propositions: a preparatory lemma and the claim on the existence of non-cut vertices.

The presentation will follow a pattern similar to Chap. 5. In analogy with Sect. 5.2, we accurately shape the selected proofs before entering into their implementation with our proof checker. This draft is offered in Sect. A.1, save for the proofs of certain accessory propositions, which the reader is invited to carry out as exercises on her/his own. Then, in Sect. A.2, in analogy with—and as a supplement to—Sect. 5.3.2, we will translate the proofs from the level of ordinary mathematical discourse down to the merciless formal idiom of Ref.

In its more advanced parts which we do not reexamine in this appendix (they have been outlined in Chap. 5), the scenario on which we report also proves that:

© Springer International Publishing AG 2017
251
E.G. Omodeo et al., *On Sets and Graphs*, DOI 10.1007/978-3-319-54981-1

(1) Every weakly extensional, acyclic digraph can be decorated *à la* Mostowski by finite sets so that: on the one hand, its arcs mimic membership; on the other hand, no collisions arise between its vertices (otherwise stated, no two vertices are sent to the same set by the decoration). This applies, of course, also to the special case of an extensional digraph. We manage to have one node sent to $\emptyset$ in the decoration.

(2) A graph whatsoever admits an orientation which is weakly extensional and acyclic; consequently, and in view of what precedes, one can regard its edges as membership arcs deprived of their natural orientation.

Then it endows each *connected claw-free graph G* with an extensional acyclic orientation so that, through such an orientation,

(3) G gets represented as a transitive set T

(4) the membership arcs between elements of T correspond to the edges of G.

To end, it exploits the representation of a connected claw-free graph as a transitive set in order to show that it always has a Hamiltonian cycle in its square and a near-perfect matching.

A.1 A Tricky Proof that Connected Graphs Have Non-cut Vertices

The part of our experiment with Ref discussed in this and the next section presupposes six definitions: we do not repeat the ones of the monadic constructs $\mathcal{P}$, Finite, and $\bigcup$ (cf. Fig. 5.2), but resume in Fig. A.1 from Chap. 5, for convenience of the reader, the ones of the relators $\mathsf{uGraph}(\cdot,\cdot)$, $\mathsf{Conn}(\cdot,\cdot)$, and $\mathsf{NonCut}(\cdot,\cdot,\cdot)$.

In preparation for the proof of Theorem 5.1, we prove a plain lemma which settles the easy case when there is a *leaf*, namely, a vertex z with exactly one incident edge (and hence with exactly one neighbor x); leaves are, in fact, non-cut vertices:

Lemma A.1 *In a connected graph with at least two edges, if there is a vertex x whose removal causes another vertex z to also disappear, then z is a non-cut vertex:*

DEF ugraph$_0$: [Undirected graph (unless null)] $\mathsf{uGraph}(V,E) \leftrightarrow_{\mathrm{Def}}$
        $\mathsf{Finite}(V)$ & $E \subseteq \{\{x,y\} : x \in V, y \in V \setminus \{x\}\}$

DEF conn$_0$: [Connectedness (typically of a graph)] $\mathsf{Conn}(V,E) \leftrightarrow_{\mathrm{Def}}$
        $V = \bigcup E$ & $\{p \subseteq E \mid \bigcup p \cap \bigcup(E \setminus p) = \emptyset\} \subseteq \{\emptyset, E\}$

DEF conn$_1$: [Non-cut vertex of a graph] $\mathsf{NonCut}(Z,V,E) \leftrightarrow_{\mathrm{Def}}$
        $Z \in V$ & $\mathsf{Conn}(V \setminus \{Z\}, \{a \in E \mid Z \notin a\})$

Fig. A.1 Three key definitions in the formal treatment of graphs

$$\Big(\mathsf{uGraph}(V,E)\,\&\,\mathsf{Conn}(V,E)\,\&\,E\nsubseteq\{\mathsf{arb}(E)\}\,\&$$
$$x\in V\,\&\ z\in V\setminus\{x\}\ \setminus\bigcup\{a\in E\,|\,x\notin a\}\Big)\to\mathsf{NonCut}(z,V,E).$$

Proof Both x and z belong to $V=\bigcup E$, but they disappear from $\bigcup\{a\in E\,|\,x\notin a\}$; this indicates that $\{x,z\}\in E$ and z belongs to no edge other than $\{x,z\}$. As for x, it must belong to some edge other than $\{x,z\}$, else $\{\{x,z\}\}$ and $E\setminus\{\{x,z\}\}$ would be nonnull and vertex disjoint, contrary to the assumption $\mathsf{Conn}(V,E)$.

Consider now the set $\{a\in E\,|\,z\notin a\}=E\setminus\{\{x,z\}\}$, whose unionset clearly equals $V\setminus\{z\}$. If this could be split into two nonnull vertex-disjoint parts, x would appear in one of the two but not the other, and, by adding edge $\{x,z\}$ to that part, we would get a decomposition of E into nonnull vertex-disjoint parts, thus contradicting $\mathsf{Conn}(V,E)$. Therefore $\mathsf{NonCut}(z,V,E)$ holds. $\dashv$

Let us now move on to the proof of our main theorem,[1] relying on various minor claims which we left as Exercises A.1, A.2, A.3, A.4, and A.5 to the reader (see Fig. A.4 below):

Proof (Theorem 5.1) Recall the claim, which was formulated as

$$\Big(\mathsf{uGraph}(V,E)\ \&\ \mathsf{Conn}(V,E)\ \&\ E\nsubseteq\{\mathsf{arb}(E)\}\Big)\to\exists z\,\mathsf{NonCut}(z,V,E).$$

Arguing by contradiction, suppose that V_0,E_0 make a counterexample to the claim; Lemma A.1 then tells us that to each $x\in V_0$, there correspond vertex-disjoint nonnull sets P_x,Q_x such that $P_x\cup Q_x=\{a\in E_0\,|\,x\notin a\}$ and $\bigcup(P_x\cup Q_x)=V_0\setminus\{x\}$, so that x must be the sole vertex w to which there correspond edges $\{w,u'\}$, $\{w,u''\}$ in E_0 that touch P_x and Q_x, respectively (see Exercise A.5).

Put for brevity $F_x=_{\mathrm{Def}}\{a\in E_0\,|\,x\notin a\}$ for each vertex x, and consider the set

$$\mathscr{C}=\{C:x\in V_0,\ C\subset F_x\,|\,\mathsf{Conn}(\textstyle\bigcup C,C)\};$$

since $\mathscr{C}$ is finite and any singleton $\{a\}$ with $a\in E_0$ belongs to it, $\mathscr{C}$ has an $\subseteq$-maximal element $C_\star$. Maximality means that there are no $y\in V_0,K\subset F_y$ satisfying $\mathsf{Conn}(\bigcup K,K)$ and $C_\star\subsetneq K$. Pick a $y_\star\in V_0$ such that $C_\star\subset F_{y_\star}$ and consider, recalling what said above, a split $F_{y_\star}=P_\star\cup Q_\star$ of $F_{y_\star}$ into vertex-disjoint nonnull sets $P_\star,Q_\star$ of edges.

In consequence of $\mathsf{Conn}(\bigcup C_\star,C_\star)$, either $C_\star\subseteq P_\star$ or $C_\star\subseteq Q_\star$ must hold (see Exercise A.1); w.l.o.g., we will assume for definiteness the inclusion $C_\star\subseteq P_\star$. Put $C^\star=\{a\in E_0\,|\,a\cap\bigcup P_\star\neq\emptyset\}$, so that $C^\star$ enlarges $P_\star$ with all edges incident to it and $C^\star\supseteq P_\star\supseteq C_\star$. We will establish three facts: (1) $y_\star\in\bigcup C^\star$ (so that the strict inclusion $C^\star\supsetneq P_\star$ holds—in fact $y_\star\notin\bigcup P_\star$); (2) $C^\star$ is connected; and (3) there is a $y^\star$ such that $C^\star\subset F_{y^\star}$. This will lead us to the sought contradiction, by conflicting with the maximality assumption.

---

[1]This is an adaptation of a proof referring to hypergraphs and due to Alberto Casagrande, see [21].

To achieve what planned, put $D^\star = \{a \in E_0 \mid a \cap \bigcup Q_\star \neq \emptyset\}$; then it turns out easily (see Exercises A.3 and A.5) that $C^\star \cup D^\star = E_0$ and $\bigcup C^\star \cap \bigcup D^\star = \{y_\star\}$. This readily gives us (1); it also implies (2) (see Exercise A.2), namely, $\mathsf{Conn}(\bigcup C^\star, C^\star)$.

In order to get (3) and thus complete this proof, it suffices to arbitrarily draw $y^\star$ from the set $\bigcup D^\star \setminus \{y_\star\}$.   ⊣

A.2   Ref's Proof that Connected Graphs Have Non-cut Vertices

Several lemmas are needed before Ref can conclusively accept formal specifications of the above proofs of Lemma A.1 and Theorem 5.1. Five of those lemmas (THMs un_0, un_2, un_4, un_8, and fin_0) have been listed in Fig. 5.18 and are not repeated here. A few more, which have to do with $\mathcal{P}$ and Finite, are listed in Fig. A.2; of these, some are proved by applying the finite induction principle of that in Fig. 5.12; the last also requires the small THEORY shown in Fig. A.3. Nine more preparatory lemmas, shown in Fig. A.4, form the rudiments on undirected graphs which we need. In the overall, the proofs of these preliminary lemmas require 563 Ref inference lines. We omit them, because none of them is demanding, albeit one—namely, the one of conn_5—is longish (100 inference lines). Here below, we show the implemented proofs of Lemma A.1 and Theorem 5.1, identified by the names conn_6 and $\mathsf{connectedness}_2$ for Ref (the former proof consists of 72 inference lines, the latter of 117).

THM pow_0: [Characterization of powerset] $X \supseteq Y \leftrightarrow Y \in \mathcal{P}X$
THM pow_1: [Monotonicity of powerset] $S \supseteq X \rightarrow \mathcal{P}X \cup \{\emptyset, X\} \subseteq \mathcal{P}S$
THM pow_2: [Powerset of null set and of singletons] $\mathcal{P}\emptyset = \{\emptyset\}$ & $\mathcal{P}\{X\} = \{\emptyset, \{X\}\}$
THM fin_1: [Finiteness of the union of a finite set with a singleton] $\mathsf{Finite}(F) \rightarrow \mathsf{Finite}(F \cup \{X\})$
THM fin_2: [Singletons are finite] $\mathsf{Finite}(\{X\})$ & $\mathsf{Finite}(\emptyset)$
THM fin_3: [Finiteness of the union of two finites sets] $\mathsf{Finite}(X)$ & $\mathsf{Finite}(Y) \rightarrow \mathsf{Finite}(X \cup Y)$
THM fin_4: [Every nonnull finite set has an inclusion-maximal element] $\mathsf{Finite}(S)$ & $S \neq \emptyset \rightarrow \langle \exists m \in S \mid \{x \in S \setminus \{m\} \mid m \subseteq x\} = \emptyset \rangle$
THM fin_5: [Powersets of finite sets are finite] $\mathsf{Finite}(F) \rightarrow \mathsf{Finite}(\mathcal{P}F)$
THM fin_6: [Any set is finite whose unionsets is finite] $\mathsf{Finite}(\bigcup F) \rightarrow \mathsf{Finite}(F)$

Fig. A.2 Additional ancillary laws on the basic set-theoretic constructs

Fig. A.3 A THEORY on
finite sets, companion of
finite induction

THEORY finiteImage$(s_0, f(X))$
   Finite(s_0)
$\Rightarrow$
   Finite$\big(\{f(x) : x \in s_0\}\big)$
END finiteImage

THM ugraph$_0$: [Finiteness of the set of edges of any graph] uGraph$(V, E) \rightarrow$ Finite(E) & $\emptyset \notin E$

THM ugraph$_1$: [Graph-induced graphs]
  uGraph(V, E) & $E \supseteq F$ & $W \supseteq V$ & $(W = V \vee$ Finite$(W)) \rightarrow$ uGraph(W, F)

THM ugraph$_2$: [Graph-induced graphs, 2]
  uGraph(V, E) & $(W \subseteq V \vee$ Finite$(W)) \rightarrow$ uGraph$(W, E \cap \{\{x, y\} : x \in W, y \in W\})$

THM conn$_0$: [Trivial cases of connectedness] Conn$(\emptyset, \emptyset)$ & Conn$(A, \{A\})$

THM conn$_1$: [See Exercise A.1]
  uGraph$(V, P \cup Q)$ & $C \subseteq P \cup Q$ & Conn$(\bigcup C, C)$ & $\bigcup P \cap \bigcup Q = \emptyset \rightarrow C \subseteq P \vee C \subseteq Q$

THM conn$_2$: [See Exercise A.2]
  uGraph$(V, R \cup S)$ & Conn$(V, R \cup S)$ & $\bigcup R \cap \bigcup S = \{X\} \rightarrow$ Conn$(\bigcup S, S)$

THM conn$_3$: [See Exercise A.3]
  uGraph(V, E) & $\bigcup (P \cup Q) = V \setminus \{X\} \rightarrow \{a \in E \mid a \cap \bigcup P \neq \emptyset\} \cup \{a \in E \mid a \cap \bigcup Q \neq \emptyset\} = E$

THM conn$_4$: [See Exercise A.4]
  uGraph(V, E) & $P \cup Q = \{a \in E \mid X \notin a\}$ & $\bigcup P \cap \bigcup Q = \emptyset$ & $Y \neq X \rightarrow$
    $Y \notin \bigcup \{a \in E \mid a \cap \bigcup P \neq \emptyset\} \cap \bigcup \{a \in E \mid a \cap \bigcup Q \neq \emptyset\}$

THM conn$_5$: [See Exercise A.5]
  uGraph(V, E) & Conn(V, E) & $P \cup Q = \{a \in E \mid X \notin a\}$ & $\bigcup P \cap \bigcup Q = \emptyset$ &
    $\bigcup (P \cup Q) = V \setminus \{X\}$ & $\emptyset \notin \{P, Q\} \rightarrow$
    $\bigcup \{a \in E \mid a \cap \bigcup P \neq \emptyset\} \cap \bigcup \{a \in E \mid a \cap \bigcup Q \neq \emptyset\} = \{X\}$

Fig. A.4 Rudiments on graphs used in the two proofs under discussion

THM conn$_6$: $\left[\begin{array}{l}\text{If removing a vertex from a connected graph causes another} \\ \text{vertex to disappear, the latter is a non-cut vertex}\end{array}\right]$

uGraph(V, E) & Conn(V, E) & $E \not\subseteq \{arb(E)\}$ & $X \in V$ & $Z \in V \setminus \{X\} \setminus \bigcup \{a \in E \mid X \notin a\} \rightarrow$
                      NonCut(Z, V, E). PROOF:
Suppose_not$(v_0, e_0, x_0, z_0) \Rightarrow$ AUTO

> Arguing by contradiction, suppose that v_0, e_0, x_0, z_0 make a counterexample. Observe that
> $\{x_0, z_0\}$ is an edge.

Use_def$(\bigcup) \Rightarrow$ $\bigcup e_0 = \{z : a \in e_0, z \in a\}$ & $z_0 \notin \{w : b \in \{a \in e_0 \mid x_0 \notin a\}, w \in b\}$
SIMPLF $\Rightarrow$ Stat1 : $z_0 \notin \{w : b \in e_0, w \in b \mid x_0 \notin b\}$
Use_def(Conn) $\Rightarrow$ $z_0, x_0 \in v_0$ & $z_0 \neq x_0$ & $v_0 = \{z : a \in e_0, z \in a\}$ &
        $\{p \subseteq e_0 \mid \bigcup p \cap \bigcup (e_0 \setminus p) = \emptyset\} \subseteq \{\emptyset, e_0\}$ & $e_0 \neq \{\{x_0, z_0\}\}$
EQUAL $\langle Stat1 \rangle \Rightarrow$ Stat2 : $z_0 \in \{z : a \in e_0, z \in a\}$ &
     Stat2a : $z_0 \notin \{w : b \in e_0, w \in b \mid x_0 \notin b\}$
$\langle a_1, z_1, a_1, z_1 \rangle \hookrightarrow Stat2(Stat2\star) \Rightarrow$ $a_1 \in e_0$ & $z_0, x_0 \in a_1$
Suppose $\Rightarrow$ $\{x_0, z_0\} \notin e_0$
    Use_def(uGraph) $\Rightarrow$ Stat3 : $a_1 \in \{\{x, y\} : x \in v_0, y \in v_0 \setminus \{x\}\}$

$\langle x_1, y_1 \rangle \hookrightarrow Stat3(Stat1\star) \Rightarrow$ false
Discharge $\Rightarrow$ AUTO

$\|$ In consequence of that, $\{a \in e_0 \mid z_0 \notin a\} = e_0 \setminus \{\{x_0, z_0\}\}$.

Suppose $\Rightarrow$ $Stat4$: $\{a \in e_0 \mid z_0 \notin a\} \neq e_0 \setminus \{\{x_0, z_0\}\}$
   $\langle a_2 \rangle \hookrightarrow Stat4(Stat4\star) \Rightarrow$ $a_2 \in \{a \in e_0 \mid z_0 \notin a\} \neq a_2 \in e_0 \setminus \{\{x_0, z_0\}\}$
   Suppose $\Rightarrow$ $Stat5$: $a_2 \notin \{a \in e_0 \mid z_0 \notin a\}$
     $\langle a_2 \rangle \hookrightarrow Stat5(Stat4\star) \Rightarrow$ $a_2 \in e_0$ & $z_0 \in a_2$ & $a_2 \neq \{x_0, z_0\}$
     Use_def(uGraph) $\Rightarrow$ $Stat6$: $a_2 \in \{\{x, y\} : x \in v_0, y \in v_0 \setminus \{x\}\}$ & $z_0 \neq x_0$
     $\langle x_2, y_2 \rangle \hookrightarrow Stat6(Stat5\star) \Rightarrow$ $x_0 \notin a_2$
     $\langle a_2, z_0 \rangle \hookrightarrow Stat2a(Stat5\star) \Rightarrow$ false
   Discharge $\Rightarrow$ AUTO
   $(Stat4\star)$ELEM $\Rightarrow$ $Stat7$: $a_2 \in \{a \in e_0 \mid z_0 \notin a\}$ & $a_2 \notin e_0 \setminus \{\{x_0, z_0\}\}$
   $\langle\ \rangle \hookrightarrow Stat7(Stat7\star) \Rightarrow$ false
Discharge $\Rightarrow$ $Stat8$: $\{a \in e_0 \mid z_0 \notin a\} = e_0 \setminus \{\{x_0, z_0\}\}$

$\|$ Moreover, x_0 belongs to some edge other than $\{x_0, z_0\}$, else $\{\{x_0, z_0\}\}$ and $e_0 \setminus \{\{x_0, z_0\}\}$
$\|$ would be nonnull and vertex disjoint, contrary to the assumption $\mathrm{Conn}(v_0, e_0)$.

$(Stat1\star)$ELEM $\Rightarrow$ $Stat9$: $\{\{x_0, z_0\}\} \notin \{p \subseteq e_0 \mid \bigcup p \cap \bigcup(e_0 \setminus p) = \emptyset\}$ & $\{\{x_0, z_0\}\} \subseteq e_0$
$\langle \{x_0, z_0\}, \{x_0, z_0\}, \{\{x_0, z_0\}\} \rangle \hookrightarrow Tun_4 \Rightarrow$ AUTO
Use_def$(\bigcup(e_0 \setminus \{\{x_0, z_0\}\})) \Rightarrow$ AUTO
$\langle \{\{x_0, z_0\}\} \rangle \hookrightarrow Stat9(Stat9\star) \Rightarrow$ $Stat10$: $\{x_0, z_0\} \cap \{z : a \in e_0 \setminus \{\{x_0, z_0\}\}, z \in a\} \neq \emptyset$
$\langle y_0 \rangle \hookrightarrow Stat10(Stat10\star) \Rightarrow$ $Stat11$: $y_0 \in \{z : a \in e_0 \setminus \{\{x_0, z_0\}\}, z \in a\}$ & $y_0 \in \{x_0, z_0\}$
$\langle a_3, y_3 \rangle \hookrightarrow Stat11(Stat8\star) \Rightarrow$ $Stat12$: $a_3 \in \{a \in e_0 \mid z_0 \notin a\}$ & $y_0 \in a_3$
$\langle\ \rangle \hookrightarrow Stat12(Stat11\star) \Rightarrow$ $Stat13$: $a_3 \in e_0$ & $x_0 \in a_3$ & $z_0 \notin a_3$

$\|$ Consider now the set $\{a \in e_0 \mid z_0 \notin a\} = e_0 \setminus \{\{x_0, z_0\}\}$, whose unionset clearly equals
$\|$ $v_0 \setminus \{z_0\}$. This set should be splittable into two nonnull vertex-disjoint parts ...

Use_def(NonCut) $\Rightarrow$ $\neg \mathrm{Conn}(v_0 \setminus \{z_0\}, \{a \in e_0 \mid z_0 \notin a\})$
EQUAL $\langle Stat2 \rangle \Rightarrow$ $\neg \mathrm{Conn}(v_0 \setminus \{z_0\}, e_0 \setminus \{\{x_0, z_0\}\})$
Use_def(Conn) $\Rightarrow$ $\neg(\{z : a \in e_0 \setminus \{\{x_0, z_0\}\}, z \in a\} = v_0 \setminus \{z_0\}$ &
     $\{p \subseteq e_0 \setminus \{\{x_0, z_0\}\} \mid \bigcup p \cap \bigcup(e_0 \setminus \{\{x_0, z_0\}\} \setminus p) = \emptyset\} \subseteq \{\emptyset, e_0 \setminus \{\{x_0, z_0\}\}\})$
Suppose $\Rightarrow$ $Stat14$: $\{z : a \in e_0 \setminus \{\{x_0, z_0\}\}, z \in a\} \neq v_0 \setminus \{z_0\}$
   $\langle z_2 \rangle \hookrightarrow Stat14(Stat14\star) \Rightarrow$ $Stat15$: $z_2 \in \{z : a \in e_0 \setminus \{\{x_0, z_0\}\}, z \in a\} \neq z_2 \in$
$v_0 \setminus \{z_0\}$
   Suppose $\Rightarrow$ $Stat16$: $z_2 \in \{z : a \in e_0 \setminus \{\{x_0, z_0\}\}, z \in a\}$
     $\langle a_4, z_4 \rangle \hookrightarrow Stat16(Stat16\star) \Rightarrow$ $a_4 \in e_0 \setminus \{\{x_0, z_0\}\}$ & $z_2 \in a_4$
     EQUAL $\langle Stat8 \rangle \Rightarrow$ $Stat17$: $a_4 \in \{a \in e_0 \mid z_0 \notin a\}$
     $\langle\ \rangle \hookrightarrow Stat17(Stat15\star) \Rightarrow$ $a_4 \in e_0$ & $z_2 \notin v_0$
     EQUAL $\langle Stat1 \rangle \Rightarrow$ $Stat18$: $z_2 \notin \{z : a \in e_0, z \in a\}$
     $\langle a_4, z_2 \rangle \hookrightarrow Stat18(Stat16\star) \Rightarrow$ false
   Discharge $\Rightarrow$ AUTO
   $(Stat15\star)$ELEM $\Rightarrow$ $Stat19$: $z_2 \notin \{z : a \in e_0 \setminus \{\{x_0, z_0\}\}, z \in a\}$ & $z_2 \in v_0 \setminus \{z_0\}$
   EQUAL $\langle Stat1 \rangle \Rightarrow$ $Stat20$: $z_2 \in \{z : a \in e_0, z \in a\}$
   $\langle a_5, z_5 \rangle \hookrightarrow Stat20(Stat20\star) \Rightarrow$ $a_5 \in e_0$ & $z_2 \in a_5$
   $\langle a_5, z_2 \rangle \hookrightarrow Stat19(Stat19\star) \Rightarrow$ $Stat21$: $x_0 \notin \{z : a \in e_0 \setminus \{\{x_0, z_0\}\}, z \in a\}$
   $\langle a_3, x_0 \rangle \hookrightarrow Stat21(Stat13, Stat13\star) \Rightarrow$ false
Discharge $\Rightarrow$ AUTO

$(Stat13\star)$ELEM $\Rightarrow$ $Stat22$: $\{z : a \in e_0 \setminus \{\{x_0, z_0\}\}, z \in a\} = v_0 \setminus \{z_0\}$ &
    $\{p \subseteq e_0 \setminus \{\{x_0, z_0\}\} \mid \bigcup p \cap \bigcup(e_0 \setminus \{\{x_0, z_0\}\} \setminus p) = \emptyset\} \not\subseteq \{\emptyset, e_0 \setminus \{\{x_0, z_0\}\}\}$

$\parallel$ ... and x_0 would appear in one of the two parts but not the other.

$\langle p_0 \rangle \hookrightarrow Stat22(Stat22\star) \Rightarrow$ $Stat23$:
$p_0 \in \{p \subseteq e_0 \setminus \{\{x_0, z_0\}\} \mid \bigcup p \cap \bigcup(e_0 \setminus \{\{x_0, z_0\}\} \setminus p) = \emptyset\}$ & $p_0 \neq \emptyset$ & $p_0 \neq e_0 \setminus \{\{x_0, z_0\}\}$
$\langle\ \rangle \hookrightarrow Stat23(Stat23\star) \Rightarrow$ $p_0 \subseteq e_0 \setminus \{\{x_0, z_0\}\}$ & $\bigcup p_0 \cap \bigcup(e_0 \setminus \{\{x_0, z_0\}\} \setminus p_0) = \emptyset$
Loc_def $\Rightarrow$ $q_0 = e_0 \setminus \{\{x_0, z_0\}\} \setminus p_0$
$(Stat9\star)$ELEM $\Rightarrow$ $Stat24$: $\bigcup(e_0 \setminus \{\{x_0, z_0\}\}) = v_0 \setminus \{z_0\}$ & $p_0 \cup q_0 = e_0 \setminus \{\{x_0, z_0\}\}$ &
    $p_0 \cap q_0 = \emptyset$
$\langle p_0, q_0, p_0 \cup q_0 \rangle \hookrightarrow Tun_8(Stat24\star) \Rightarrow$ $\bigcup p_0 \cup \bigcup q_0 = \bigcup(p_0 \cup q_0)$
EQUAL $\langle Stat23 \rangle \Rightarrow$ $Stat25$: $\bigcup p_0 \cap \bigcup q_0 = \emptyset$ & $\bigcup p_0 \cup \bigcup q_0 = v_0 \setminus \{z_0\}$
$(Stat1\star)$ELEM $\Rightarrow$ $x_0 \in \bigcup p_0 \cup \bigcup q_0$ & $z_0 \in v_0$ & $\{x_0, z_0\} \in e_0$
Suppose $\Rightarrow$ $x_0 \in \bigcup p_0$
    $\langle p_0, \{x_0, z_0\}, p_0 \cup \{\{x_0, z_0\}\} \rangle \hookrightarrow Tun_2 \Rightarrow$ AUTO
    $\langle p_0 \cup \{\{x_0, z_0\}\}, q_0, p_0 \cup \{\{x_0, z_0\}\} \cup q_0 \rangle \hookrightarrow Tun_8(Stat25\star) \Rightarrow$
        $\bigcup(p_0 \cup \{\{x_0, z_0\}\} \cup q_0) = v_0$
    Suppose $\Rightarrow$ $Stat26$: $p_0 \cup \{\{x_0, z_0\}\} \notin \{p \subseteq e_0 \mid \bigcup p \cap \bigcup(e_0 \setminus p) = \emptyset\}$
        $\langle p_0 \cup \{\{x_0, z_0\}\} \rangle \hookrightarrow Stat26(Stat24\star) \Rightarrow$ $(\bigcup p_0 \cup \{z_0\}) \cap \bigcup(e_0 \setminus (p_0 \cup \{\{x_0, z_0\}\}$
$)) \neq \emptyset$
        $\langle e_0 \setminus (p_0 \cup \{\{x_0, z_0\}\}), q_0, q_0 \rangle \hookrightarrow Tun_8(Stat24\star) \Rightarrow$ false
    Discharge $\Rightarrow$ AUTO
    $(Stat1\star)$ELEM $\Rightarrow$ $p_0 \cup \{\{x_0, z_0\}\} = e_0$
    $(Stat23\star)$Discharge $\Rightarrow$ AUTO
$(Stat25\star)$ELEM $\Rightarrow$ $x_0 \in \bigcup q_0$
$\langle q_0, \{x_0, z_0\}, q_0 \cup \{\{x_0, z_0\}\} \rangle \hookrightarrow Tun_2 \Rightarrow$ AUTO
$\langle q_0 \cup \{\{x_0, z_0\}\}, p_0, q_0 \cup \{\{x_0, z_0\}\} \cup p_0 \rangle \hookrightarrow Tun_8(Stat25\star) \Rightarrow$ $\bigcup(q_0 \cup \{\{x_0, z_0\}\} \cup$
$p_0) = v_0$
Suppose $\Rightarrow$ $Stat27$: $q_0 \cup \{\{x_0, z_0\}\} \notin \{p \subseteq e_0 \mid \bigcup p \cap \bigcup(e_0 \setminus p) = \emptyset\}$
    $\langle q_0 \cup \{\{x_0, z_0\}\} \rangle \hookrightarrow Stat27(Stat24\star) \Rightarrow$ $(\bigcup q_0 \cup \{z_0\}) \cap \bigcup(e_0 \setminus (q_0 \cup \{\{x_0, z_0\}\}))$
$\neq \emptyset$
    $\langle e_0 \setminus (q_0 \cup \{\{x_0, z_0\}\}), p_0, p_0 \rangle \hookrightarrow Tun_8(Stat24\star) \Rightarrow$ false
Discharge $\Rightarrow$ AUTO
$(Stat1\star)$ELEM $\Rightarrow$ $q_0 \cup \{\{x_0, z_0\}\} = e_0$
$(Stat23\star)$Discharge $\Rightarrow$ QED

THM connectedness$_2$: [Every connected graph with at least two edges has a non-cut vertex]
    uGraph(V, E) & Conn(V, E) & E $\not\subseteq$ $\{arb(E)\}$ $\rightarrow$ $\langle \exists z \mid NonCut(z, V, E) \rangle$. PROOF:
Suppose_not(v_0, e_0) $\Rightarrow$ AUTO

$\parallel$ Arguing by contradiction, suppose that v_0, e_0 make a counterexample to the claim.
$\parallel$ For each $x \in v_0$, put $f(x) = \{a \in e_0 \mid x \notin a\}$. Then, by Theorem conn$_6$, the equality
$\parallel$ $v_0 \setminus \{x\} = \bigcup f(x)$ holds for every $x \in v_0$.

Loc_def $\Rightarrow$ $a_0 = arb(e_0)$ & $b_0 = arb(e_0 \setminus \{a_0\})$
ELEM $\Rightarrow$ $Stat0$: $\neg\langle \exists z \mid NonCut(z, v_0, e_0) \rangle$ & $a_0, b_0 \in e_0$ & $a_0 \neq b_0$
Suppose $\Rightarrow$ $Stat1$: $\neg\langle \forall x, \exists f \mid f = \{a \in e_0 \mid x \notin a\} \rangle$
    $\langle x_0 \rangle \hookrightarrow Stat1 \Rightarrow$ $Stat2$: $\neg\langle \exists f \mid f = \{a \in e_0 \mid x_0 \notin a\} \rangle$
    $\langle \{a \in e_0 \mid x_0 \notin a\} \rangle \hookrightarrow Stat2(Stat2\star) \Rightarrow$ false
Discharge $\Rightarrow$ $\langle \forall x, \exists f \mid f = \{a \in e_0 \mid x \notin a\} \rangle$
APPLY $\langle v1_\Theta : f \rangle$ Skolem $\Rightarrow$ $Stat3$: $\langle \forall x \mid f(x) = \{a \in e_0 \mid x \notin a\} \rangle$

Suppose $\Rightarrow$ $Stat4:$ $\neg\langle\forall x \in v_0 \mid v_0\setminus\{x\} = \bigcup f(x)\ \&\ f(x) \subseteq e_0\rangle$

   $\langle x_1\rangle\hookrightarrow Stat4 \Rightarrow$ $\neg(v_0\setminus\{x_1\} = \bigcup f(x_1)\ \&\ f(x_1) \subseteq e_0)\ \&\ x_1 \in v_0$

   $\langle x_1\rangle\hookrightarrow Stat3(Stat3\star) \Rightarrow$ $f(x_1) = \{a \in e_0 \mid x_1 \notin a\}$

   Set_monot $\Rightarrow$ $\{a \in e_0 \mid x_1 \notin a\} \subseteq \{a \in e_0 \mid \text{true}\}$

   Suppose $\Rightarrow$ $Stat5:\ v_0\setminus\{x_1\} \not\subseteq \bigcup f(x_1)$

     $\langle z_1\rangle\hookrightarrow Stat5(Stat5\star) \Rightarrow$ $z_1 \in v_0\setminus\{x_1\}\setminus\bigcup f(x_1)$

     EQUAL $\langle Stat4\rangle \Rightarrow$ $z_1 \in v_0\setminus\{x_1\}\setminus\bigcup\{a \in e_0 \mid x_1 \notin a\}$

     $\langle v_0, e_0, x_1, z_1\rangle\hookrightarrow Tconn_6(\star) \Rightarrow$ $NonCut(z_1, v_0, e_0)$

     $\langle z_1\rangle\hookrightarrow Stat0(Stat0\star) \Rightarrow$ false

Discharge $\Rightarrow$ AUTO

$(Stat4\star)$ELEM $\Rightarrow$ $Stat6:\ \bigcup f(x_1) \not\subseteq v_0\setminus\{x_1\}$

   $\langle x_2\rangle\hookrightarrow Stat6 \Rightarrow$ $x_2 \in \bigcup f(x_1)\ \&\ x_2 \notin v_0\setminus\{x_1\}$

   Use_def$(\bigcup\{a \in e_0 \mid x_1 \notin a\}) \Rightarrow$ AUTO

   EQUAL $\langle Stat4\rangle \Rightarrow$ $x_2 \in \{x : y \in \{a \in e_0 \mid x_1 \notin a\}, x \in y\}$

   SIMPLF $\Rightarrow$ $Stat7: x_2 \in \{x : a \in e_0, x \in a \mid x_1 \notin a\}$

   $\langle a_3, x_3\rangle\hookrightarrow Stat7(Stat6\star) \Rightarrow$ $a_3 \in e_0\ \&\ a_3 \not\subseteq v_0$

   Use_def(uGraph) $\Rightarrow$ $Stat8:\ a_3 \in \{\{x, y\} : x \in v_0, y \in v_0\setminus\{x\}\}$

   $\langle x_4, y_4\rangle\hookrightarrow Stat8(Stat8\star) \Rightarrow$ $a_3 \subseteq v_0$

$(Stat7\star)$Discharge $\Rightarrow$ $Stat9:\ \langle\forall x \in v_0 \mid v_0\setminus\{x\} = \bigcup f(x)\ \&\ f(x) \subseteq e_0\rangle$

Moreover, since we are supposing that $Conn(v_0\setminus\{x\}, f(x))$ holds for no $x \in v_0$, we can associate with each $x \in v_0$ vertex-disjoint nonnull sets $p(x)$, $f(x)\setminus p(x)$ with $p(x) \subseteq f(x)$.

Suppose $\Rightarrow$ $Stat10: \neg\langle\forall x, \exists p \mid x \in v_0 \to p \subseteq f(x)\ \&\ p \neq \emptyset\ \&\ p \neq f(x)\ \&\ \bigcup p \cap \bigcup(f(x)\setminus p) = \emptyset\rangle$

   $\langle x_5\rangle\hookrightarrow Stat10(Stat10\star) \Rightarrow$

     $Stat11: \neg\langle\exists p \mid x_5 \in v_0 \to p \subseteq f(x_5)\ \&\ p \neq \emptyset\ \&\ p \neq f(x_5)\ \&\ \bigcup p \cap \bigcup(f(x_5)\setminus p) = \emptyset\rangle$

   $\langle\emptyset\rangle\hookrightarrow Stat11(\star) \Rightarrow$ $x_5 \in v_0$

   Use_def$(NonCut(x_5, v_0, e_0)) \Rightarrow$ AUTO

   $\langle x_5\rangle\hookrightarrow Stat0 \Rightarrow$ $\neg Conn(v_0\setminus\{x_5\}, \{a \in e_0 \mid x_5 \notin a\})$

   $\langle x_5\rangle\hookrightarrow Stat3 \Rightarrow$ $f(x_5) = \{a \in e_0 \mid x_5 \notin a\}$

   EQUAL $\langle Stat11\rangle \Rightarrow$ $\neg Conn(v_0\setminus\{x_5\}, f(x_5))$

   Use_def$\big(Conn(v_0\setminus\{x_5\}, f(x_5))\big) \Rightarrow$ AUTO

   $\langle x_5\rangle\hookrightarrow Stat9(Stat9\star) \Rightarrow$ $Stat12: \{p \subseteq f(x_5) \mid \bigcup p \cap \bigcup(f(x_5)\setminus p) = \emptyset\} \not\subseteq \{\emptyset, f(x_5)\}$

   $\langle p_5\rangle\hookrightarrow Stat12 \Rightarrow$ $Stat13: p_5 \in \{p \subseteq f(x_5) \mid \bigcup p \cap \bigcup(f(x_5)\setminus p) = \emptyset\}\ \&\ p_5 \neq \emptyset\ \&\ p_5 \neq$
$f(x_5)$

   $\langle\ \rangle\hookrightarrow Stat13 \Rightarrow$ $p_5 \subseteq f(x_5)\ \&\ \bigcup p_5 \cap \bigcup(f(x_5)\setminus p_5) = \emptyset$

   $\langle p_5\rangle\hookrightarrow Stat11(Stat13\star) \Rightarrow$ false

Discharge $\Rightarrow$ $\langle\forall x, \exists p \mid x \in v_0 \to p \subseteq f(x)\ \&\ p \neq \emptyset\ \&\ p \neq f(x)\ \&\ \bigcup p \cap \bigcup(f(x)\setminus p) = \emptyset\rangle$

APPLY $\langle v1_\Theta : p\rangle$ Skolem $\Rightarrow$

$Stat14:\ \langle\forall x \mid x \in v_0 \to p(x) \subseteq f(x)\ \&\ p(x) \neq \emptyset\ \&\ p(x) \neq f(x)\ \&\ \bigcup p(x) \cap \bigcup(f(x)\setminus p(x)) = \emptyset\rangle$

Define cc to be the set of those connected sets of edges each of which is included in some set of the form $f(x)$ with $x \in v_0$.

Loc_def $\Rightarrow$ cc $= \{c : x \in v_0, c \subseteq f(x) \mid Conn(\bigcup c, c)\}$

In order to see that the set cc just defined is nonnull, we prove next that $\{a_0\} \in$ cc, where a_0 has been arbitrarily selected from the sets of edges.

Suppose $\Rightarrow$ $Stat15$: $\{c : x \in v_0, c \subseteq f(x) \mid \text{Conn}(\bigcup c, c)\} = \emptyset$
Use_def(uGraph) $\Rightarrow$ $Stat16$: $e_0 \subseteq \{\{x, y\} : x \in v_0, y \in v_0 \setminus \{x\}\}$
EQUAL $\langle Stat0, Stat16 \rangle \Rightarrow$ $Stat17$: $a_0, b_0 \in \{\{x, y\} : x \in v_0, y \in v_0 \setminus \{x\}\}$ & $a_0 \in e_0$
$\langle x_6, y_6, x_7, y_7 \rangle \hookrightarrow Stat17(Stat16, Stat0\star) \Rightarrow$ $Stat18$: $b_0 \not\subseteq a_0$ & $b_0 \subseteq v_0$
$\langle y_5 \rangle \hookrightarrow Stat18(Stat18\star) \Rightarrow$ $y_5 \in v_0$ & $y_5 \notin a_0$
$\langle a_0 \rangle \hookrightarrow Tconn_0(Stat18\star) \Rightarrow$ $\text{Conn}(a_0, \{a_0\})$
$\langle a_0, a_0, \{a_0\} \rangle \hookrightarrow Tun_4 \Rightarrow$ $\bigcup \{a_0\} = a_0$
EQUAL $\langle Stat18 \rangle \Rightarrow$ $\text{Conn}(\bigcup \{a_0\}, \{a_0\})$
$\langle y_5 \rangle \hookrightarrow Stat3 \Rightarrow$ AUTO
$\langle y_5, \{a_0\} \rangle \hookrightarrow Stat15(Stat18\star) \Rightarrow$ $Stat19$: $a_0 \notin \{a \in e_0 \mid y_5 \notin a\}$
$\langle a_0 \rangle \hookrightarrow Stat19(Stat17\star) \Rightarrow$ false
Discharge $\Rightarrow$ AUTO

> The finiteness of v_0 yields the finiteness of e_0, hence of $\mathcal{P}e_0$ and, consequently, the
> finiteness of any subset of $\mathcal{P}e_0$: in particular, the set cc of subsets of e_0 introduced above
> is finite; therefore it has an inclusion-maximal element c_0. From the definition of cc, it
> follows that $\text{Conn}(\bigcup c_0, c_0)$ and $c_0 \subseteq f(y_0)$ holds for some vertex y_0.

Suppose $\Rightarrow$ $\neg \langle \exists m \in cc \mid \{x \in cc \setminus \{m\} \mid m \subseteq x\} = \emptyset \rangle$
$\langle cc \rangle \hookrightarrow Tfin_4(Stat14\star) \Rightarrow$ $Stat20$: $\neg\text{Finite}(cc)$
$\langle v_0, e_0 \rangle \hookrightarrow Tugraph_0 \Rightarrow$ $\text{Finite}(e_0)$
$\langle e_0 \rangle \hookrightarrow Tfin_5(Stat20\star) \Rightarrow$ $Stat21$: $cc \not\subseteq \mathcal{P}e_0$
$\langle k_0 \rangle \hookrightarrow Stat21(Stat14\star) \Rightarrow$ $Stat22$: $k_0 \in \{c : x \in v_0, c \subseteq f(x) \mid \text{Conn}(\bigcup c, c)\}$ & $k_0 \notin$
$\mathcal{P}e_0$
$\langle e_0, k_0 \rangle \hookrightarrow Tpow_0 \Rightarrow$ AUTO
$\langle x_8 \rangle \hookrightarrow Stat3 \Rightarrow$ AUTO
Set_monot $\Rightarrow$ $\{a \in e_0 \mid x_8 \notin a\} \subseteq \{a \in e_0 \mid \text{true}\}$
$\langle x_8, c_8 \rangle \hookrightarrow Stat22(Stat22\star) \Rightarrow$ false
Discharge $\Rightarrow$ $Stat24$: $\langle \exists m \in cc \mid \{x \in cc \setminus \{m\} \mid m \subseteq x\} = \emptyset \rangle$
$\langle c_0 \rangle \hookrightarrow Stat24(Stat14\star) \Rightarrow$ $Stat25$: $c_0 \in \{c : x \in v_0, c \subseteq f(x) \mid \text{Conn}(\bigcup c, c)\}$ &
        $Stat25a$: $\{x \in cc \setminus \{c_0\} \mid c_0 \subseteq x\} = \emptyset$
$\langle y_0 \rangle \hookrightarrow Stat9 \Rightarrow$ AUTO
$\langle y_0, c_9 \rangle \hookrightarrow Stat25(Stat25\star) \Rightarrow$
        $Stat26$: $y_0 \in v_0$ & $c_0 \subseteq f(y_0)$ & $\text{Conn}(\bigcup c_9, c_9)$ & $c_0 = c_9$ & $v_0 \setminus \{y_0\} = \bigcup f(y_0)$

> Let us call p_0 the one, between $p(y_0)$ and $f(y_0) \setminus p(\emptyset)$, which satisfies the inclusion $c_0 \subseteq p_0$
> (one of the two in fact does, by Theorem $conn_1$ and because of $\text{Conn}(\bigcup c_0, c_0)$). Also, put
> $q_0 = f(y_0) \setminus p_0$, $c_1 = \{a \in e_0 \mid a \cap \bigcup p_0 \neq \emptyset\}$, and $d_1 = \{a \in e_0 \mid a \cap \bigcup q_0 \neq \emptyset\}$.

$\langle y_0 \rangle \hookrightarrow Stat14(Stat26\star) \Rightarrow$ $p(y_0) \subseteq f(y_0)$ & $p(y_0) \neq \emptyset$ &
        $p(y_0) \neq f(y_0)$ & $\bigcup p(y_0) \cap \bigcup (f(y_0) \setminus p(y_0)) = \emptyset$
$\langle v_0, e_0, p(y_0) \cup (f(y_0) \setminus p(y_0)), v_0 \rangle \hookrightarrow Tugraph_1(\star) \Rightarrow$ $\text{uGraph}\big(v_0, p(y_0) \cup (f(y_0) \setminus p(y_0))\big)$
Loc_def $\Rightarrow$ $p_0 = \text{if } c_9 \subseteq p(y_0) \text{ then } p(y_0) \text{ else } f(y_0) \setminus p(y_0) \text{ fi}$
$\langle v_0, p(y_0), f(y_0) \setminus p(y_0), c_9 \rangle \hookrightarrow Tconn_1(Stat25\star) \Rightarrow$ $Stat27$: $c_0 \subseteq p_0$ & $p_0 \subseteq e_0$
Loc_def $\Rightarrow$ $Stat28$: $c_1 = \{a \in e_0 \mid a \cap \bigcup p_0 \neq \emptyset\}$
Suppose $\Rightarrow$ $Stat29$: $p_0 \not\subseteq c_1$
$\langle u_1 \rangle \hookrightarrow Stat29(Stat27\star) \Rightarrow$ $Stat30$: $u_1 \notin \{a \in e_0 \mid a \cap \bigcup p_0 \neq \emptyset\}$ & $u_1 \in p_0$
$\langle p_0, u_1, p_0 \rangle \hookrightarrow Tun_2(Stat30\star) \Rightarrow$ $u_1 \subseteq \bigcup p_0$
$\langle v_0, e_0 \rangle \hookrightarrow Tugraph_0 \Rightarrow$ $\emptyset \notin e_0$

$\langle u_1 \rangle \hookrightarrow Stat30(Stat27\star) \Rightarrow$ false

Discharge $\Rightarrow$ $Stat31 : p_0 \subseteq c_1$

Loc_def $\Rightarrow$ $Stat32 : q_0 = f(y_0)\backslash p_0 \ \& \ d_1 = \{a \in e_0 \mid a \cap \bigcup q_0 \neq \emptyset\}$

$(Stat26\star)$ELEM $\Rightarrow$ $p_0 \cup q_0 = f(y_0)$

EQUAL $\langle Stat26 \rangle \Rightarrow$ $Stat33 : \bigcup(p_0 \cup q_0) = v_0 \backslash \{y_0\}$

> We will establish three facts: 1) $y_0 \in \bigcup c_1$ (so that the strict inclusion $c_1 \supseteq p_0$ holds, since $y_0 \notin \bigcup p_0$); 2) c_1 is connected; and 3) there is a y_1 such that $c_1 \subseteq f(y_1)$. This will lead us to the sought contradiction, by conflicting with the maximality assumption.

$\langle v_0, e_0, p_0, q_0, y_0 \rangle \hookrightarrow Tconn_3(\star) \Rightarrow$ $\{a \in e_0 \mid a \cap \bigcup p_0 \neq \emptyset\} \cup \{a \in e_0 \mid a \cap \bigcup q_0 \neq \emptyset\}$
$= e_0$

$\langle y_0 \rangle \hookrightarrow Stat3 \Rightarrow$ $f(y_0) = \{a \in e_0 \mid y_0 \notin a\}$

> Note that both of Theorem $conn_3$ and Theorem $conn_5$ can be applied to c_1 and d_1, since p_0 and q_0 are nonnull and vertex disjoint. Therefore, $c_1 \cup d_1 = e_0$ and $\bigcup c_1 \cap \bigcup d_1 = \{y_0\}$. This readily gives us 1); ...

$(Stat26\star)$ELEM $\Rightarrow$ $\emptyset \notin \{p_0, q_0\}$

Suppose $\Rightarrow$ $\bigcup p_0 \cap \bigcup q_0 \neq \emptyset$

   Suppose $\Rightarrow$ $p_0 = p(y_0)$

   EQUAL $\langle Stat26 \rangle \Rightarrow$ false;   Discharge $\Rightarrow$ AUTO

   $(Stat26\star)$ELEM $\Rightarrow$ $p_0 = f(y_0)\backslash p(y_0) \ \& \ q_0 = p(y_0) \ \& \ \bigcup(f(y_0)\backslash p(y_0)) \cap \bigcup p(y_0) = \emptyset$

   EQUAL $\langle Stat26 \rangle \Rightarrow$ false

Discharge $\Rightarrow$ AUTO

$\langle v_0, e_0, p_0, q_0, y_0 \rangle \hookrightarrow Tconn_5(\star) \Rightarrow$
      $Stat34 : \bigcup\{a \in e_0 \mid a \cap \bigcup p_0 \neq \emptyset\} \cap \bigcup\{a \in e_0 \mid a \cap \bigcup q_0 \neq \emptyset\} = \{y_0\}$

> ... it also implies 2), thanks to Theorem $conn_2$.

EQUAL $\Rightarrow$ $Stat35 : uGraph(v_0, c_1 \cup d_1) \ \& \ Conn(v_0, c_1 \cup d_1) \ \& \ \bigcup c_1 \cap \bigcup d_1 = \{y_0\}$

$\langle v_0, d_1, c_1, y_0 \rangle \hookrightarrow Tconn_2(Stat34\star) \Rightarrow$ $Stat36 : Conn(\bigcup c_1, c_1)$

> In order to get 3) and thus complete this proof, we will draw y_1 from the set $\bigcup d_1 \backslash \{y_0\}$. To see that this is doable, notice that the inclusion $c_1 \subseteq p_0$ must be strict, else $y_0 \in \bigcup p_0$ and hence $y_0 \in \bigcup(p_0 \cup q_0)$ would hold, whereas we know from Stat33 that this is not the case. But then, since $c_0 \subseteq p_0$, we must have $c_1 \notin cc$ by the assumed maximality of c_0.

$\langle c_1, q_0, p_0 \cup q_0 \rangle \hookrightarrow Tun_8(Stat33\star) \Rightarrow$ $Stat37 : c_1 \neq p_0$

$\langle c_1 \rangle \hookrightarrow Stat25a(Stat27, Stat31, Stat37\star) \Rightarrow$ $c_1 \notin cc$

> Also notice that the set $\{a \in d_1 \mid a\backslash\{y_0\} \neq \emptyset\}$ is nonnull ...

Suppose $\Rightarrow$ $Stat38 : \{a \in d_1 \cap e_0 \mid a \cap v_0 \backslash \{y_0\} \neq \emptyset\} = \emptyset$

   Use_def($\bigcup d_1$) $\Rightarrow$ AUTO

   $(Stat35\star)$ELEM $\Rightarrow$ $Stat39 : y_0 \in \{y : a \in d_1, y \in a\}$

   $\langle a_9, y_9 \rangle \hookrightarrow Stat39(Stat39, Stat32\star) \Rightarrow$ nothing

      $Stat40 : a_9 \in \{a \in e_0 \mid a \cap \bigcup q_0 \neq \emptyset\} \ \& \ a_9 \in d_1 \ \& \ y_0 \in a_9$

   $\langle \ \rangle \hookrightarrow Stat40 \Rightarrow$ $a_9 \in e_0$

   Use_def(uGraph) $\Rightarrow$ $Stat41 : a_9 \in \{\{x, y\} : x \in v_0, y \in v_0 \backslash \{x\}\}$

   $\langle x_{10}, y_{10} \rangle \hookrightarrow Stat41(Stat41\star) \Rightarrow$ $a_9 \cap v_0 \backslash \{y_0\} \neq \emptyset$

   $\langle a_9 \rangle \hookrightarrow Stat38(Stat38\star) \Rightarrow$ false

Discharge $\Rightarrow$ $Stat42 : \{a \in d_1 \cap e_0 \mid a \cap v_0 \backslash \{y_0\} \neq \emptyset\} \neq \emptyset$

> ... and hence $\bigcup d_1 \backslash \{y_0\}$ is indeed nonnull and we can draw a y_1 from it.

$\langle b_1 \rangle \hookrightarrow Stat42(Stat42\star) \Rightarrow \quad Stat43 : b_1 \cap v_0 \setminus \{y_0\} \neq \emptyset \ \& \ b_1 \in d_1$

$\langle y_1 \rangle \hookrightarrow Stat43(Stat43\star) \Rightarrow \quad Stat44 : y_1 \in b_1 \ \& \ y_1 \in v_0 \ \& \ b_1 \in d_1 \ \& \ y_1 \neq y_0$

‖ To get the desired contradiction, it now suffices to derive that $c_1 \subseteq f(y_1)$.

Suppose $\Rightarrow \quad Stat45 : c_1 \nsubseteq f(y_1)$

$\quad \langle d_1, b_1, d_1 \rangle \hookrightarrow Tun_2(Stat43\star) \Rightarrow \quad y_1 \in \bigcup d_1$

$\quad \langle y_1 \rangle \hookrightarrow Stat3 \Rightarrow \quad Stat46 : f(y_1) = \{a \in e_0 \mid y_1 \notin a\}$

$\quad \langle a_1 \rangle \hookrightarrow Stat45(Stat28, Stat46\star) \Rightarrow$

$\qquad\qquad Stat47 : a_1 \in \{a : a \in e_0 \mid a \cap \bigcup p_0 \neq \emptyset\} \ \& \ a_1 \notin \{a \in e_0 \mid y_1 \notin a\} \ \& \ a_1 \in c_1$

$\quad \langle a_2, a_1 \rangle \hookrightarrow Stat47(Stat47\star) \Rightarrow \quad y_1 \in a_1$

$\quad \langle c_1, a_1, c_1 \rangle \hookrightarrow Tun_2(Stat47\star) \Rightarrow \quad y_1 \in \bigcup c_1$

$(Stat35\star)$Discharge $\Rightarrow \quad Stat48 : c_1 \subseteq f(y_1)$

EQUAL $\langle Stat14 \rangle \Rightarrow \quad Stat49 : c_1 \notin \{c : x \in v_0, c \subseteq f(x) \mid \text{Conn}(\bigcup c, c)\}$

$\langle y_1, c_1 \rangle \hookrightarrow Stat49(Stat48, Stat36, Stat44\star) \Rightarrow \quad$ false

Discharge $\Rightarrow$ QED

Exercises

A.1 Prove that *if a graph $G = (V, E)$ has a connected set C of edges and vertex-disjoint sets P, Q such that $P \cup Q = E$, then either $C \subseteq P$ or $C \subseteq Q$ must hold:*

$$\left(\text{uGraph}(V, P \cup Q) \ \& \ C \subseteq P \cup Q \ \& \ \text{Conn}(\bigcup C, C) \ \& \ \bigcup P \cap \bigcup Q = \emptyset \right) \rightarrow$$
$$\left(C \subseteq P \vee C \subseteq Q \right).$$

A.2 Prove that *if two sets of edges form a connected graph and they share exactly one vertex, then each of them is connected:*

$$\left(\text{uGraph}(V, R \cup S) \ \& \ \text{Conn}(V, R \cup S) \ \& \ \bigcup R \cap \bigcup S = \{x\} \right) \rightarrow$$
$$\text{Conn}(\bigcup S, S).$$

A.3 Prove that *if a graph $G = (V, E)$ has sets P, Q of edges such that $\bigcup(P \cup Q) = V \setminus \{x\}$, then any of its edges touches either P or Q:*

$$\left(\text{uGraph}(V, E) \ \& \ \bigcup(P \cup Q) = V \setminus \{x\} \right) \rightarrow$$
$$\{a \in E \mid a \cap \bigcup P \neq \emptyset\} \cup \{a \in E \mid a \cap \bigcup Q \neq \emptyset\} = E.$$

262 A Excerpts from a Referee-Checked Proof-Script

A.4 Prove that *if a graph $G = (V, E)$ has vertex-disjoint sets P, Q such that $P \cup Q = \{ a \in E \mid x \notin a \}$, then no vertex other than x has incident edges, some of which touch P and some of which touch Q:*

$$\left(\mathsf{uGraph}(V, E) \,\&\, P \cup Q = \{ a \in E \mid x \notin a \} \,\&\, \bigcup P \cap \bigcup Q = \emptyset \,\&\, y \neq x \right) \rightarrow$$
$$y \notin \bigcup \{ a \in E \mid a \cap \bigcup P \neq \emptyset \} \cap \bigcup \{ a \in E \mid a \cap \bigcup Q \neq \emptyset \}.$$

A.5 Prove that *if a connected graph $G = (V, E)$ has vertex-disjoint nonnull sets P, Q such that $P \cup Q = \{ a \in E \mid x \notin a \}$ and $\bigcup(P \cup Q) = V \setminus \{x\}$, then x is the sole vertex w which has edges $\{w, u_P\}, \{w, u_Q\} \in E$ touching P and Q, respectively:*

$$\left(\mathsf{uGraph}(V, E) \,\&\, \mathsf{Conn}(V, E) \,\&\, P \cup Q = \{ a \in E \mid x \notin a \} \,\&\, \right.$$
$$\bigcup P \cap \bigcup Q = \emptyset \,\&\, \left. \bigcup(P \cup Q) = V \setminus \{x\} \,\&\, \emptyset \notin \{P, Q\} \right) \rightarrow$$
$$\bigcup \{ a \in E \mid a \cap \bigcup P \neq \emptyset \} \cap \bigcup \{ a \in E \mid a \cap \bigcup Q \neq \emptyset \} = \{x\}.$$

List of Symbols

$\varphi \& \psi$	Conjunction of the formulae φ, ψ (alias of $\varphi \wedge \psi$). See p. 32
$\varphi \vee \psi$	Disjunction of the formulae φ, ψ. See p. 32
$\varphi \wedge \psi$	Conjunction of the formulae φ, ψ (also written as $\varphi \ \& \ \psi$). See p. 32
$\bigcap s$	Set of the elements shared by all elements of the set s. See p. 45
$\bigcup s$	Set of all elements of elements of the set s. See p. 45
$\mathcal{P}(s)$	Set of all subsets of the set s. See p. 45
$x \times y$	Cartesian product of the sets x, y. See p. 41
$R \circ S$	Composition of relations R and S. See p. 45
$R \vec{\Gamma} D$	(Multi)-image of R with respect to D. See p. 45
$R^{\triangleleft} D$	Reverse image of R with respect to D. See p. 45
$\prec$	Ackermann order. See p. 70
$\beth(k)$	Function for the number of hereditarily finite sets at rank k. See p. 71
$\iota\iota(a,b)$	Formula describing a double-stranded infinite spiral. See p. 224
$|s|$	Cardinality, i.e., number of elements, of a set s. See p. 47
$\iota\iota^{[n]}(a,b)$	Formula describing an n-stranded infinite spiral. See p. 227
$\mathbb{N}_A(h)$	Ackermann number of the set h. See p. 70
$\mathbb{R}_A(h)$	Ackermann-like number of the rational hyperset h. See p. 197
HF_n	The n-th level of HF. See p. 64
HF	The hereditarily finite sets. See p. 64
$\mathcal{V}_n$	n-th level of $\mathcal{V}$. See p. 66
$\mathcal{V}$	von Neumann's cumulative hierarchy of sets. See p. 66
Ω	Self-singleton hyperset. See p. 6
$D[W]$	Vertex-induced (directed) subgraph. See p. 51
$G[W]$	Vertex-induced (undirected) subgraph. See p. 54
$N_D^-(v)$	In-neighborhood of vertex v in the digraph D. See p. 51
$N_D^+(v)$	Out-neighborhood of vertex v in the digraph D. See p. 51
$N_G(v)$	Neighborhood of vertex v in the undirected graph G. See p. 54
$d_D^-(v)$	In-degree of vertex v in the digraph D. See p. 51

© Springer International Publishing AG 2017
E.G. Omodeo et al., *On Sets and Graphs*, DOI 10.1007/978-3-319-54981-1

$d_D^+(v)$	Out-degree of vertex v in the digraph D. See p. 51
$d_G(v)$	Degree of vertex v in the undirected graph G. See p. 54
K_n	The complete undirected graph on n vertices. See p. 55
$K_{n,m}$	The complete bipartite graph with parts of sizes n and m. See p. 55
P_n	Peddicord set systems for transitive sets with n elements. See p. 188
$c_{n,s}$	The number of non-isomorphic extensional acyclic graphs with n vertices, out of which exactly s are sources. See p. 180
e_n	The number of extensional acyclic graphs with vertex set $\{1, \ldots, n\}$. See p. 179
s_n	The number of transitive well-founded sets of cardinality n. See p. 176
s_n^r	The number of transitive well-founded sets of cardinality n with exactly r elements of maximum rank. See p. 177
$\mathbf{arb}(w)$	Arbitrary selection operator (for Ref). See p. 157
$\mathbf{dom}(R)$	Domain of R. See p. 45
$\mathbf{img}(R)$	Overall (multi-)image of R. See p. 44
Conn(V, E)	Connectedness of a graph (for Ref). See p. 133
Finite(F)	Property of being finite of a set F. See p. 60
NonCut(z, V, E)	Non-cut vertex (for Ref). See p. 134
trCl(s)	Transitive closure of the set s. See p. 46
uGraph(V, E)	Undirected graph (for Ref). See p. 132

References

1. Ackermann, W.: Die Widerspruchfreiheit der allgemeinen Mengenlehre. Mathematische Annalen **114**, 305–315 (1937)
2. Aczel, P.: Non-Well-Founded Sets. CSLI Lecture Notes, vol. 14. Center for the Study of Language and Information, Stanford (1988)
3. Aho, A.V., Hopcroft, J.E., Ullman, J.D.: The Design and Analysis of Computer Algorithms. Addison-Wesley, Reading (1976)
4. Andersson, S.A., Madigan, D., Perlman, M.D.: A characterization of Markov equivalence classes for acyclic digraphs. Ann. Stat. **25**, 502–541 (1997)
5. Audrito, G., Tomescu, A.I., Wagner, S.G.: Enumeration of the adjunctive hierarchy of hereditarily finite sets. J. Log. Comput. **25**(3), 943–963 (2015)
6. Bandelt, H.J., Mulder, H.M.: Distance-hereditary graphs. J. Comb. Theory Ser. B **41**(2), 182–208 (1986)
7. Bang-Jensen, J., Gutin, G.: Digraphs: Theory, Algorithms and Applications. Springer Monographs in Mathematics, 2nd edn. Springer, London (2008)
8. Barwise, J., Moss, L.: Hypersets. Math. Intell. **13**(4), 31–41 (1991)
9. Barwise, J., Moss, L.S.: Vicious Circles. CSLI Lecture Notes, Stanford (1996)
10. Behzad, M., Chartrand, G., Lesniak-Foster, L.: Graphs & Digraphs. Prindle, Weber & Schmidt, Boston (1979)
11. Belinfante, J.G.F.: On computer-assisted proofs in ordinal number theory. J. Autom. Reason. **22**(2), 341–378 (1999)
12. Bellè, D., Parlamento, F.: Truth in $\mathbf{V}$ for $\exists^* \forall \forall$-sentences is decidable. J. Symb. Log. **71**(4), 1200–1222 (2006)
13. Boyer, R.S., Lusk, E.L., McCune, W., Overbeek, R.A., Stickel, M.E., Wos, L.: Set theory in first-order logic: clauses for Gödel's axioms. J. Autom. Reason. **2**(3), 287–327 (1986)
14. Brandstädt, A., Klembt, T., Lozin, V.V., Mosca, R.: On independent vertex sets in subclasses of apple-free graphs. Algorithmica **56**, 383–393 (2010)
15. Brandstädt, A., Le, V.B., Spinrad, J.P.: Graph Classes: A Survey. Monographs on Discrete Mathematics and Applications, vol. 3. SIAM Society for Industrial and Applied Mathematics, Philadelphia (1999)
16. Brandstädt, A., Lozin, V.V., Mosca, R.: Independent sets of maximum weight in apple-free graphs. SIAM J. Discret. Math. **24**, 239–254 (2010)
17. Burstall, R., Goguen, J.: Putting theories together to make specifications. In: Reddy, R. (ed.) Proceedings of 5th International Joint Conference on Artificial Intelligence, Cambridge, MA, pp. 1045–1058 (1977)

18. Calligaris, P., Omodeo, E.G., Tomescu, A.I.: A proof-checking experiment on representing graphs as membership digraphs. In: Cantone, D., Asmundo, M.N. (eds.) CILC 2013, 28th Italian Conference on Computational Logic, CEUR Workshop Proceedings, Catania, pp. 227–233 (2013)

19. Cantone, D., Omodeo, E.G., Policriti, A.: Set Theory for Computing. From Decision Procedures to Declarative Programming with Sets. Texts and Monographs in Computer Science. Springer, New York (2001)

20. Cantone, D., Omodeo, E.G., Schwartz, J.T., Ursino, P.: Notes from the logbook of a proof-checker's project. In: Dershowitz, N. (ed.) Verification: Theory and Practice, Essays Dedicated to Zohar Manna on the Occasion of his 64th Birthday. Lecture Notes in Computer Science, vol. 2772, pp. 182–207. Springer, Berlin Heidelberg (2003)

21. Casagrande, A., Omodeo, E.G.: Reasoning about connectivity without paths. In: Bistarelli, S., Formisano, A. (eds.) ICTCS'13, Fifteenth Italian Conference on Theoretical Computer Science, CEUR Workshop Proceedings, pp. 93–108 (2014)

22. Casagrande, A., Piazza, C., Policriti, A.: Is hyper-extensionality preservable under deletions of graph elements? Electron. Notes Theor. Comput. Sci. **322**, 103–118 (2016)

23. Chudnovsky, M., Robertson, N., Seymour, P., Thomas, R.: The strong perfect graph theorem. Ann. Math. **164**(1), 51–229 (2006)

24. Cohen, P.J.: Set Theory and the Continuum Hypothesis. Mathematics Lecture Note Series. W. A. Benjamin, Inc., Reading, MA(1966)

25. D'Agostino, G., Omodeo, E.G., Policriti, A., Tomescu, A.I.: Mapping sets and hypersets into numbers. Fundam. Inform. **140**(3–4), 307–328 (2015)

26. Davis, M.: Applied Nonstandard Analysis. John Wiley & Sons, New York (1977)

27. Davis, M., Schonberg, E. (eds.): From Linear Operators to Computational Biology: Essays in Memory of Jacob T. Schwartz. Springer, London (2012)

28. Dewar, R.: SETL and the Evolution of Programming. In: [27, pp. 39–46] (2012)

29. Dovier, A., Omodeo, E.G., Pontelli, E., Rossi, G.: {log}: A language for programming in logic with finite sets. J. Log. Program. **28**(1), 1–44 (1996)

30. Dovier, A., Piazza, C., Policriti, A.: An efficient algorithm for computing bisimulation equivalence. Theor. Comput. Sci. **311**(1–3), 221–256 (2004)

31. Dovier, A., Piazza, C., Pontelli, E., Rossi, G.: Sets and constraint logic programming. ACM Trans. Program. Lang. Syst. **22**(5), 861–931 (2000)

32. Dyer, M., Frieze, A., Jerrum, M.: Approximately counting Hamilton paths and cycles in dense graphs. SIAM J. Comput. **27**(5), 1262–1272 (1998)

33. Enderton, H.B.: A Mathematical Introduction to Logic, 2nd edn. Hartcourt/Academic Press, Burlington (2001)

34. Erdős, P., Moser, L.: On the representation of directed graphs as unions of orderings. Publ. Math. Inst. Hung. Acad. Sci. Ser. A **9**, 125–132 (1964)

35. Euler, L.: Solutio problematis ad geometriam situs pertinentis. Commentarii academiae scientiarum Petropolitanae **8**, 128–140 (1741)

36. Felgner, U.: Comparison of the axioms of local and universal choice. Fundamenta mathematicae **71**(1), 43–62 (1971)

37. Ferro, A., Omodeo, E., Schwartz, J.: Decision procedures for some fragments of set theory. In: Bibel, W., Kowalski, R. (eds.) Proceedings of the 5th Conference on Automated Deduction. LNCS, vol. 87, pp. 88–96. Springer, Berlin/New York (1980)

38. Ferro, A., Omodeo, E.G., Schwartz, J.T.: Decision procedures for elementary sublanguages of set theory. I. Multi-level syllogistic and some extensions. Commun. Pure Applied Math. **33**(5), 599–608 (1980)

39. Formisano, A., Omodeo, E.G.: Theory-specific automated reasoning. In: Dovier, A., Pontelli, E. (eds.) A 25-Year Perspective on Logic Programming: Achievements of the Italian Association for Logic Programming, GULP. Lecture Notes in Computer Science, vol. 6125, pp. 37–63. Springer, Berlin/Heidelberg (2010)

40. Forti, M., Honsell, F.: Set theory with free construction principles. Annali Scuola Normale Superiore di Pisa, Classe di Scienze **IV**(10), 493–522 (1983)

41. Fraenkel, A.A.: The notion of "definite" and the independence of the axiom of choice. In: [46, pp. 284–289]. (The original publication, in German language, is of 1922)
42. Garey, M.R., Johnson, D.S.: Computers and Intractability; A Guide to the Theory of NP-Completeness. W. H. Freeman & Co., New York (1990)
43. Gentilini, R., Piazza, C., Policriti, A.: From bisimulation to simulation: Coarsest partition problems. J. Autom. Reason. **31**(1), 73–103 (2003)
44. Gödel, K.: The Consistency of the Axiom of Choice and of the Generalized Continuum-Hypothesis. Proc. Natl. Acad. Sci. U.S.A. **24**(12), 556–557 (1938)
45. Harary, F., Palmer, E.: Graphical Enumeration. Academic Press, New York (1973)
46. van Heijenoort, J. (ed.): From Frege to Gödel – A Source Book in Mathematical Logic, 1879–1931, 3rd printing edn. Source books in the history of the sciences. Harvard University Press, Cambridge MA (1977)
47. Hendry, G., Vogler, W.: The square of a connected $S(K_{1,3})$-free graph is vertex pancyclic. J. Graph Theory **9**(4), 535–537 (1985)
48. Hopcroft, J.E.: An $n \log n$ algorithm for minimizing states in a finite automaton. In: Kohavi, Z., Paz, A. (eds.) Theory of Machines and Computations, pp. 189–196. Academic Press, New York (1971)
49. Howorka, E.: On metric properties of certain clique graphs. J. Comb. Theory Ser. B **27**(1), 67–74 (1979)
50. Jech, T.: Set Theory. Springer Monographs in Mathematics, 3rd Millennium edn. Springer, Berlin/Heidelberg (2003)
51. Keller, J.P., Paige, R.: Program derivation with verified transformations – a case study. Commun. Pure Appl. Math. **48**(9–10), 1053–1113 (1995). Special issue in honor of J.T. Schwartz
52. Kirby, L.: A hierarchy of hereditarily finite sets. Arch. Math. Log. **47**(2), 143–157 (2008)
53. König, D.: Theorie der endlichen und unendlichen graphen. Akademische Verlagsgesellschaft Leipzig (1936)
54. Kunen, K.: Set Theory. Studies in Logic. College Publications, London (2011)
55. Levin, D.A., Peresand, Y., Wilmer, E.: Markov Chains and Mixing Times. AMS, Providence (2009)
56. Lin, R., Olariu, S., Pruesse, G.: An optimal path cover algorithm for cographs. Comput. Math. Appl. **30**(8), 75–83 (1995)
57. Lisitsa, A., Sazonov, V.: Linear ordering on graphs, anti-founded sets and polynomial time computability. Theor. Comput. Sci. **224**, 173–213 (1999)
58. Lovász, L., Plummer, M.D.: Matching Theory. North-Holland Mathematics Studies. North-Holland/Amsterdam, New York (1986). Includes indexes
59. Lozin, V.V., Milanič, M., Purcell, C.: Graphs without large apples and the maximum weight independent set problem. Graphs Comb. **30**(2), 395–410 (2014)
60. Matthews, M.M., Sumner, D.P.: Hamiltonian Results in $K_{1,3}$-Free Graphs. J. Graph Theory **8**, 139–146 (1984)
61. Melançon, G., Philippe, F.: Generating connected acyclic digraphs uniformly at random. Inf. Process. Lett. **90**(4), 209–213 (2004)
62. Melançon, G., Dutour, I., Bousquet-Mélou, M.: Random generation of directed acyclic graphs. In: Nesetril, J., Noy, M., Serra, O. (eds.) Euroconference on Combinatorics, Graph Theory and Applications (Comb01). Electronic Notes in Discrete Mathematics, vol. 10, pp. 202–207 (2001)
63. Milanič, M., Rizzi, R., Tomescu, A.I.: Set graphs. II. Complexity of set graph recognition and similar problems. Theor. Comput. Sci. **547**, 70–81 (2014)
64. Milanič, M., Tomescu, A.I.: Set graphs. I. Hereditarily finite sets and extensional acyclic orientations. Discret. Appl. Math. **161**(4–5), 677–690 (2013)
65. Milanič, M., Tomescu, A.I.: Set graphs. IV. Further connections with claw-freeness. Discret. Appl. Math. **174**, 113–121 (2014)
66. Minty, G.J.: On maximal independent sets of vertices in claw-free graphs. J. Comb. Theory Ser. B **28**(3), 284–304 (1980)

67. Mostowski, A.: An undecidable arithmetical statement. Fund. Math. **36**, 143–164 (1949)
68. von Neumann, J.: Zur Einführung der trasfiniten Zahlen. Acta Sci. Math. (Szeged) **1**(4–4), 199–208 (1922–23). Available in English translation in [46, pp. 346–354]
69. von Neumann, J.: Über eine Widerspruchsfreiheitsfrage in der axiomatischen Mengenlehre. J. für die reine und angewandte Mathematik (160), 227–241, reprinted in [70, pp. 494–508] (1929)
70. von Neumann, J.: Collected Works, vol. I: Logic, Theory of Sets and Quantum Mechanics. Pergamon Press, New York (1961)
71. Olariu, S.: The strong perfect graph conjecture for pan-free graphs. J. Comb. Theory Ser. B **47**, 187–191 (1989)
72. Omodeo, E.G., Cantone, D., Policriti, A., Schwartz, J.T.: A computerized referee. In: Stock, O., Schaerf, M. (eds.) Reasoning, Action and Interaction in AI Theories and Systems – Essays Dedicated to Luigia Carlucci Aiello. LNAI, vol. 4155, pp. 117–139. Springer, Berlin Heidelberg (2006)
73. Omodeo, E.G., Parlamento, F., Policriti, A.: Decidability of $\exists^*\forall$-sentences in membership theories. Math. Log. Q. **42**(1), 41–58 (1996)
74. Omodeo, E.G., Policriti, A.: The Bernays-Schönfinkel-Ramsey class for set theory: semidecidability. J. Symb. Logic **75**(2), 459–480 (2010)
75. Omodeo, E.G., Policriti, A.: The Bernays-Schönfinkel-Ramsey class for set theory: decidability. J. Symb. Logic **77**(3), 896–918 (2012)
76. Omodeo, E.G., Policriti, A., Tomescu, A.I.: Statements of ill-founded infinity in set theory. In: Cégielski, P. (ed.) Studies in Weak Arithmetics. Lecture Notes, vol. 196, pp. 173–199. Center for the Study of Language and Information, Stanford University, Stanford CA (2009)
77. Omodeo, E.G., Policriti, A., Tomescu, A.I.: Infinity, in short. J. Logic Comput. **22**(6), 1391–1403 (2012)
78. Omodeo, E.G., Policriti, A., Tomescu, A.I.: Set-syllogistics meet combinatorics. Math. Struct. Comput. Sci. **27**(2), 296–310 (2017)
79. Omodeo, E.G., Schwartz, J.T.: A 'theory' mechanism for a proof-verifier based on first-order set theory. In: Kakas, A.C., Sadri, F. (eds.) Computational Logic: Logic Programming and Beyond, Essays in Honour of Robert A. Kowalski, Part II. Lecture Notes in Computer Science, vol. 2408, pp. 214–230. Springer, London (2002)
80. Omodeo, E.G., Tomescu, A.I.: Set graphs. III. Proof Pearl: claw-free graphs mirrored into transitive hereditarily finite sets. J. Automat. Reason. **52**(1), 1–29 (2014)
81. Omodeo, E.G., Tomescu, A.I.: Set graphs. V. On representing graphs as membership digraphs. J. Log. Comput. **25**(3), 899–919 (2015)
82. Paige, R., Tarjan, R.E.: Three partition refinement algorithms. SIAM J. Comput. **16**(6), 973–989 (1987)
83. Paige, R., Tarjan, R.E., Bonic, R.: A linear time solution to the single function coarsest partition problem. Theoret. Comput. Sci. **40**, 67–84 (1985)
84. Papadimitriou, C.H.: Computational Complexity. Addison-Wesley, Reading (1994)
85. Parlamento, F., Policriti, A.: Decision procedures for elementary sublanguages of set theory. IX. Unsolvability of the decision problem for a restricted subclass of the Δ_0-formulas in set theory. Commun. Pure Appl. Math. **XLI**, 221–251 (1988)
86. Parlamento, F., Policriti, A.: The logically simplest form of the infinity axiom. Proc. Am. Math. Soc. **103**(1), 274–276 (1988)
87. Parlamento, F., Policriti, A.: Note on "The logically simplest form of the infinity axiom". Proc. Am. Math. Soc. **108**(1), 285–286 (1990)
88. Parlamento, F., Policriti, A.: Decision procedures for elementary sublanguages of set theory: XIII. Model graphs, reflection and decidability. J. Automat. Reason. **7**(2), 271–284 (1991)
89. Parlamento, F., Policriti, A.: Expressing infinity without foundation. J. Symb. Logic **56**(4), 1230–1235 (1991)
90. Parlamento, F., Policriti, A.: Undecidability results for restricted universally quantified formulae of set theory. Commun. Pure Appl. Math. **XLVI**(1), 57–73 (1993)

91. Paulson, L.C.: A simple formalization and proof for the mutilated chess board. Logic J. IGPL
    9(3), 499–509 (2001)
92. Peddicord, R.: The number of full sets with n elements. Proc. Am. Math. Soc. **13**, 825–828
    (1962)
93. Piazza, C., Policriti, A.: Ackermann encoding, bisimulations, and OBDDs. Theory Pract.
    Logic Programm. **4**(5–6), 695–718 (2004)
94. Policriti, A., Tomescu, A.I.: Counting extensional acyclic digraphs. Inf. Process. Lett. **111**(3),
    787–791 (2011)
95. Policriti, A., Tomescu, A.I.: Markov chain algorithms for generating sets uniformly at
    random. Ars Math. Contemp. **6**(57–68) (2013)
96. Prüfer, H.: Neuer Beweis eines Satzes über Permutationen. Arch. Math. Phys. **27**, 742–744
    (1918)
97. Quaife, A.: Automated deduction in von Neumann-Bernays-Gödel set theory. J. Autom.
    Reason. **8**(1), 91–147 (1992)
98. Ramsey, F.P.: On a problem of formal logic. Proc. Lond. Math. Soc. **30**, 264–286 (1930).
    Read on 13 Dec 1928
99. Rizzi, R., Tomescu, A.I.: Ranking, unranking and random generation of extensional acyclic
    digraphs. Inf. Process. Lett. **113**(5–6), 183–187 (2013)
100. Robinson, R.M.: The theory of classes, a modification of von Neumann's system. J. Symb.
     Logic **2**, 29–36 (1937)
101. Robinson, R.W.: Enumeration of acyclic digraphs. In: Proceedings of Second Chapel
     Hill Conference on Combinatorial Mathematics and its Applications. University of North
     Carolina, Chapel Hill (1970)
102. Robinson, R.W.: Counting labeled acyclic digraphs. In: Harary, F. (ed.) New Directions in
     the Theory of Graphs, pp. 239–273. Academic Press, New York (1973)
103. Sbihi, N.: Algorithme de recherche d'un stable de cardinalite maximum dans un graphe sans
     etoile. Discret. Math. **29**(1), 53–76 (1980)
104. Schwartz, J., Dewar, R., Dubinsky, E., Schonberg, E.: Programming with Sets: An Introduc-
     tion to SETL. Texts and Monographs in Computer Science. Springer, New York (1986)
105. Schwartz, J.T., Cantone, D., Omodeo, E.G.: Computational Logic and Set Theory. Springer,
     London (2011). Foreword by Martin Davis
106. Steinsky, B.: Efficient coding of labeled directed acyclic graphs. Soft Comput. **7**, 350–356
     (2003)
107. Steinsky, B.: Enumeration of labelled chain graphs and labelled essential directed acyclic
     graphs. Discret. Math. **270**(1–3), 266–277 (2003)
108. Sumner, D.: Graphs with 1-factors. Proc. Am. Math. Soc. **42**, 8–12 (1974)
109. Tarski, A.: Sur les ensembles fini. Fundamenta Mathematicae **VI**, 45–95 (1924)
110. Tomescu, A.I.: A simpler proof for vertex-pancyclicity of squares of connected claw-free
     graphs. Discret. Math. **312**(15), 2388–2391 (2012)
111. Vergnas, M.L.: A note on matchings in graphs. Cahiers Centre Etudes Rech. Opér. **17**, 257–
     260 (1975)
112. Wagner, S.: Asymptotic enumeration of extensional acyclic digraphs. In: Proceedings of the
     ANALCO12 Meeting on Analytic Algorithmics and Combinatorics, pp. 1–8 (2012)
113. West, D.B.: Introduction to Graph Theory. Prentice-Hall, New Jersey (2001)
114. Zermelo, E.: Untersuchungen über die Grundlagen der Mengenlehre I. In: [46, pp. 199–215]
     (1908)

Index

© Springer International Publishing AG 2017   271
E.G. Omodeo et al., *On Sets and Graphs*, DOI 10.1007/978-3-319-54981-1

Printed in the United States
By Bookmasters